Birkhoff Interpolation

GIAN-CARLO ROTA, *Editor*
ENCYCLOPEDIA OF MATHEMATICS AND ITS APPLICATIONS

GIAN-CARLO ROTA, *Editor*
ENCYCLOPEDIA OF MATHEMATICS AND ITS APPLICATIONS

ENCYCLOPEDIA
OF MATHEMATICS
and Its Applications

GIAN-CARLO ROTA, Editor
Department of Mathematics
Massachusetts Institute of Technology
Cambridge, Massachusetts

Editorial Board

GIAN-CARLO ROTA, *Editor*

ENCYCLOPEDIA OF MATHEMATICS AND ITS APPLICATIONS

Volume 19

Section: Interpolation and Approximation
G. G. Lorentz, *Section Editor*

Birkhoff Interpolation

G. G. Lorentz
Department of Mathematics
The University of Texas
Austin, Texas

K. Jetter
Mathematics Institute
University of Duisburg
Duisburg, Federal Republic of Germany

S. D. Riemenschneider
Department of Mathematics
University of Alberta
Edmonton, Alberta, Canada

1983

Addison-Wesley Publishing Company
Advanced Book Program/World Science Division
Reading, Massachusetts

London • Amsterdam • Don Mills, Ontario • Sydney • Tokyo

CAMBRIDGE UNIVERSITY PRESS
Cambridge, New York, Melbourne, Madrid, Cape Town, Singapore, São Paulo, Delhi

Cambridge University Press
The Edinburgh Building, Cambridge CB2 8RU, UK

Published in the United States of America by Cambridge University Press, New York

www.cambridge.org
Information on this title: www.cambridge.org/9780521104043

First published 1983 by Addison Wesley
First published by Cambridge University Press 1984
This digitally printed version 2009

A catalogue record for this publication is available from the British Library

Library of Congress Cataloguing in Publication data

Lorentz, G. G.
  Birkhoff interpolation.

  (Encyclopedia of mathematics and its applications; v. 19)
  Bibliography: p.
  Includes indexes.
  1. Interpolation.   2. Spline theory.   I. Jetter, K.
II. Riemenschneider, S. D.   III. Title.   IV. Series.
QA281.L67 1983   511'.4   82-18422

ISBN 978-0-521-30239-5 hardback
ISBN 978-0-521-10404-3 paperback

Contents

Editor's Statement

A large body of mathematics consists of facts that can be presented and described much like any other natural phenomenon. These facts, at times explicitly brought out as theorems, at other times concealed within a proof, make up most of the applications of mathematics, and are the most likely to survive change of style and of interest.

This ENCYCLOPEDIA will attempt to present the factual body of all mathematics. Clarity of exposition, accessibility to the non-specialist, and a thorough bibliography are required of each author. Volumes will appear in no particular order, but will be organized into sections, each one comprising a recognizable branch of present-day mathematics. Numbers of volumes and sections will be reconsidered as times and needs change.

It is hoped that this enterprise will make mathematics more widely used where it is needed, and more accessible in fields in which it can be applied but where it has not yet penetrated because of insufficient information.

GIAN - CARLO ROTA

Preface

Birkhoff, or lacunary, interpolation appears whenever observation gives scattered, irregular information about a function and its derivatives. First discovered by G. D. Birkhoff in 1906, it received little attention until I. J. Schoenberg revived interest in the subject in 1966. Lacunary interpolation differs radically from the more familiar Lagrange and Hermite interpolation in both its problems and its methods. It could even be described as "non-Hermitian" interpolation. The name *Birkhoff interpolation* is justified also from a historical point of view.

At present, the main definitions and theorems for polynomial Birkhoff interpolation seem to have been found, while the theory for other systems of functions, most notably splines, is in healthy development. Since this material can be found only in research periodicals and proceedings of conferences, it is time for a comprehensive exposition of the material. We have gone to great lengths to unify, simplify, and improve the information already published or in press, and to set the stage for further developments.

The book should be of interest to approximation theorists, numerical analysts, and analysts in general, as well as to computer specialists and engineers who need to analyze functions when their values and those of their derivatives are given in an erratic way. The book could be used as a text for a graduate course, requiring little more than an undergraduate mathematics background.

Many novel ideas and tools have been developed in this theory; interpolation matrices, coalescence of rows in matrices, independent knots, probabilistic methods, diagrams of splines, and the Rolle theorem for splines. There are applications to approximation with constraints, to quadrature formulas, to splines and their zeros, and to the theory of Chebyshev systems.

The book is largely self-contained, at least in its central parts (Chapters 1–8 and 13–14). We begin with the basic definitions and elementary properties of Birkhoff interpolation by linear combinations of smooth functions in Chapters 1 and 2. In Chapters 3 and 4 we introduce coalescence of rows of matrices, and obtain many applications. Rolle theorem methods and independent knots are discussed in Chapter 5; these methods work for interpolation by polynomials in quite general systems of functions. Chapter 6 is concerned with singularity theorems; conditions are given under which the Birkhoff interpolation problem is not solvable.

Chapter 7 returns to the original problem of Birkhoff—to describe the remainders of interpolation formulas by means of an integral involving a kernel function. This naturally leads to the introduction of Birkhoff splines and to the study of their zeros.

In Chapter 8 we investigate a special case that illustrates the complexity of the Birkhoff interpolation problem in even a very simple situation. Selected applications of Birkhoff interpolation to approximation theory and Chebyshev systems are presented in Chapter 9. Many related applications had to be omitted, but they can be traced through the literature cited in the notes. Birkhoff interpolation of functions of several variables—a subject that needs much further investigation—also must have useful applications.

In Chapter 10 we deal with quadrature formulas based on general Birkhoff interpolation matrices. This relatively new theory culminates in theorems about the existence of formulas of Gaussian type.

Lacunary interpolation at special knots has received considerable attention since the work of Turán and his associates in the 1950s. In Chapters 11 and 12 we give a sample of these results; much related material had to be omitted for lack of space.

Chapters 13 and 14 offer an introduction to Birkhoff interpolation by splines. This subject is inherently complicated, and considerable effort has been made to simplify and unify the theory by means of new notation and methods. Applications abound here, and are presented in the last sections of Chapter 14. For the convenience of the reader, we include a Bibliography (with the indication of the sections where different papers are quoted), a Symbol Index, and a Subject Index.

The book has been developed from the report of one of us to the Center of Numerical Analysis, University of Texas, in 1975, as well as from lectures by each of us at our universities and from our recent publications. The authors gratefully acknowledge support of their research activities during the time of writing this book by the National Science Foundation of the United States and the National Science and Engineering Research Council of Canada. We are grateful for generous advice and comments from C. de Boor, T. N. T. Goodman, M. Marsden, C. Micchelli, P. Nevai, A. Sharma, P. W. Smith, and others. We are especially thankful to G. C. Rota for encouraging our participation in the *Encyclopedia*.

G. G. Lorentz
K. Jetter
S. D. Riemenschneider

Introduction

Approximation and Interpolation in the Last 20 Years

This is the first in a series of books dealing with approximation and interpolation of functions. Many changes have occurred in this theory during the last decades. In what follows, we shall try to describe some of the problems and achievements of this period.

Until about 1955, the leading force in approximation was the Russians, in particular, Bernstein and his school (Ahiezer), Chebyshev, Kolmogorov, and Markov. The development of the subject in Germany, Hungary, and the United States occurred later. The West certainly leads in the number of papers published—see the bulky *Journal of Approximation Theory*. The twelve sections that follow review the newer developments.

The two classical books dealing with approximation and interpolation are those of Natanson [0-N] and Ahiezer [0-A]. Important recent books include two Russian works devoted to special problems: Korneichuk [0-K$_2$] (see also [0-K$_3$]) deals with best constants in the trigonometric approximation, while Tihomirov [0-T$_1$] treats extremal problems, particularly widths and optimization. The book of Butzer and Berens [0-B$_2$] introduced functional analytic methods into the field; the two books by de Boor [0-B$_1$] and Schumaker [0-S] deal with splines, an American development rich in practical applications. Karlin and Studden [0-K$_1$] treat Chebyshev systems exhaustively. Books on general approximation theory are those of Rice [0-R], Lorentz [0-L], Dzyadyk [0-D], and Timan [0-T$_2$]; the last book contains a wealth of material. Several books will be mentioned in later sections.

What might one recommend to someone who wants to begin a study of the subject? My choice would be the books of Natanson [0-N] and Cheney [0-C], and perhaps also my own [0-L], for the real approximation, and that of Gaier [0-G] for the complex approximation.

REFERENCES

[0-A]   N. I. Ahiezer, *Lectures on the Theory of Approximation*, 2nd ed., Izd. Nauka, Moscow, 1965.

[0-B₁]   C. de Boor, *A Practical Guide to Splines* (Applied Mathematical Sciences, Vol. 27), Springer-Verlag, New York, 1978.

[0-B₂]   P. L. Butzer and H. Berens, *Semi-Groups of Operators and Approximation* (Grundlehren, Vol. 145), Springer-Verlag, Berlin, 1967.

[0-C]   E. W. Cheney, *Introduction to Approximation Theory*, 2nd ed., Chelsea, New York, 1981.

[0-D]   V. K. Dzyadyk, *Introduction to Uniform Approximation of Functions by Polynomials*, Izd. Nauka, Moscow, 1977.

[0-G]   D. Gaier, *Vorlesungen über Approximation im Komplexen*, Birkhäuser-Verlag, Basel, 1980.

[0-K₁]   S. Karlin and W. J. Studden, *Tchebycheff Systems: With Applications in Analysis and Statistics*, Wiley-Interscience, New York, 1966.

[0-K₂]   N. P. Korneichuk, *Extremal Problems of Approximation Theory*, Izd. Nauka, Moscow, 1976.

[0-K₃]   N. P. Korneichuk, A. A. Ligun and V. G. Doronin, *Approximation with Constraints*, Naukova Dumka, Kiev, 1982.

[0-L]   G. G. Lorentz, *Approximation of Functions*, Holt, Rinehart and Winston, New York, 1966.

[0-N]   I. P. Natanson, *Constructive Function Theory*, Vols. I–III, Ungar, New York, 1964–1965.

[0-R]   J. R. Rice, *The Approximation of Functions*, Vols. 1, 2, Addison-Wesley, Reading, MA, 1964, 1969.

[0-S]   L. L. Schumaker, *Spline Functions: Basic Theory*, Wiley, New York, 1981.

[0-T₁]   V. M. Tihomirov, *Some Problems of Approximation Theory*, Izd. Mosk. Univ., Moscow, 1976.

[0-T₂]   A. F. Timan, *Theory of Approximation of Functions of a Real Variable*, Izd. Fiz. Mat., Moscow, 1966.

§1.   BEST APPROXIMATION; CHEBYSHEV SYSTEMS

We denote by $\mathscr{P}_n$ (or by $\mathscr{T}_n$) the set of all algebraic polynomials P_n (or all trigonometric polynomials T_n) of degree $\leqslant n$ on an interval $[a, b]$ (or the circle $\mathbb{T}$). The degree of approximation of a function f in the L_p norm is

$$E_n(f)_p = \min\|f - P_n\|_p, \qquad E_n^*(f)_p = \min\|f - T_n\|_p \qquad (1.1)$$

(the space L_∞ is here interpreted to be C). The polynomial of best approximation to f is one that realizes the minimum in (1.1). The problem of finding or describing polynomials of best approximation is the problem of the qualitative theory of approximation, which has been overshadowed in recent decades by the quantitative theory.

In the case $1 < p < +\infty$, the uniqueness of the polynomial of best approximation follows from the convexity of the space L_p. For spaces C and L_1, we also have unicity, but not trivially. We have:

Let $P_n \neq f \in C[a, b]$. Then P_n is the best approximation to f in $\mathscr{P}_n$ if there exists an alternance, that is, points $a \leqslant x_1 < \cdots < x_{n+2} \leqslant b$ for which

$$f(x_i) - P_n(x_i) = \varepsilon(-1)^i \|f - P_n\|_\infty, \qquad i = 1, \ldots, n+2, \quad \varepsilon = +1 \text{ or } -1.$$
$$(1.2)$$

This is Chebyshev's theorem. Actually, Chebyshev only proved, using calculus, the existence of $n + 2$ points, which are either $x = a$ or $x = b$, or satisfy $f'(x) = P_n'(x)$ where $|f(x) - P_n(x)|$ attains its maximum (see [1-C, p. 284, Theorem 2]). Theorem (1.2) has been gradually developed since then. There is also a different condition of Kolmogorov (see [0-L, p. 18]) which characterizes the polynomial of best approximation. It is much less concrete than (1.2), but implies this condition. Its advantage is high flexibility: It can be adjusted to characterize the best approximation in many other cases. For the practical determination of the polynomial of best approximation we have the famous Remez algorithm.

It is very hard to pinpoint the polynomials of best approximation of a given function. The best example that we have is this. If

$$f(x) = \sum_{k=1}^{\infty} a_k C_{3^k}(x), \qquad \sum |a_k| < +\infty, \qquad (1.3)$$

where the C_n are Chebyshev polynomials, then all polynomials of best uniform approximation of f are the partial sums of the series (1.3).

Because of this difficulty, many concrete questions about polynomials P_n of best approximation have not been completely answered. For example, is it true that $f \in C[-1, +1]$ is odd if $P_n(0) = 0$, $n = 0, 1, \ldots$? (Partial affirmative answers are given by Saff and Varga [1-S₁].) Or: What can be the highest multiplicity of 0 as a root of P_n? (It can be $\geqslant \text{const.} \log n$ for infinitely many n; see [1-L].) A peculiar question has been answered by Borosh, Chui, and Smith [1-B]. We approximate x^{n+1} by a polynomial

$$P_n(x) = \sum_{k=1}^{s} a_k x^{\lambda_k}$$

of length $s \leqslant n$, where the λ_k are integers $\leqslant n$. Which selection of the λ_k produces best results? The answer is that one should take the λ_k as close as possible to n; $\lambda_{s-i} = n - i$, $i = 0, \ldots, s - 1$.

Sometimes best approximation is unique even for infinite-dimensional subspaces. Diliberto and Strauss show this for $f(x, y) \in C(I^2)$, $I = [0, 1]$, approximated by functions $g_1(x) + g_2(y)$, $g_i \in C(I)$, $i = 1, 2$. However, generalizations of this theorem proved to be difficult (see papers of Cheney, v Golitschek).

It is impossible to achieve the exact degree of approximation $E_n^*(f)$ by means of a linear polynomial operator on C^*. But one can hope for approximation of order $\leqslant$ const. $E_n^*(f)$. This can be realized, both for C^{*r}, $r = 0, 1, \ldots$ (by means of the de la Vallée–Poussin sums) and for the space A_σ^*, $\sigma > 0$, of periodic functions $f(z)$, analytic in $|\mathrm{Re}\, z| < \sigma$ and continuous on the boundary (by means of Fourier partial sums). However, it cannot be done by the same operators for both spaces at once (Dahmen and Görlich [1-D]).

In 1937, Bernstein coined the term *Chebyshev system* for a set of linearly independent functions $g_0, \ldots, g_n$ on $[a, b]$ or $\mathbb{T}$, which has the property that polynomials $P_n = \Sigma_0^n a_k g_k$ interpolate arbitrary data c_i at any set of distinct points x_i, $i = 0, \ldots, n$. In other words, the determinant $\det[g_k(x_i)]_{i, k=0}^n$ must be $\neq 0$, let us say > 0, for each selection of points $x_0 < \cdots < x_n$. Instead of the Lagrange interpolation, we can take Hermite interpolation here, and obtain the extended Chebyshev systems [0-K$_1$]. Very important is Haar's theorem (1911) that $\{g_k\}$ is a Chebyshev system if and only if each continuous function has a unique polynomial of best approximation. The book of Karlin and Studden [0-K$_1$] treats Chebyshev systems exhaustively; Zalik and Zielke (see [1-Z]) also allow discontinuous Chebyshev systems. Newman and Shapiro [1-N$_1$] show that for each Chebyshev system and each $f \in C$ one has, for the polynomial P_n of best approximation to f, and any other polynomial Q_n,

$$\|f - Q_n\| \geqslant \|f - P_n\| + \gamma \|Q_n - P_n\|, \qquad (1.4)$$

where $\gamma = \gamma(f) > 0$ is a constant. This is the so-called strong uniqueness theorem. Many authors (Bartelt, Henry, Roulier) have studied the behavior of the constants γ (see, e.g., [1-H]).

A *weak Chebyshev system* is defined by means of the inequality $\det[g_k(x_i)] \geqslant 0$, $x_0 < \cdots < x_n$. In important papers, Nürnberger and Sommer have studied these systems in great detail. We give one of their results [1-N$_2$]. For a weak Chebyshev system, some functions $f \in C$ may have several P_n of best approximation. These P_n form a compact convex subset $\pi(f)$ of the $(n + 1)$-dimensional space G spanned by the g_k; the map $f \to \pi(f)$ is called the metric projection. The property of being a weak Chebyshev system is necessary for the existence of a *continuous selection*, that is, of a continuous map $f \to P_n(f)$ of C onto G, with $P_n(f) \in \pi(f)$ for all f. See [1-S$_2$] for literature and other problems.

A related notion is that of *total positivity*. A function $K(x, y)$, $x \in X$, $y \in Y$, where X, Y are linearly ordered, is called totally positive of order r (Karlin [1-K]) if

$$\det\left[K(x_i, y_j)\right] > 0, \qquad x_1 < \cdots < x_k, \qquad y_1 < \cdots < y_k, \qquad 1 \leqslant k \leqslant r.$$

$$(1.5)$$

Properties of total positivity are often very useful (used, e.g., in [1-B], [10-M$_2$], [10-M$_3$]); at other times it is difficult to compute all the determinants contained in (1.5).

REFERENCES

[1-B]   I. Borosh, C. K. Chui, and P. W. Smith, Best uniform approximation from a collection of subspaces, *Math. Z.* **156** (1977), 13–18.

[1-C]   P. L. Tchebychef, *Oeuvres*, Vol. 1, Chelsea, New York, 1961.

[1-D]   W. Dahmen and E. Görlich, Asymptotically optimal linear approximation processes and a conjecture of Golomb. In *Linear Operators and Approximation*, II (P. L. Butzer and B. Sz.-Nagy, eds.), pp. 327–335, Birkhäuser-Verlag, 1974 (ISNM Vol. 25).

[1-H]   M. S. Henry and J. A. Roulier, Lipschitz and strong unicity constants for changing dimension, *J. Approx. Theory* **22** (1978), 85–94.

[1-K]   S. Karlin, *Total Positivity*, Stanford University Press, Stanford, CA, 1968.

[1-L]   G. G. Lorentz, Incomplete polynomials of best approximation, *Israel J. Math.* **29** (1978), 132–140.

[1-N$_1$] D. J. Newman and H. S. Shapiro, Some theorems on Čebyšev approximation, *Duke Math. J.* **30** (1963), 673–681.

[1-N$_2$] G. Nürnberger and M. Sommer, Weak Chebyshev subspaces and continuous selections for the metric projection, *Trans. Amer. Math. Soc.* **238** (1978), 129–138.

[1-S$_1$] E. B. Saff and R. S. Varga, Remarks on some conjectures of G. G. Lorentz, *J. Approx. Theory* **30** (1980), 29–36.

[1-S$_2$] M. Sommer, Existence and pointwise-Lipschitz continuous selections of the metric projection for a class of Z-spaces, *J. Approx. Theory* **34** (1982), 131–145.

[1-Z]   R. Zielke, *Discontinuous Čebyšev Systems* (Lecture Notes in Mathematics, Vol. 707), Springer-Verlag, Berlin, 1979.

§2. MODULI OF SMOOTHNESS; SPACES OF FUNCTIONS; DEGREE OF APPROXIMATION

2.1. Classical Quantitative Theorems

The quantitative theory is the main part of approximation. Here we want good approximation, not the best, mainly because we are seldom able to find the elements of best approximation. In this sense, "best is the enemy of the good." The two main theorems of the quantitative theory are those of Jackson and Bernstein. The first asserts that the degree of approximation $E_n^*(f) = \min_{T_n \in \mathcal{T}_n} \|f - T_n\|_\infty$ satisfies

$$E_n^*(f) \leqslant C_r n^{-r} \omega(f^{(r)}, 1/n) \qquad (2.1)$$

if $f \in C^{*r}$. [C^r is the space of r times continuously differentiable functions, $\omega(g, t)$ is the modulus of continuity of g; the asterisk means that the functions are on the circle $\mathbb{T}$.] The theorem of Bernstein states that conversely, $E_n^*(f) \leqslant Cn^{-r-\alpha}$, $0 < \alpha < 1$, implies that $f \in C^{*r}$ and that $f^{(r)} \in$ Lip α. To these two, Butzer likes to add as a third element the theorem of Zamansky, namely, that for $f \in C^*$

$$\|f - T_n\|_\infty = \mathcal{O}(n^{-\beta}), \beta > 0, \text{ implies } \|T_n^{(k)}\|_\infty = \mathcal{O}(n^{k-\beta}), k > \beta.$$

Since about 1960, new fields of quantitative theory have flourished—constrained approximation, Korovkin theorems, Müntz approximation, incomplete polynomials; these will be reviewed in later sections.

But even in this section progress will be evident. New features are (a) the use of functional analysis (interpolation of linear operators), particularly the use of the K-functionals of Lions and Peetre; (b) some new spaces (e.g., Besov spaces); and (c) problems that refer to two different spaces or to two norms. In §10, some problems of this type can be solved by embedding theorems, but the best of them do not allow this reduction.

2.2. Moduli of Smoothness

We say that $f(x)$, $x \in [a, b]$, has rth derivative $f^{(r)}$ if $f', \ldots, f^{(r-1)}$ exist and are absolutely continuous; then $f^{(r)}$ exists a.e. For functions of several variables, one uses distributional derivatives. The moduli of smoothness of f are given by

$$\omega_r(f, h) = \sup_{t, h} |\Delta_h^r f(t)|, \qquad r = 1, 2, \ldots, \qquad (2.2)$$

where $\Delta_h^r f(t)$ is the rth difference of f with step h. The main point in this definition is that Jackson's theorem (2.1) allows an improvement: $E_n^*(f) \leqslant C_r \omega_r(f, 1/n)$. In the space L_p one puts $\omega_r(f, h)_p = \sup_{u \leqslant h} \|\Delta_u^r f(\cdot)\|$, and also has (2.1).

Other moduli of smoothness have been used by Popov [2-P]; for instance, $\tau_r(f, h)_p = \|\omega_r(f, x; \delta)\|_p$, where in $[a, b]$,

$$\omega_r(f, x; \delta) = \sup \left\{ |\Delta_h^r f(t)|, t, t + rh \in \left[x - \frac{r\delta}{2}, x + \frac{r\delta}{2} \right] \right\}. \qquad (2.3)$$

This τ_r plays the same role in the one-sided approximation as ω_r in the ordinary Jackson theorem. The one-sided degree of approximation of f is

$$\mathscr{E}_n(f) = \inf\{\|P_n - Q_n\|: P_n(x) \leqslant f(x) \leqslant Q_n(x), a \leqslant x \leqslant b\} \quad (2.4)$$

where P_n, Q_n are two polynomials of degree $\leqslant n$.

2.3. Basic Function Spaces

The basic spaces of functional analysis are C, $L_p (1 \leqslant p \leqslant +\infty)$, the Orlicz spaces and the Lorentz space L_{pq}. For f defined on a finite or infinite interval, let f^* be the decreasing rearrangement of $|f|$, and let $f^{**}(t) = t^{-1} \int_0^t f^*$. The space L_{pq} is given by the norm

$$\|f\|_{pq} = \left[\int_0^\infty (t^{1/p} f^{**})^q \frac{dt}{t} \right]^{1/q}, \quad 1 \leqslant q < +\infty, \quad (2.5)$$

(with the q norm replaced by sup for $q = +\infty$). This is Calderón's defini-tion. The original definition [2-L] has f^{**} replaced by f^* in (2.5). For $1 \leqslant q \leqslant p < +\infty$ this is equivalent to (2.5), but is not a norm for other values of p, q.

2.4. Besov Spaces

The spaces needed in the approximation theory are derived from these by applying their norm to the important quantities $f^{(r)}$, $\omega(f, h)/h^\alpha$, $\omega_r(f, h)_p$. In this way one obtains spaces Lip$(\alpha, p) = H_p^\alpha$ for which $\omega(f, h)_p \leqslant Ch^\alpha$, spaces H^ω with $\omega(f, h) \leqslant C\omega(h)$, where ω is some fixed (often concave) modulus of continuity. By $W^r E$ we mean the space of functions f with $f^{(r)} \in E$. The spaces W_p^r (often called Sobolev spaces) consist of functions f with $f^{(r)} \in L_p$. Finally, if we apply to $t^{-1}\omega_r(f, t)_p$ the L_{pq} norm, taking $1 - (1/s) = \theta$, and the original definition (we almost have that $t^{-1}\omega_r$ is positive decreasing, for ω_r is often concave), we obtain the Besov spaces $B_p^{\theta, q}$, $\theta > 0$, with the norm

$$\|f\|_{B_p^{\theta, q}} = \|f\|_p + \left[\int_0^{+\infty} (t^{-\theta}\omega_r(f, t)_p)^q \frac{dt}{t} \right]^{1/q}, \quad r = [\theta] + 1. \quad (2.6)$$

(If $q = +\infty$, the q norm is replaced by the supremum norm.)

We shall mention a few facts about Besov spaces. The space $B_p^{\theta, \infty}$, with $\theta = r + \alpha$, is equivalent to $W^r H^\alpha$; and if $0 < \theta < 1$, then $B_p^{\theta, \infty}$ is the space Lip(θ, p). One also has the embedding for $1 \leqslant s \leqslant +\infty$,

$$B_p^{\alpha+\theta, s} \hookrightarrow B_{p_1}^{\alpha, s} \quad \text{if} \quad 1 \leqslant p < p_1 \leqslant +\infty, \quad \theta := \frac{1}{p} - \frac{1}{p_1}. \quad (2.7)$$

The Hardy spaces H_p, and BMO, are also important in functional analysis. The latter has not yet been discovered by approximation theorists.

2.5. K-functionals

Interpolation of operators appears in approximation theory mainly in the Peetre form, which is based on K-functionals. A definitive first exposition of this theory was presented in the important book of Butzer and Berens [0-B$_2$]. Based on this, and the semigroups of operators, the Aachen school of approximation flourished. Compare Butzer [2-B$_3$] for an exposition of their results until 1973.

Taking the special case, let $X_1 \hookrightarrow X$ be two Banach spaces with a continuous embedding. The K-functional of the spaces X_1, X is the function of t,

$$K(f, t, X_1, X) := \inf_{g \in X_1} \{\|f - g\|_X + t\|g\|_{X_1}\}, \qquad t \geqslant 0. \qquad (2.8)$$

There is a standard way of generating intermediate spaces between X_1, X by applying to $K(f, t)$ one of the norms of type (2.5). In this scheme, the Besov space $B_p^{\theta, q}$ is intermediate between W_p^r and L_p.

One of the reasons why the K-functionals are important is the fact, discovered by Johnen [2-J] and Peetre, that they are equivalent to the smoothness moduli:

$$C_1 \omega_r(f, t)_p \leqslant K(f, t^r; W_p^r, L_p) \leqslant C_2 \omega_r(f, t)_p. \qquad (2.9)$$

This allows us, in proving theorems of the Jackson type with ω_r, to assume f to have several derivatives. See DeVore [2-D] for an exposition of the approximation theorems from this point of view.

2.6. Jackson-type Theorems

What is new in theorems of the Jackson type? A function f on $\mathbb{T}$ belongs to $B_p^{\theta, q}$, $\theta < r$, if and only if (Besov [2-B$_1$])

$$\sum_{n=0}^{\infty} \left(2^{\theta n} E_{2^n}^*(f)\right)^q < +\infty.$$

The influence of the end points in polynomial approximation has been clarified. Timan and Dzyadyk show that $f \in W'H^\alpha[-1, +1]$ if and only if for some $P_n \in \mathscr{P}_n$,

$$|f(x) - P_n(x)| \leqslant \text{const.} \Delta_n(x)^{r+\alpha}, \qquad \Delta_n(x) := n^{-2} + n^{-1}\sqrt{1-x^2}.$$
$$(2.10)$$

There is a corresponding result with ω_r. Here, the difficult inverse theorem is due to Brudnyi [2-B$_2$]. The peculiar fact that (2.10) remains a necessary and sufficient condition if $\Delta_n(x)$ is replaced simply by $n^{-1}\sqrt{1-x^2}$ has been observed by Teljakovskii (see [2-T]). Theorems like (2.10) are also valid in the L_p norm (Oswald [2-O]). In the paper of Butzer and Scherer [2-B$_4$], analogues to the theorems of Jackson, Bernstein, and Zamansky for an arbitrary Banach space are given and their equivalence proved for elements of best approximation.

REFERENCES

[2-B₁] O. V. Besov, Investigation of a family of function spaces in connection with theorems of imbedding and extension, *Trudy Mat. Inst. Steklov* **60** (1961), 42–81.

[2-B₂] Yu. A. Brudnyi, Approximation of functions by algebraic polynomials, *Izv. Akad. Nauk SSSR* **32** (1968), 780–787.

[2-B₃] P. L. Butzer, A survey of work on approximation in Aachen, 1968–72. In *Approximation Theory*, Vol. I (G. G. Lorentz, ed.), pp. 31–100, Academic Press, New York, 1973.

[2-B₄] P. L. Butzer and K. Scherer, On the fundamental approximation theorems of D. Jackson, S. N. Bernstein and theorems of M. Zamansky and S. B. Stečkin, *Aequationes Math.* **3** (1969), 170–185.

[2-D] R. A. DeVore, Degree of approximation. In *Approximation Theory*, Vol. II (G. G. Lorentz et al., eds.), pp. 117–161, Academic Press, New York, 1976.

[2-J] H. Johnen, Inequalities connected with moduli of smoothness, *Mat. Vesnik* **9** (24) (1972), 289–303.

[2-L] G. G. Lorentz, Some new functional spaces, *Ann. of Math.* **51** (1950), 37–55.

[2-O] P. Oswald, Ungleichungen vom Jackson-Typ für die algebraische beste Approximation in L_p, *J. Approx. Theory* **23** (1978), 113–136.

[2-P] V. Popov, Averaged local moduli and their applications. In *Approximation and Function Spaces* (Z. Ciesielski, ed.), pp. 572–583, North Holland, New York, 1981.

[2-T] S. A. Teljakovskii, Two theorems on the approximation of functions by algebraic polynomials, *Mat. Sb.* **70**(112) (1966), 252–266.

§3. APPROXIMATION WITH CONSTRAINTS

Here the approximating polynomials, and sometimes the approximated function, are in some way restricted. For a nice review, see that of Chalmers and Taylor [3-C]. See also the book [0-K₃], discussed in §5.

3.1. Müntz Approximation

We approximate here by polynomials $P_n(x) = \sum_{k=0}^{n} a_k x^{\lambda_k}$, with given $0 = \lambda_0 < \cdots < \lambda_n < \cdots$. Weierstrass' theorem for any $f \in C[0,1]$ is true for this case if and only if $\sum_{1}^{\infty} \lambda_n^{-1} = +\infty$.

What has been investigated recently are Jackson-type theorems. Many important results are due to Newman [3-F$_1$]. We restrict ourselves to the case when f belongs to the unit ball $\mathscr{B}$ of Lip 1. Then

$$\rho_n = \sup_{f \in \mathscr{B}} E_n(f) \approx \varepsilon_n, \qquad \text{where} \quad \varepsilon_n = \sup_{\operatorname{Re} z = 1} \left| \frac{B(z)}{z} \right|, \qquad (3.1)$$

and the $B(z)$ are Blaschke products $B(z) = \prod_{k=1}^{n}(z - \lambda_k)/(z + \lambda_k)$. One proves that $\rho_n = \sup_\mu \int_0^\infty e^{-t}|\mu(t)|\,dt$, where μ are all measures with $\mu(0) = 0$, $\int_0^\infty |d\mu| \leqslant 1$, for which the Laplace transform $F(z) = \int_0^\infty e^{-zt}\,d\mu(t)$ vanishes at the points λ_k, $k = 0,\ldots,n$. The main difficulty is in the proof of $\rho_n \leqslant \mathrm{const.}\ \varepsilon_n$. To overcome it, one takes a concrete μ and estimates $\int |F|^2$ by means of the Paley–Wiener theorem.

From this one obtains concrete results: (a) In the "separated case," $\lambda_{n+1} - \lambda_n \geqslant 2$, $n = 1, 2, \ldots, \rho_n \approx \exp(-2\sum_{k=1}^{n} \lambda_k^{-1})$; (b) in the "unseparated case," $\lambda_{n+1} - \lambda_n \leqslant 2$ for all n, $\rho_n \approx (\sum_1^n \lambda_k)^{-1/2}$.

This method of proof is very nice, but v. Golitschek obtained many of these results by elementary methods.

3.2. Monotone and Co-Monotone Approximation

Monotone approximation is defined in §9.2 of this book; unicity of the polynomials of best approximation is proved there. Here we discuss the degree of monotone approximation. We restrict ourselves to the most important case when $f \in C[a, b]$ and the approximating polynomials are increasing on $[a, b]$. We have a strong result of DeVore [3-D]:

$$E_n(f) \leqslant C_r n^{-r} \omega(f^{(r)}, 1/n), \qquad f \in C^r, \quad r = 0, 1, \ldots. \qquad (3.2)$$

For $r = 0, 1$ this is proved in [3-L$_1$] by means of an integral operator with smoothed-out Jackson kernel. For reasons of saturation, such an operator cannot exist for $r \geqslant 2$. Perhaps this explains why DeVore's methods are so difficult. He also has a similar theorem for splines with equidistant knots, but a very simple proof for this case has been found by Beatson [3-B$_1$].

Let $a = x_0 < x_1 < \cdots < x_n = b$, let y_k be arbitrary real numbers for which $y_k \neq y_{k+1}$, $k = 0, \ldots, n-1$. Wolibner and S. Young proved the existence of a polynomial P_N satisfying

P_N increases (decreases) on $[x_k, x_{k+1}]$ whenever $\Delta y_k > 0$ (or < 0).

$$(3.3)$$

For a piecewise monotone function $f \in C[a, b]$ this shows the existence of a co-monotone polynomial P_N which is monotone in the same sense as f on each subinterval where f is monotone. Two questions present themselves: What is an upper bound for the degree N of P_N? What is the degree of approximation of f by P_N? Iliev [3-I] gives a nice answer to the first question [if all $\Delta y_k > 0$ in (3.3)]: One can take $N \leqslant$ const. $\log(e + A/B)n$ if $A = \max \Delta y_k$, $B = \min \Delta y_k$, and $\Delta x_k \geqslant$ const. n^{-1}. For the second question one has the Jackson estimate with the error of order $n^{-r}\omega(f^{(r)}, 1/n)$, $r = 0, 1$ (see Newman [3-N]; Beatson and Leviatan [3-B$_2$]).

3.3. Restricted Polynomials; Improved Inequalities

Without caring about constants, we state the inequalities of Markov and of Bernstein for P_n on $[-1, +1]$:

$$\|P_n^{(r)}\|_\infty \leqslant C_r(n^2)^r \|P_n\|_\infty; \qquad |P_n^{(r)}(x)| \leqslant C_r \left(\frac{n}{\sqrt{1 - x^2} + n^{-1}} \right)^r \|P_n\|_\infty.$$

$$(3.4)$$

The inequalities can be improved if we restrict the P_n in some way. We can restrict (a) zeros of P_n; (b) coefficients of the polynomials.

(a)   Long ago, Erdös proved that if P_n has only real roots, none of which are in $(-1, +1)$, then on $[-1, +1]$, $\|P_n'\|_\infty \leqslant \frac{1}{2}en\|P_n\|_\infty$. There have been several related results (Szabados; Varma, e.g., [3-V]; also Buslaev, see 4.3(A), belongs here).

(b)   Polynomials with positive coefficients in $x, 1 - x$ on $[0, 1]$ are given by

$$P_n(x) = \sum_{k=0}^{n} a_{n,k} x^k (1 - x)^{n-k}, \qquad a_{n,k} \geqslant 0. \qquad (3.5)$$

For these polynomials, Lorentz and Schieck prove inequalities that are much better than (3.4); they can replace the exponent r in these inequalities by $r/2$. Erdös and Varma [3-E] even have exact inequalities of this type in the L_1 metric.

One remark should be made here. The second inequality (3.4) is quite prominent in the proof of the inverse theorems of approximation; but nobody has proved inverse theorems for the restricted polynomials discussed here. For example, one can assume that the best polynomials (3.5) for a function $f \geqslant 0$ on $[0, 1]$ are its Bernstein polynomials. I offer the conjecture: If there is a sequence of polynomials (3.5) with $\|f - P_n\| \leqslant Cn^{-\alpha/2}$, $0 < \alpha < 1$, then $f \in \mathrm{Lip}\,\alpha$.

3.4. Integral Coefficients

Let $\mathcal{P}^*$ be the set of all polynomials with real *integral coefficients*. Which functions $f \in C[a, b]$ belong to $\overline{\mathcal{P}}^*$, the uniform closure of $\mathcal{P}^*$? This depends on the interval $[a, b]$. If $b - a \geqslant 4$, then $\overline{\mathcal{P}}^* = \mathcal{P}^*$. In the case where $b - a < 4$, there exists a finite set of points $J \subset [a, b]$ so that $f \in \overline{\mathcal{P}}^*$ if and only if the Lagrange interpolation polynomial of f at the points of J has integral coefficients. (Sets J can be described in an elementary way.) In the L_p norm the situation is simpler: $\overline{\mathcal{P}}* = \mathcal{P}*$ if $b - a \geqslant 4$, and $\overline{\mathcal{P}}* = L_p$ if $b - a < 4$ (see L. B. O. Ferguson [3-F$_2$]). It should be noted that these theorems require serious algebraic tools. The most natural approach to them is in the complex plane; then the length of an interval should be replaced by Fekete's transfinite diameter of a compact set.

3.5. Incomplete Polynomials

These are polynomials of the form $P_n(x) = \sum_{k=n_1}^{n} a_k x^k$, where $n_1 \geqslant \theta n$, $0 < \theta < 1$. The main theorem here belongs to the author (see Kemperman and Lorentz [3-K]):

$$|P_n(x)| \leqslant 1 \text{ on } [\theta^2, 1] \text{ implies } P_n(x) \to 0 \text{ on } [0, \theta^2). \qquad (3.6)$$

(This actually happens inside a certain lemniscate in the complex plane surrounding $[0, \theta^2)$.) This is best possible (Saff and Varga [3-S]). For a detailed review of this field see my article [3-L$_2$]. I will mention only two new developments. Statements like (3.6) also hold for trigonometric polynomials $T_n(t)$ with a zero of high order at $t = 0$, or with many zeros in an interval $|t| < \delta$. The second result leads to the following. Let T_n be trigonometric polynomials, let N_n be the number of real zeros of T_n, and assume that $|T_n(t)| \leqslant 1$ on $\mathbb{T}$ and that $T_n(t) \to f(t)$ a.e. Then the function f vanishes on a set of measure $\geqslant 2 \limsup(N_n/n)$, and the constant 2 is the best possible [3-G].

In [3-M], Mhaskar and Saff employ methods of incomplete polynomials to estimate polynomials on $(-\infty, +\infty)$.

REFERENCES

[3-B$_1$] R. K. Beatson, Monotone and convex approximation by splines: Error estimates and a curve fitting algorithm. To appear in *SIAM J. Numer. Anal.*

[3-B$_2$] R. K. Beatson and D. Leviatan, On co-monotone approximation, *Canad. Math. Bull.* (in press).

[3-C]   B. L. Chalmers and G. D. Taylor, Uniform approximation with constraints, *Jahresber. Deutsch. Math.-Verein.* **81** (1978/79), 49–86.

[3-D]   R. A. DeVore, Monotone approximation by polynomials, *SIAM J. Math. Anal.* **8** (1977), 906–921.

[3-E]   P. Erdös and A. K. Varma, An extremum problem concerning algebraic polynomials; in print in *Pacific J. Math.*

[3-F$_1$] R. P. Feinerman and D. J. Newman, *Polynomial Approximation*, Williams & Wilkins, Baltimore, 1974.

[3-F$_2$] L. B. O. Ferguson, *Approximation by Polynomials with Integral Coefficients* (Math. Surveys No. 17), Amer. Math. Soc., Providence, RI, 1980.

[3-G]   M. v. Golitschek and G. G. Lorentz, Trigonometric polynomials with many real zeros; preprint.

[3-I]   G. L. Iliev, Exact estimates for monotone interpolation, *J. Approx. Theory* **28** (1980), 101–112.

[3-K]   J. H. B. Kemperman and G. G. Lorentz, Bounds for polynomials with applications, *Indagationes Math.* **88** (1979), 13–26.

[3-L$_1$]   G. G. Lorentz, Monotone approximation. In *Inequalities*, Vol. III, pp. 201–215, Academic Press, New York, 1972.

[3-L$_2$]   G. G. Lorentz, Problems for incomplete polynomials. In *Approximation Theory*, Vol. III (E. W. Cheney, ed.), pp. 41–74, Academic Press, 1980.

[3-M]   H. N. Mhaskar and E. Saff, Extremal problems for polynomials with exponential weight; preprint.

[3-N]   D. J. Newman, Efficient co-monotone approximation, *J. Approx. Theory* **25** (1979), 189–192.

[3-S]   E. B. Saff and R. S. Varga, The sharpness of Lorentz' theorem on incomplete polynomials, *Trans. Amer. Math. Soc.* **249** (1975), 163–186.

[3-V]   A. K. Varma, Some inequalities of algebraic polynomials having all zeros inside $[-1, 1]$, *Proc. Amer. Math. Soc.* **75** (1979), 243–250.

§4.   POSITIVE OPERATORS; KOROVKIN THEOREMS; SPECIAL RESULTS

4.1.   Positive Operators

The most important special positive polynomial operators on $C[0,1]$ remain the Bernstein polynomials:

$$B_n(f, x) = \sum_{k=0}^{n} f\left(\frac{k}{n}\right) p_{nk}(x), \qquad p_{nk}(x) = \binom{n}{k} x^k (1-x)^{n-k}. \quad (4.1)$$

One obtains the Kantorovich polynomials K_n by replacing $f(k/n)$ by the average $(n+1)\int_{k/(n+1)}^{(k+1)/(n+1)} f$. Main results obtained are saturation theorems of B_n, K_n.

We say that a sequence of operators U_n on a Banach space E for which $U_n f \rightarrow f, f \in E$, has saturation if there exists a function $\varphi(n) \downarrow 0$ with the property that $\|U_n f - f\| = \mathcal{O}(\varphi(n))$ is true for a substantial set $S \subset E$, but $\|U_n f - f\| = o(\varphi(n))$ is true only for certain trivial functions f. Then S is called the saturation class. A sequence $\langle U_n \rangle$ of operators need not have saturation. If it does, the problem is to determine $\varphi(n)$ and S. For Bernstein polynomials, because of the influence of end points of $[0,1]$ the saturation condition must be modified to $|U_n f(x) - f(x)| \leqslant C\varphi(n, x)$. The optimal saturation theorem for B_n given in [O-L] now has much simpler proofs. One of them can be based on the curious formula of Averbach (see [1-K, p. 306]) for the difference $B_n f - B_{n+1} f$. The intermediate saturation theorem is as follows: f belongs to Lip* α, $0 < \alpha \leqslant 2$, if and only if for some M,

$|B_n(f, x) - f(x)| \leqslant M\delta_n(x)^{\alpha/2}$, $0 < \alpha \leqslant 2$, where $\delta_n(x) = x(1-x)/n$, and Lip* α is given by $\omega_2(f, h) \leqslant$ const. h^α. (See Berens and Lorentz [4-B$_3$]; Becker [4-B$_2$].)

Saturation theorems for K_n are due to Maier [4-M] for L_1 and to Riemenschneider [4-R] for L_p, $1 < p < +\infty$. They obtain that $\|f - K_n f\| = \mathcal{O}(n^{-1})$ holds for $f \in L_p$ if and only if one has the representation $f(x) = \text{const.} + \int_c^x t^{-1}(1-t)^{-1}h(t)\,dt$, where $h(0) = h(1) = 0$ and $h' \in L_p$, $1 < p < +\infty$, or $h \in V$ for $p = 1$. Many other saturation theorems for positive operators are given in the book of DeVore [4-D].

4.2. Korovkin's Theorems

In 1957 Korovkin discovered that for the convergence of a sequence U_n of positive linear operators to the identity operator, $U_n f \to f, f \in C[a, b]$, it is sufficient that this be true for the set of only three functions, $f(x) = 1$, x, and x^2. A similar set in $L_1[a, b]$ is given even by two functions, 1 and x. In general, let G be the linear hull of elements g_j, $j = 0, 1, \ldots, m$ (infinite sequences are also possible), in a Banach lattice E. Then the g_j form a Korovkin set for a class $\mathcal{U}$ of operators U_n if convergence $U_n g_j \to g_j$, $j = 0, 1, \ldots$, implies $U_n f \to f$ for all $f \in E$. If this is not the case, the *shadow* $\mathcal{S}(G)$ of G for operators $\mathcal{U}$ consists of all $f \in E$ with this property. The most important paper here belongs to Shashkin [4-S$_2$] (for its modern exposition, see [4-B$_6$]). Here $E = C(X)$, where X is a compact Hausdorff space. For a finite $\langle g_j \rangle_0^m$, Shashkin considers the embedding $\Phi: X \to \mathbb{R}^{m+1}$ given by $y_j = g_j(x), j = 0, \ldots, m$. Let K be the convex cone with vertex in the origin, spanned by $\Phi(X)$. Roughly speaking, $\{g_j\}$ is a Korovkin set for positive operators on $C(X)$ if and only if each $x \in X$ defines an extreme ray of K. In this geometric way one obtains many beautiful examples (for a list, see [4-B$_6$, §2]).

Let ε_x for $x \in X$ be the evaluation functional $\varepsilon_x(f) = f(x)$ on $C(X)$, let $\mathcal{L}_x(G)$ stand for all linear functionals L on $C(X)$ with the property $L(g) = g(x), g \in G$. Then G is a Korovkin set if and only if

$$\mathcal{L}_x(G) \text{ consists only of } \varepsilon_x. \tag{4.2}$$

Shashkin found that the definitions of this theory parallel those of Choquet boundaries. In fact, more general definitions are needed—for example, "quasi-peak points" instead of the more familiar peak points [4-K].

Another method was given in [4-B$_1$] and [4-B$_4$]. For $f \in C(X)$, its upper and lower envelopes $\bar{f}, \underline{f}$ are defined by

$$\bar{f}(x) = \inf\{g(x): g \in G, g \geqslant f\}; \qquad \underline{f}(x) = -\overline{(-f)}(x). \tag{4.3}$$

Then f belongs to $\mathcal{S}(G)$ if and only if $f = \bar{f}$. This has the advantage of having an extension to positive operators from $C(X)$ into Banach lattices E

of functions on X (see [4-B$_4$]). However, for positive operators mapping $L_p(X)$ into itself, the solution has been given only by Donner ([4-D$_1$, 4-D$_2$]). It is far from being simple; the necessity part depends on extension theorems for linear operators in L_p.

Still another method is lattice theoretic. The shadow of a set $G \subset L_p(X)$ with respect to all *positive contractions* on L_p is the linear closed lattice $\hat{G}$ in $L_p(X)$ generated by G (Berens and Lorentz [4-B$_5$]): $\mathcal{S}(G) = \hat{G}$.

Bernau [4-B$_7$] characterizes also the shadows of *contractions* of L_p in lattice-theoretic terms. All of these results are possible because Andô has described the range of a positive contraction in the spaces L_p.

4.3. Special Results

Approximation theory abounds in small brilliances that cannot be properly classified, some of which we give next. My selection is, of course, necessarily very personal; readers may find many other equally scintillating gems.

(A) From Bernstein's inequality it follows that for a polynomial P_n of degree n, in the uniform norm, $|P_n'(0)| \leq n\|P_n\|_{[-1,+1]}$. Buslaev [4-B$_8$] assumes that the real polynomial P_n has r zeros in the unit disk and $n - r$ zeros a_i outside. Then, with an absolute constant C,

$$|P_n'(0)| \leq C\left(r + 1 + \sum_{i=1}^{n-r} |a_i|^{-2}\right)\|P_n\|. \tag{4.4}$$

(B) Newman and Rivlin [4-N] approximate x^n uniformly in $C[0,1]$ by polynomials P_t of degree $t = t(n) < n$. The degree of approximation tends to zero for $n \to \infty$ if and only if $t(n)/\sqrt{n} \to +\infty$.

(C) Let $f \in C[a, b]$, and let $a \leq x_1^{(n)} < \cdots < x_{n+2}^{(n)} \leq b$ be the points of Chebyshev alternation for f and its polynomial of best approximation P_n. If $\Delta_n(f) = \max_k(x_{k+1}^{(n)} - x_k^{(n)})$, then

$$\liminf \Delta_n(f)\frac{n}{\log n} \leq 8(b - a). \tag{4.5}$$

In other words, for some arbitrarily large n, the points are at most $[8(b - a) + \varepsilon](\log n)/n$ apart (Tashev [4-T]).

(D) For Bernstein polynomials, $\|f - B_n f\|_\infty \leq C\omega(f, n^{-1/2})$. Here, Esseen [4-E] finds the asymptotically best constant C, that is, the number

$$C = \limsup_{n \to \infty} \left(\frac{\|f - B_n f\|}{\omega(f, n^{-1/2})}\right).$$

It is equal to

$$2 \sum_{\nu=0}^{\infty} (\nu+1)(\Phi_{2\nu+2} - \Phi_{2\nu})$$

where

$$\Phi_{\nu} = (2\pi)^{-1/2} \int_{-\infty}^{\nu} e^{-x^2/2} \, dx.$$

(E) Fiedler and Jurkat [4-F] deduce the exact formula $E_n(f)_1 = \int_{-1}^{+1} V_n(t) f^{(n)}(t) \, dt$, where $V_n(t)$ is expressible by means of the sign of the Chebyshev polynomial of the second kind; by means of an asymptotic formula for $V_n(t)$, they estimate $E_n(f)_1$. For example,

$$E_{2n-1}(e^{x^2})_1 = \frac{2\sqrt{e}}{4^n n!} (1 + \mathcal{O}(n^{-1})).$$

(F) If $\mathcal{T}_n^*$ are trigonometric polynomials that have all real zeros, then (Saff and Sheil-Small [4-S$_1$]) $\|T_n\|_1 \leqslant 4\|T_n\|_\infty$, $T_n \in \mathcal{T}_n^*$. For a similar inequality see Szabados [4-S$_3$].

REFERENCES

[4-B$_1$] H. Bauer, Theorems of Korovkin type for adopted spaces, *Ann. Inst. Fourier* **23** (1973), 245–260.

[4-B$_2$] M. Becker, An elementary proof of the inverse theorem for Bernstein polynomials, *Aequationes Math.* **19** (1979), 145–150.

[4-B$_3$] H. Berens and G. G. Lorentz, Inverse theorems for Bernstein polynomials, *Indiana Univ. Math. J.* **21** (1972), 693–708.

[4-B$_4$] H. Berens and G. G. Lorentz, Theorems of Korovkin type for positive linear operators on Banach lattices. In *Approximation Theory*, Vol. I (G. G. Lorentz, ed.), pp. 1–30, Academic Press, New York, 1973.

[4-B$_5$] H. Berens and G. G. Lorentz, Korovkin theorems for sequences of contractions on L^p-spaces. In *Linear Operators and Approximation* (P. L. Butzer and B. Sz.-Nagy, eds.), pp. 367–375, Birkhäuser-Verlag, Basel, 1974 (ISNM Vol. 25).

[4-B$_6$] H. Berens and G. G. Lorentz, Geometric theory of Korovkin sets, *J. Approx. Theory* **15** (1975), 161–189.

[4-B$_7$] S. J. Bernau, Theorems of Korovkin type for L_p spaces, *Pacific J. Math.* **53** (1974), 11–19.

[4-B$_8$] V. I. Buslaev, Estimation of the derivative of a polynomial with real coefficients, *Izv. Akad. Nauk SSSR* **39** (1975), 413–417.

[4-D] R. A. DeVore, *The Approximation of Continuous Functions by Positive Operators* (Lecture Notes in Mathematics, Vol. 293), Springer-Verlag, Berlin, 1972.

[4-D$_1$] K. Donner, Korovkin theorems in L^p-spaces, *J. Functional Anal.* **42** (1981), 12–28.

[4-D$_2$] K. Donner, *Extension of Positive Operators and Korovkin Theorems* (Lecture Notes in Mathematics, Vol. 904), Springer-Verlag, Berlin, 1982.

[4-E] C. G. Esseen, Über die asymptotisch beste Approximation stetiger Funktionen mit Hilfe von Bernstein-Polynomen, *Numer. Math.* **2** (1960), 206–213.

[4-F] H. Fiedler and W. B. Jurkat, Best L^1-approximation by polynomials, *J. Approx. Theory* (in press).

[4-K] S. S. Kutateladze and A. M. Rubinov, Minkowski duality and its applications, *Uspehi Mat. Nauk* **27** (1972), 127–176.

[4-M] V. Maier, The L_1-saturation class of the Kantorovič operator, *J. Approx. Theory* **22** (1978), 223–232.

[4-N] D. J. Newman and T. J. Rivlin, Approximation of monomials by lower degree polynomials, *Aequationes Math.* **14** (1976), 451–455.

[4-R] S. D. Riemenschneider, The L^p-saturation of the Bernstein–Kantorovitch polynomials, *J. Approx. Theory* **23** (1978), 158–162.

[4-S$_1$] E. B. Saff and T. Sheil-Small, Coefficient and integral mean estimates for algebraic and trigonometric polynomials with restricted zeros, *J. London Math. Soc.* **9** (1974), 16–22.

[4-S$_2$] Yu.A. Shashkin, Korovkin systems in spaces of continuous functions, *Izv. Akad. Nauk SSSR, Ser. Mat.* **26** (1962), 495–512.

[4-S$_3$] J. Szabados, On some extremum problems for polynomials. In *Approximation and Function Spaces* (Z. Ciesielski, ed.), pp. 731–748, North Holland, New York, 1981.

[4-T] S. Tashev, On the distribution of points of maximal deviation. In *Approximation and Function Spaces* (Z. Ciesielski, ed.), pp. 791–799, North Holland, New York, 1981.

§5. EXACT CONSTANTS

Particularly attractive are results in approximation theory that not only give asymptotic expressions of basic quantities, but also determine the exact constants. Many theorems of this type are known, particularly for functions on the circle.

An example is the inequality of Bernstein: For a trigonometric polynomial T_n of order $\leqslant n$, and the L_p norm, $1 \leqslant p \leqslant +\infty$, according to Zygmund,

$$\|T_n'\|_p \leqslant Cn\|T_n\|_p, \qquad (5.1)$$

and the best possible constant is $C = 1$. With some absolute constant, (5.1) was known also for $0 < p < 1$ (Máté and Nevai [5-M]), but Arestov [5-A] shows that here too we can take $C = 1$.

Another famous example is the theorem of Favard (see [0-L]). Let $W^rB^*(1) = W_r^*$, $r = 1, 2, \ldots$, be the class of functions on $\mathbb{T}$ with $|f^{(r)}(x)| \leqslant 1$ a.e. One of the forms of Jackson's theorem (2.1) is that $E_n^*(f) \leqslant C_r n^{-r}$ for

$f \in W_r^*$. Constants C_r that are computed by means of Jackson's integral are not the best ones; they increase exponentially to $+\infty$ with r. It is not possible to determine $E_n^*(f)$ exactly for an individual function $f \in W_r^*$. But we can try to find the degree of approximation of the whole class W_r^*, that is, the degree of approximation $E_n^*(f)$ of the worst function $f \in W_r^*$. Favard finds that

$$E_n^*(W_r^*) = \sup_{f \in W_r^*} E_n^*(f) = \mathcal{K}_r n^{-r}, \qquad r = 1, 2, \ldots, \qquad (5.2)$$

where

$$\mathcal{K}_r = \frac{4}{\pi} \sum_{k=0}^{\infty} \frac{(-1)^{k(r+1)}}{(2k+1)^{r+1}}. \qquad (5.3)$$

The proof of (5.2) is instructive: One represents $f \in W_r^*$ in the form of convolution

$$f = h * D_r,$$

$$D_r(x) = \frac{1}{\pi} \sum_{k=1}^{\infty} k^{-r} \cos\left(kx - \frac{r\pi}{2}\right), \qquad |h(x)| \leqslant 1 \text{ a.e.,} \qquad (5.4)$$

and then approximates h in the L_1 norm. Here, the D_r are periodic splines on $\mathbb{T}$.

    Intermediate between W_r^* and W_{r+1}^* is $W^r \text{ Lip}_1 \alpha$, or more generally, the sets $W^r H^\omega(1)$, where ω is a convex modulus of continuity. In his remarkable investigations, Korneichuk [0-K$_2$] finds the degree of approximation by trigonometric polynomials (in the L_∞ and L_1 norms) for these classes. He has, for instance,

$$E_n^*(W^r H^\omega(1))_\infty = n^{-r-1} \int_0^\pi \Phi_r(t) \omega'\left(\frac{t}{n}\right) dt$$

$$\leqslant \frac{1}{2} \mathcal{K}_r n^{-r} \omega\left(\frac{\eta_r}{n}\right), \qquad (5.5)$$

where Φ_r is a known polynomial for each r, and η_r are constants, $\eta_r \to 2$ for $r \to \infty$, with $\eta_1 = \frac{1}{2}\pi$, $\eta_2 = \frac{2}{3}\pi, \ldots$.

    The book [0-K$_3$] returns to the same subject with some even better proofs; the main point of it are analogues of his theorems (to (5.5), for example) for one-sided approximation in the L_1 norm.

    The original elementary proofs of Korneichuk applied only for $r = 1, 2, 3$. For $r > 3$, he uses his own "Σ-rearrangement" $\Sigma(f)(t)$ of a function f. This is first done for continuously differentiable $f \geqslant 0$. (See [0-K$_2$] for the definition.) At the Oberwolfach meeting in November 1981, Riemenschneider remarked that for such f, $\Sigma(f)$ can be defined by means of the K-functional of the spaces L_1 and V:

$$K\left(f, \frac{t}{2}; V, L_1\right) = \int_0^t \Sigma(f)(s) \, ds; \qquad (5.6)$$

the implications of this are not yet known. Korneichuk uses Σ-rearrangements to estimate integrals containing splines of the type $D_r(x)$.

There are useful inequalities that estimate the norm of an intermediate derivative $f^{(k)}$ in terms of the norms of the function f and of $f^{(r)}$, $0 < k < r$. If the functions are bounded on $(-\infty, +\infty)$, one has an inequality with the best possible constant $c_{r,k}$ (Kolmogorov, 1939):

$$\|f^{(k)}\|_{\infty} \leq c_{r,k} \|f\|_{\infty}^{1-k/r} \|f^{(r)}\|_{\infty}^{k/r}, \qquad c_{r,k} = \frac{\mathcal{K}_{r-k}}{\mathcal{K}_r^{1-k/r}}. \qquad (5.7)$$

(For a simple proof see Cavaretta [5-C].) There are several related results: Stein [5-S$_2$] proves (5.7) in the metric of L_1; Schoenberg and Cavaretta [5-S$_1$] establish (5.7), with different constants, for functions $f^{(k)}$, $k = 1, \ldots, r-1$, bounded on $(0, +\infty)$.

See also several results in §10 with exact constants.

REFERENCES

[5-A]   V. V. Arestov, On integral inequalities for trigonometric polynomials and their derivatives, *Izv. Akad. Nauk SSSR, Ser. Math.* **45** (1981), 3–22.

[5-C]   A. S. Cavaretta, Jr., An elementary proof of Kolmogorov's theorem, *Amer. Math. Monthly* **81** (1974), 480–486.

[5-M]   A. Máté and P. Nevai, Bernstein inequality in L^p for $0 < p < 1$ and $(C,1)$ bounds of orthogonal polynomials, *Ann. of Math.* **111** (1980), 145–154.

[5-S$_1$]   I. J. Schoenberg and A. Cavaretta, Solution of Landau's problem concerning higher derivatives on a halfline. In *Proceedings of the International Conference on Constructive Function Theory*, pp. 297–308, Bulgarian Acad. Sci., Sofia, 1972.

[5-S$_2$]   E. M. Stein, Functions of exponential type, *Ann. of Math.* **65** (1957), 582–592.

§6. SPLINES

Splines S are piecewise polynomials. Let the knots $X: -1 < x_1 < \cdots < x_n < 1$ lie in $[-1, +1]$. Then a spline S of order r (degree $r-1$) and the multiplicities k_i at the knots x_i has the form

$$S(x) = P_{r-1}(x) + \sum_{i=1}^{n} \sum_{k=1}^{k_i} a_{k,i}(x - x_i)_+^{r-k}. \qquad (6.1)$$

Thus, S is a polynomial of degree $\leq r-1$ on each interval (x_i, x_{i+1}), and is $r - k_i - 1$ times continuously differentiable at x_i. See the books of Schumaker [0-S] and de Boor [0-B$_1$]. [For the purpose of Birkhoff interpolation, more general splines are needed, with arbitrarily distributed powers of $(x - x_i)_+$.] The space $\mathcal{S}_r(X)$ of splines (6.1) has dimension $r + k_1 + \cdots + k_n$.

Systematic use of splines in approximation theory was begun around 1950 by Schoenberg; splines were known, however, much earlier. The Euler–Maclaurin formula contains splines. The solutions of many extremal problems are splines. This is true for the theorems of Favard and Korneichuk of §5, for the inequality of Kolmogorov (5.8). The extremal spaces in the definition of the widths $d_n(B)$ of the unit ball $B \subset C^r$ in the uniform norm also consist of splines. Another source of splines is kernels in functionals L in the space C^r. If this functional vanishes for all P_{r-1}, then one can often prove that for $f \in C^r[a, b]$,

$$L(f) = \frac{1}{(r-1)!} \int_a^b f^{(r)}(t) L\left([\cdot - t]_+^{r-1}\right) dt. \tag{6.2}$$

[Here L acts on $(x - t)_+^{r-1}$ as a function of x.]

Very useful are the *basic splines* (or B-splines) of Curry and Schoenberg. They provide a basis of $\bar{S}_r(X)$ of minimal support.

It is convenient to consider infinite sets of knots $\{x_i\}$, $x_{i+1} - x_i \geqslant 0$, $\lim_{i = \pm \infty} x_i = \pm \infty$, each x_i repeated k_i times. The basic splines of order r are the kernels $M_{i,r}(x)$ of the divided differences in the formula

$$[x_i, \ldots, x_{i+r}] f = \frac{1}{r!} \int_{x_i}^{x_{i+r}} M_{i,r}(t) f^{(r)}(t) \, dt; \tag{6.3}$$

they have the support $[x_i, x_{i+r}]$. They provide the representation

$$S(x) = \sum c_j(S) \frac{(x_{i+r} - x_i)}{r} M_{i,r}(x), \tag{6.4}$$

and there is a useful formula (based on an identity of Marsden and derived by de Boor and Fix) for the coefficients $c_j(S)$:

$$c_j(S) = \sum_{k=0}^{r-1} (-1)^{r-k-1} \psi_i^{(r-k-1)}(\xi_i) S^{(k)}(\xi_i), \tag{6.5}$$

where ξ_i is any point in (x_i, x_{i+r}), and

$$\psi_i(x) = \frac{1}{(r-1)!} (x_{i+1} - x) \cdots (x_{i+r-1} - x).$$

Formula (6.4) can be extended to functions f other than splines, for instance, to $f \in L_p$. It then gives a projection with uniformly bounded norm of the space L_p into $\bar{S}_r(X)$. See de Boor [6-B$_1$, 0-B$_1$]. The "fundamental theorem of algebra for monosplines" [6-K] gives an upper bound for the number of zeros of a spline. There are many theorems of this type; see this volume and the book [0-S].

Cardinal splines are defined on $(-\infty, +\infty)$ and have knots $\nu = 0, \pm 1, \ldots$. Schoenberg [6-S] assumes that the data y_ν satisfy $y_\nu = \mathcal{O}(|\nu|^\gamma)$, $\nu \to \pm \infty$, where $\gamma \geqslant 0$ is given, and looks for a spline of order r that solves the Lagrange interpolation problem $S(\nu) = y_\nu$, $\nu = 0, \pm 1, \ldots$. The main result is that there is a unique solution with the growth condition $S(x) = \mathcal{O}(|x|^\gamma)$, $x \to \pm\infty$; S decays exponentially if only finitely many y_ν are different from zero. There are many papers continuing this development.

Splines, in particular splines of low degree, provide a *very useful practical tool* of approximation, for instance, in curve fitting and finite element approximations. This gave rise to an enormous literature. Theoretical justification of the usefulness of splines is provided by, for instance, the Markov-type inequality $|S'(x)| \leqslant Mn\|S\|_\infty$ for equally spaced knots and a fixed order r. Here n is the number of knots. The corresponding inequality for polynomials of degree n has the factor n^2 instead of n.

In determining the degree of approximation of a smooth function f by splines, one often uses a theorem of Whitney, which estimates the degree of approximation $E_n(f)$ of f on the interval I by algebraic polynomials. Whereas Jackson's estimate (2.1) is useful when I is fixed and n is large, Whitney's theorem presupposes a fixed n and small interval I:

$$E_n(f)_p \leqslant C_n \omega_n(f, |I|)_p. \tag{6.6}$$

The approximation theorems that one obtains for spline approximation have the same order of convergence as the theorems (of Jackson type) for trigonometric approximation.

One obtains a better degree of approximation (which we denote by $\sigma_n(f)$), if one allows the n knots of the approximating spline to be *optimal*, best suited for a given function. Freud and Popov [6-F] show that $\sigma_n(f) = o(n^{-1})$ if $f \in \text{Lip } 1$. By a remark of Kahane, f is of bounded variation if and only if $\sigma_n(f) = O(n^{-1})$. Burchard and Hale [6-B$_3$, 6-B$_4$] and Peetre have tried to characterize functions f for which $\sigma_n(f) = O(n^{-\alpha})$.

There seems to exist a relation between the degrees of approximation by splines with variable knots and by rational functions. However, only partial results are known at present. See DeVore [6-D], where approximation of the first type is used to obtain results about the latter.

A famous problem of Banach was the question whether each separable Banach space possesses a basis $\{x_n\}_1^\infty$, that is, a sequence of elements so that each $x \in E$ has a unique representation of the form $x = \Sigma a_n x_n$; the basis is unconditional if this holds for any rearrangement of the series.

Since the negative answer to this by Enflo, the problem of finding bases in concrete spaces has become even more interesting. Ciesielski [6-C] finds bases of this type in spaces W_p^m (Sobolev spaces), Besov spaces, Hardy spaces, spaces of analytic functions of arbitrary dimension m. They can be obtained by successive integrations, starting with the Haar and the Franklin orthogonal systems or their multivariate analogues. Thus, these bases consist of splines.

Related is an unresolved conjecture of de Boor. In the space $\mathcal{S}_r(X)$ we introduce some orthonormal basis $\varphi_1, \ldots, \varphi_N$ and for $f \in L_2[-1, +1]$ the projector $P_X(f) = \Sigma_1^N(f, \varphi_j)\varphi_j$. The conjecture is that the norm of P_X as a map from $C[-1, +1]$ into L_∞ is uniformly bounded for all X. This is known for $r \leqslant 4$ (see de Boor [6-B$_2$]).

REFERENCES

[6-B$_1$]   C. de Boor, Splines as linear combinations of B-splines: A survey. In *Approximation Theory*, Vol. II (G. G. Lorentz, et al., eds.), pp. 1–47, Academic Press, New York 1976.

[6-B$_2$]   C. de Boor, On a max-norm bound for the least squares approximant. In *Approximation and Function Spaces* (Z. Ciesielski, ed.), pp. 163–175, North Holland, New York, 1981.

[6-B$_3$]   H. G. Burchard and D. F. Hale, Piecewise polynomial approximation on optimal meshes, *J. Approx. Theory* **14** (1975), 128–147.

[6-B$_4$]   H. G. Burchard, On the degree of convergence of piecewise polynomial approximation on optimal meshes, *Trans. Amer. Math. Soc.* **234** (1977), 531–559.

[6-C]   Z. Ciesielski, to appear in *Proceedings* of the Edmonton Conference, 1982.

[6-D]   R. A. DeVore, Maximal functions measuring smoothness and their role in approximation, to appear in *Proceedings* of the Edmonton Conference, 1982.

[6-F]   G. Freud and V. A. Popov, Certain questions connected with approximation by spline-functions and polynomials, *Studia Sci. Math. Hungar.* **5** (1970), 161–171.

[6-K]   S. Karlin and L. L. Schumaker, The fundamental theorem of algebra for Tchebycheffian monosplines, *J. Analyse Math.* **20** (1967), 233–270.

[6-S]   I. J. Schoenberg, Cardinal spline interpolation, *SIAM Reg. Conf. Ser. Appl. Math.* **12**, 1973.

§7. RATIONAL APPROXIMATION

The qualitative theory of rational approximation started early. The class $\mathcal{R}_{m,n} := \mathcal{R}_{m,n}[a, b]$ of rational functions consists of all $r = P/Q$ that, in the irreducible form, have $P \in \mathcal{P}_m$, $Q \in \mathcal{P}_n$, and $Q(x) > 0$ on $[a, b]$. We put $\mathcal{R}_n := \mathcal{R}_{n,n}$. Walsh proved in 1935 that each $f \in C[a, b]$ has a function $r \in \mathcal{R}_{m,n}$ of best uniform approximation. This is not quite obvious, for the familiar compactness argument cannot be used here. Ahiezer (1930) proved the uniqueness and alternation theorem: A function r is best approximation to $f \in C[a, b]$ if and only if $f - r$ has a Chebyshev alternance in $[a, b]$ consisting of $m + n - d(r) + 2$ points, where $d(r)$, the defect of r, is given by

$$d(r) = \min(m - \deg P, n - \deg Q).$$

A function $f \in C[a, b]$ is normal if $d(r) = 0$ for $r \in \mathcal{R}_{m,n}$ of best approximation.

Maehly and Witzgall (1960) proved that the operator $r = r(f)$ of best approximation is continuous at f if f is normal, and Werner (1964, [7-W]) showed that it is discontinuous at f in the opposite case.

An important source of functions in $\mathscr{R}_{m,n}$ that may approximate $f(x) = \sum_{k=0}^{\infty} c_k x^k$ are the Padé approximants $r_{m,n}(x) = P_m(x)/Q_n(x)$. They are uniquely determined by the condition that the first $n + m + 1$ coefficients of the power expansion of $f(x)Q_n(x) - P_m(x)$ should be zero (see the two volumes of this series [7-B$_1$]). Saff and Varga have investigated in depth the Padé approximants of the exponential, e^x.

In the quantitative theory, one tries to determine the degree of rational approximation,

$$R_n(f) = \min_{r_n \in \mathscr{R}_n} \|f - r_n\|.$$

Almost all of the following estimations are for $R_n(f)$ in the uniform norm. In the fifties and sixties there appeared several papers of Dolženko and Gončar devoted to the properties of rational functions (on the line $\mathbb{R}^1$ or in the complex plane) that are useful in approximation. For instance, if $\Sigma R_n(f, A) < +\infty$, $A \subset \mathbb{R}^1$, then $f'(x)$ exists a.e. on A (taken along A) [7-D$_1$]; if $|r_n(x)| \leqslant M$ on A, then for each $\delta > 0$ there is a subset $e \subset A$, $me < \delta$, for which $|r_n'(x)| \leqslant 2Mn/\delta$, $x \in A \backslash e$ [7-D$_2$]; for $A = [0,1]$, $\|r_n'\|_1 \leqslant 2n\|r\|_\infty$. The latest paper in this direction is by Fuchs and Gončar [7-F$_3$]. An important contribution of Newman [7-N$_1$] is an estimation of the degree of rational approximation of $|x|$ on $[-1, +1]$:

$$\tfrac{1}{2} e^{-9\sqrt{n}} \leqslant R_n[|x|] \leqslant 3e^{-\sqrt{n}} \tag{7.1}$$

(by comparison, for approximation by polynomials, $E_n(|x|) \approx n^{-1}$). The method of proof is to represent the approximating rational function by means of certain polynomials whose zeros are explicitly given. Newman's theorem has been improved by the Russian mathematicians Bulanov [7-B$_4$] and Vjačeslavov [7-V] to

$$e^{-\pi\sqrt{n+1}} \leqslant R_n(|x|) \leqslant Ane^{-\pi\sqrt{n}}, \tag{7.2}$$

where A is a constant.

Also, the approximation of x^α on $[0,1]$, where $\alpha > 0$ is not an integer, has been investigated with similar methods. Freud and Szabados [7-F$_2$] proved

$$e^{-c_1(\alpha)\sqrt{n}} \leqslant R_n(x^\alpha) \leqslant e^{-c_2(\alpha)\sqrt{n}}.$$

For polynomial approximation, one can estimate $E_n(f)$ in the following way. One approximates f by a polygonal line g; one expresses g as a linear combination of linear functions and functions $|x - c|$; one approximates $|x - c|$ by polynomials of degree $\leqslant n$ with error comparable to $E_n(|x|)$. In this way one obtains a single polynomial approximation to f. This does not work as well for rational approximation, for a sum of m rational functions of degrees $\leqslant n$ can be of degree mn.

Nevertheless, using Newman's estimate (7.1), Freud [7-F$_1$] obtained for $f \in \text{Lip } \alpha \cap V$ (where V are the functions of bounded variation)

$$R_n(f) \leqslant Cn^{-1}\log^2 n, \qquad (7.3)$$

which should be compared with $Cn^{-\alpha}$ for $E_n(f)$. The situation for Lip α is different—here both $E_n(f)$ and $R_n(f)$ can be of order $\geqslant n^{-\alpha}$ (Szabados, 1967 [7-S$_3$]). Apparently, good rational approximation is possible for a function $f(x)$ that behaves well at most x, badly only at exceptional points. This idea is also the method of proof of most of the following theorems. However, the selection of "good intervals" for a function of a given class is different in different cases, and depends on the oscillation properties, or the differentiability properties, or the associated maximal functions [6-D], and so on. Single points with bad behavior of f do not impair $R_n(f)$ too much; hence there are no inverse theorems of rational approximation.

Using deep and difficult methods, the Bulgarian mathematician Popov obtained important results for many function classes: (1) $R_n(f) = o(n^{-1})$ for each $f \in \text{Lip } 1$; (2) $R_n(f) = \mathcal{O}(n^{-2})$ if $f' \in V$ [7-P$_2$]; (3) $R_n(f) \leqslant$ const. $n^{-2}E_n(f'')_1$ if $f'' \in L_1$ [7-P$_3$]. Relation (1) could perhaps be expected, for functions $f \in \text{Lip } 1$ are good (= differentiable) almost everywhere. Statement (1) means that $R_n(f) = \varepsilon_n n^{-1}$ for some unknown sequence $\varepsilon_n \to 0$; for the numerical computation results of this type are useless. Statement (2) has been improved by Petrushev [7-P$_1$] to $R_n(f) = o(n^{-2})$. Recently Brudnyi [7-B$_3$] announced theorems that "almost" characterize the classes $R_{p,\alpha}$ of functions f for which $R_n(f)_p = \mathcal{O}(n^{-\alpha})$, $\alpha > 0$. They are related to the generalized Lipschitz spaces

$$\Lambda_{p,\alpha} = \left\{ f: \sup_{h>0}\left(h^{-\alpha}\|\Delta_h^k f\|_p \right) < +\infty \right\}, \qquad k = [\alpha]+1.$$

If one puts

$$\Lambda_{p-0,\alpha} = \bigcap_{p_1 < p} \Lambda_{p_1,\alpha}, \qquad \Lambda_{p+0,\alpha} = \bigcup_{p_2 > p} A_{p_2,\alpha},$$

then for $q = (\alpha + 1/p)^{-1}$,

$$\Lambda_{q+0,\alpha} \subset R_{p,\alpha} \subset \Lambda_{q-0,\alpha} \qquad (7.4)$$

(compare this with Vituškin's theorems in §8).

Finally, we note the amusing fact that there does not exist a Müntz theorem for rational functions: If $\lambda_i \geqslant 0$, $i = 1, 2, \ldots$, then each $f \in C[a, b]$ is uniformly approximable by a quotient of two polynomials of the form $\sum_{i=1}^n a_i x^{\lambda_i}$ (see Somorjai [7-S$_2$] and [7-N$_3$]).

Among special functions, approximation of the exponential e^{-x} attracted particular attention. Let $\lambda_{m,n}$ (or $\rho_{m,n}$) stand for the degree of approximation of e^{-x} on $[0, +\infty)$ (or $[-1, +1]$, respectively) by functions of the class $\mathfrak{R}_{m,n}$. One has

$$C_1^n \leqslant \lambda_{n,n} \leqslant C_2^n, \qquad 0 < C_1 < C_2 < +\infty$$

(Newman [7-N$_2$]; Rahman and Schmeisser [7-R]), while Schönhage [7-S$_1$] has $\sqrt[n]{\lambda_{0,n}} \to \frac{1}{3}$. The constants $\rho_{m,n}$ are known more precisely; one has the weak equivalence

$$\rho_{m,n} \approx \frac{m!n!2^{-m-n}}{(m+n)!\,(m+n+1)!} \quad \text{for} \quad m+n \to \infty.$$

The last result (Newman [7-N$_4$]; Braess [7-B$_2$]) has been achieved by studying the Padé approximant $r_{m,n}$ to e^{-x} in the complex plane.

REFERENCES

[7-B$_1$]   G. A. Baker, Jr., and P. Graves-Morris, *Padé Approximants*, Parts I and II, Vol. 13 and 14 of this *Encyclopedia*.

[7-B$_2$]   D. Braess, On the conjecture of Meinardus on rational approximation; preprint, Univ. of Bochum.

[7-B$_3$]   Yu. A. Brudnyi, Rational approximation and exotic Lipschitz spaces. In *Quantitative Approximation* (R. A. DeVore and K. Scherer, eds.), pp. 25–30, Academic Press, New York, 1980.

[7-B$_4$]   A. P. Bulanov, The asymptotics of the maximum derivation of $|x|$ from rational functions, *Mat. Sb.* **76** (1968), 288–303.

[7-D$_1$]   E. P. Dolženko, The rapidity of approximation by rational functions and properties of functions, *Mat. Sb.* **56** (1962), 403–432.

[7-D$_2$]   E. P. Dolženko, Estimates of derivatives of rational functions, *Izv. Akad. Nauk SSSR, Ser. Mat.* **27** (1963), 9–28.

[7-F$_1$]   G. Freud, Über die Approximation reeller Funktionen durch rationale gebrochene Funktionen, *Acta Math. Acad. Sci. Hungar.* **17** (1966), 313–324.

[7-F$_2$]   G. Freud and J. Szabados, Rational approximation to x^α, *Acta Math. Acad. Sci. Hungar.* **18** (1967), 393–399.

[7-F$_3$]   W. H. J. Fuchs and A. A. Gončar, Functions admitting very good rational approximations, to appear in the *Proceedings* of the Edmonton Conference, 1982.

[7-N$_1$]   D. J. Newman, Rational approximation to $|x|$, *Michigan Math. J.* **11** (1964), 11–14.

[7-N$_2$]   D. J. Newman, Rational approximation to e^{-x}, *J. Approx. Theory* **10** (1974), 301–303.

[7-N$_3$]   D. J. Newman, Approximation with rational functions, *SIAM Reg. Conf. Ser. Math.* **41**, 1978.

[7-N$_4$]   D. J. Newman, Rational approximation to e^x, *J. Approx. Theory* **27** (1979), 234–235.

[7-P$_1$]   P. P. Petrushev, Uniform rational approximation of the functions of the class V_r, *Mat. Sb.* **108** (1979), 418–432.

[7-P$_2$]   V. A. Popov, Uniform rational approximation of the class V_r and its applications, *Acta Math. Acad. Sci. Hungar.* **29** (1977), 119–129.

[7-P$_3$]   V. A. Popov, On the connection between rational uniform approximation and polynomial L_p approximation of functions. In *Quantitative Approximation* (R. A. DeVore and K. Scherer, eds.), pp. 267–278, Academic Press, New York, 1980.

[7-R]   Q. J. Rahman and G. Schmeisser, Rational approximation to e^x: II, *Trans. Amer. Math. Soc.* **235** (1978), 395–402.

[7-S$_1$]   A. Schönhage, Zur rationalen Approximierbarkiet von e^{-x} über $[0, \infty)$, *J. Approx. Theory* **7** (1973), 395–398.

[7-S$_2$]   G. Somorjai, A Müntz-type problem for rational approximation, *Acta Math. Acad. Sci. Hungar.* **27** (1976), 197–199.

[7-S$_3$]   J. Szabados, Negative results in the theory of rational approximation, *Studia Sci. Math. Hungar.* **2** (1967), 385–390.

[7-V]   N. S. Vyačeslavov, On the approximation of x^α by rational functions, *Izv. Akad. Nauk SSSR, Ser. Mat.* **44** (1980), 92–109.

[7-W]   H. Werner, On the rational Tchebycheff operator, *Math. Z.* **86** (1964), 317–326.

§8. NONLINEAR APPROXIMATION

8.1. General Theorems

Vituškin [8-V$_1$] has developed theorems that estimate from below the degree of approximation of functions by means of quite complicated expressions. Let

$$R(x, t) = \frac{P(x, t)}{Q(x, t)}, \qquad x \in [-1, +1], \quad t = (t_1, \ldots, t_n) \in \mathbb{R}^n, \quad (8.1)$$

where P, Q are polynomials in t of total degree $\leqslant p$, with coefficients that are given functions of x. One also considers piecewise expressions of this type, given by different formulas (8.1) in several regions $\Gamma_j(x)$ of the space of $(t_1, \ldots, t_n)$ for each x. We assume that the $\Gamma_j(x)$ are the components into which the surface $\tilde{P}(t, x) = 0$ decomposes $\mathbb{R}^n$, and that $\tilde{P}$ is a polynomial of degree q in t.

The approximate content of Vituškin's theorems is that for large classes of functions, approximation of this type is only marginally better than that by polynomials $t_1 + \cdots + t_n x^{n-1}$. The original proofs of Vituškin used his notion of the multidimensional variation of sets and were very complicated. Simple proofs and improvements of some (but not all) of his theorems were found by H. S. Shapiro and the present author (see Lorentz [8-L]).

One of his theorems concerns the approximation of the ball of the space of functions $f \in W_r^m$ (m is the dimension) with $D'f \in \text{Lip}\,\alpha$. If approximation by expressions (8.1) is possible with error $< \varepsilon$, then

$$n \log\left(p + q + \frac{1}{\varepsilon} \right) \geq \text{const.}\ \varepsilon^{-(r+\alpha)/m}. \qquad (8.2)$$

To give a special example, there is a function f of one variable with $\omega(f, h)_p \leq h^\alpha$ whose approximation by the expressions

$$M_{n,m}(x) := \max_{j=1,\ldots,m} \frac{t_{0j} + t_{1j}x + \cdots + t_{nj}x^n}{s_{0j} + s_{1j}x + \cdots + s_{nj}x^n} \qquad (8.3)$$

satisfies [8-L]

$$E_{n,m}(f)_p := \min_{M_{n,m}} \|f - M_{n,m}\|_p \geq Cn^{-\alpha}\log^{-\alpha}N, \qquad N = \max(m, n). \tag*{(8.4)}$$

8.2. Approximation in the Hausdorff Metric

If one interprets the functions f and its approximating polynomial P_n to be curves $C\colon y = f(x)$ and $C_n\colon y = P_n(x)$ in the plane, then the most natural measure of the degree of approximation is the Hausdorff distance $h(f, P_n) := L(C, C_n)$ between the curves. One can also approximate discontinuous functions by polynomials.

This point of view has been developed mainly by Bulgarian mathematicians (in particular, by Popov and Sendov; see Sendov [8-S]). Some of the theorems here parallel those of the uniform approximation; others are quite different. For instance, if $\tau(f, \delta)$ is the Hausdorff distance of the curves $y = f(x + \delta/2)$ and $y = f(x - \delta/2)$, then one has for the Bernstein polynomial B_n: If $\tau(f, \delta) \leq C\delta^\alpha$, then $h[f, B_n(f)] = \mathcal{O}(n^{-1}\log n)^{\alpha/2}$, with best order of approximation (Veselinov). Compare this with the fact that $\omega(f, \delta) \leq C\delta^\alpha$ implies $|f(x) - B_n(f)| \leq \text{const.}\ n^{-\alpha/2}$. For approximation with trigonometric polynomials, one has $h(f, T_n) \leq C\delta^\alpha$ if $\tau(f, \delta) \leq \text{const.}\ \delta^\alpha$, $0 < \alpha \leq 1$. But the inverse theorems have a different character. It is the constant in the inequality $h(f, T_n) \leq Cn^{-1}$ that determines the class of the function f. (Dolženko and Sevastjanov, 1976.)

8.3. Varisolvent Families; Exponential Sums

In 1960, Rice introduced the notion of *varisolvent families* $\mathcal{V}$ (see [8-B₃]), which were a generalization of the solvent families of Tornheim of 1950. The family of rational functions $\mathcal{R}_{m,n}$ is one of them. It was believed that there is a characterization of best approximation $v \in \mathcal{V}$ to $f \in C[a, b]$

based on sign alternation of $f - v$. This proved to be not quite true (Dunham; Bartke [8-B$_3$]), because it may happen that $f - v = \text{const.} \neq 0$, and some mild additional assumptions are necessary.

The *exponential sums* form the family

$$\mathcal{E}_n^0 = \left\{ h : h(x) = \sum_{k=1}^{n} a_k e^{\lambda_k x} \right\},$$

where a_k, λ_k are arbitrary reals. They have been investigated in several papers of Braess (for instance, [8-B$_1$, 8-B$_2$]). The set $\mathcal{E}_n^0$ is not closed—for instance, $x e^{\lambda x} = \lim_{h \to 0} [h^{-1}(e^{(\lambda + h)x} - e^{\lambda x})]$ does not belong to $\mathcal{E}_n^0$; the closure of $\mathcal{E}_n^0$ is

$$\mathcal{E}_n = \left\{ h : h(x) = \sum_{j=1}^{l} p_j(x) e^{\lambda_j x}, \Sigma \deg p_j \leqslant n \right\},$$

where p_j are arbitrary polynomials. Each $f \in C[a, b]$ has an $h \in \mathcal{E}_n$ of uniform best approximation, but it is not necessarily unique. Moreover, there may exist local best approximations. Braess has found $\frac{1}{2}(n - 1)!$ to be an upper bound of the number of the latter (for given n, and each $f \in C[a, b]$). According to Verführth [8-V$_2$], some $f \in C[a, b]$ can have as much as const. $n^{-1/2} 2^{(2/3)n}$ local best approximants. Thus, the qualitative theory of approximation for $\mathcal{E}_n$ is not as simple as that for $\mathcal{P}_n$. Strangely enough, nothing seems to be known from the quantitative theory.

REFERENCES

[8-B$_1$] D. Braess, Chebyshev approximation by γ-polynomials: I, II, *J. Approx. Theory* **9** (1973), 20–43; **11** (1974), 16–37.

[8-B$_2$] D. Braess, Global analysis in nonlinear approximation and its application to exponential approximation, 1, 2. In *Approximation Theory and Applications* (Z. Ziegler, ed.), pp. 23–38 and 39–64, Academic Press, New York, 1981.

[8-B$_3$] K. Bartke, Eine varisolvente Familie, welche das Phänomen der konstanten Fehlerkurve zulässt, *J. Approx. Theory* **24** (1978), 324–329.

[8-L] G. G. Lorentz, Metric entropy and approximation, *Bull. Amer. Math. Soc.* **72** (1966), 903–937.

[8-S] B. Sendov, *Hausdorff Approximations* (in Russian), Bulgarian Acad. Sci., Sofia, 1979. (A German translation will appear in the Springer Lecture Notes series.)

[8-V$_1$] A. G. Vituškin, *Theory of Transmission and Processing of Information*, Pergamon Press, New York, 1961. (In spite of the title, this book is only about approximation.)

[8-V$_2$] R. Verführth, On the number of local best approximations by exponential sums, *J. Approx. Theory* **34** (1982), 306–323.

§9. INTERPOLATION; QUADRATURE FORMULAS; COMPLEX APPROXIMATION

In *interpolation* (other than Birkhoff interpolation), we have two important developments. Let $-1 \leqslant x_1^{(n)} < \cdots < x_n^{(n)} \leqslant 1$, $n = 1, 2, \ldots$, be a matrix of knots of Lagrange interpolation. In 1936–1937, Grünwald and Marcinkiewicz proved independently that in the case when the knots are zeros of the nth Chebyshev polynomial, there exists a function $f \in C[-1, +1]$ whose Lagrange interpolation polynomial $L_n(f)$ diverges everywhere. Since the Chebyshev knots are in many respects the best possible, it has been expected that the result must be true for any sets of knots. To establish this proved to be very difficult.

Only in 1980 did Erdös and Vértesi [9-E] show that for any set of knots, $L_n(f, x)$ will diverge a.e. for a proper $f \in C[-1, +1]$. One cannot assert divergence everywhere; if the knots repeat themselves, $x_k^{(n)} = x_k$, $k = 1, \ldots, n$, $n = 1, \ldots$, one has the identity $L_n(f, x_k) = f(x_k)$ for each k and each large n.

This paper proved to have a predecessor in that of N. B. Tihomirov (not V. M. Tihomirov!) [9-T$_1$]. This 1977 paper, well hidden in a Russian collection of articles on differential geometry (!), was not yet available to me when this review was written. Reportedly, Tihomirov proves the Erdös–Vértesi theorem under the assumption that all $x_k^{(n)}$ are different.

Fixing n, let $X: -1 = x_0 < \cdots < x_{n-1} < x_n = 1$ be the knots of a Lagrange interpolation formula $L_n(X)(f)(x) = \sum_0^n f(x_i) l_i(x)$, where $l_i(x)$ are the fundamental polynomials of this interpolation. Then $L_n(X)$ is a linear operator of $C[-1, +1]$ into itself.

By a reasonable definition, the best formula corresponds to a set of knots X^* for which the norm of the operator, $\|L_n(X)\|$, is minimal. It is well known that this norm is equal to $\Lambda_n(X) = \max_x \sum_{i=0}^n |l_i(x)|$ and hence to $\max_i \lambda_{n,i}$, where

$$\lambda_{n,i} := \lambda_{n,i}(X) = \max_{x_i \leqslant x \leqslant x_{i+1}} \sum_0^n |l_i(x)|.$$

A very old conjecture that the best set X^* is unique and that it is characterized by the conditions $\lambda_{n,i} = \Lambda_n$, $i = 1, \ldots, n$, has been proved by Kilgore [9-K] and de Boor and Pinkus [9-B]. The nature of the optimal set X^* remains a mystery. The roots of the Chebyshev polynomials provide only asymptotically the smallest norm.

We mention two review articles: by Nevai [9-N], devoted to the divergence and convergence of interpolation at roots of orthogonal polynomials; and by Turán [9-T$_2$], which contains the history of the subject and some interesting open problems. We can leave the questions of interpolation at this point, since P. Nevai is writing a monograph on this subject for the *Encyclopedia*.

As regards *quadrature formulas*, interest has recently shifted to the Birkhoff quadrature formulas (see Chapter 11 of the present volume) and to the multidimensional or cubature formulas. Compare the paper of Schmid [9-S$_1$] and the recent book of Mysovskih [9-M].

We cannot discuss the subject of *complex approximation* here; we shall mention only the excellent books of Gaier [0-G], Smirnov and Lebedev [9-S$_2$], and Varga [9-V], and Chapter 9 of the book of Dzyadyk [0-D]. Most of the papers of Saff and Varga also belong to this field.

REFERENCES

[9-B]   C. de Boor and A. Pinkus, Proof of the conjectures of Bernstein and Erdös concerning the optimal nodes for polynomial interpolation, *J. Approx. Theory* **24** (1978), 289–303.

[9-E]   P. Erdös and P. Vértesi, On the almost everywhere divergence of Lagrange interpolatory polynomials for arbitrary system of nodes, *Acta Math. Acad. Sci. Hungar.* **36** (1980), 71–89; **38** (1981), 263

[9-K]   T. A. Kilgore, A characterization of Lagrange interpolating projections with minimal Tchebycheff norm, *J. Approx. Theory* **24** (1978), 273–288.

[9-M]   I. P. Mysovskih, *Interpolatory Quadrature Formulas*, Izd. Nauka, Moscow, 1981.

[9-N]   P. Nevai, Lagrange interpolation at zeros of orthogonal polynomials. In *Approximation Theory*, Vol. II (G. G. Lorentz et al., eds.), pp. 163–201, Academic Press, New York, 1976.

[9-S$_1$]   H. L. Schmid, Interpolatory cubature formulae and real ideals. In *Quantitative Approximation* (R. A. DeVore and K. Scherer, eds.), pp. 245–254, Academic Press, New York, 1980.

[9-S$_2$]   V. I. Smirnov and N. A. Lebedev, *Constructive Theory of Functions of a Complex Variable*, Izd. Nauka, Moscow, 1964.

[9-T$_1$]   N. B. Tihomirov, On the divergence of interpolation processes. In *Differential Geometry*, pp. 104–111, Kalinin, 1977.

[9-T$_2$]   P. Turán, On some open problems in approximation theory, *J. Approx. Theory* **29** (1980), 23–89.

[9-T$_3$]   A. Kh. Tureckiĭ, *Theory of Interpolation in Problems*, Vols. 1 and 2, Izd. Vyš. Škola, Minsk, 1968 and 1977.

[9-V]   R. S. Varga, *Topics in Polynomial and Rational Interpolation and Approximation*, Université de Montréal, Montreal, 1982.

§10. WIDTHS AND ENTROPY

Kolmogoroff introduced these two notions in 1936 and 1956 in order to measure the "size," the "massiviness' of relatively small (for instance, compact) sets A in a Banach space E. Tihomirov in his book [0-T$_1$] gives some fifteen definitions of n-width (where $n = 1, 2, \ldots$). We give four of them, for a central-symmetric set $A \subset E$ (for which $f \in A$ implies $-f \in A$). Kolmogorov's n-width is given by

$$d_n(A, E) = \inf_{X_n} \sup_{f \in A} \rho(f, X_n), \qquad (10.1)$$

where the X_n are all possible linear subspaces of E of dimension n. Here $\rho(f, X_n) = E_{X_n}(f)$ is the degree of approximation of f by X_n, and $\sup_{f \in A} E_{X_n}(f) = E_{X_n}(A)$ is the degree of approximation of the set A. A linear subspace X_n is extreme if the infimum in (10.1) is attained for it.

The linear n-width $\lambda_n(A, E)$ is defined by means of linear bounded operators L that map E into a subspace X_n of dimension n:

$$\lambda_n(A, E) = \inf_{X_n} \inf_{L(E) \subset X_n} \sup_{f \in A} \| f - Lf \|. \qquad (10.2)$$

Then we have the entropy width:

$$\varepsilon_n(A, E) = \inf_{\Sigma_n} \sup_{f \in A} \inf_{g \in \Sigma_n} \| f - g \|, \qquad (10.3)$$

where Σ_n is any set of n points of E. Relation $\varepsilon_n(A, E) < \varepsilon$ means that A can be covered by n balls of radii $\leqslant \varepsilon$. The inverse of $\varepsilon_n(A, E)$ is the function $N_\varepsilon(A)$, and $\log N_\varepsilon(A)$ is the entropy of A (see later).

Finally,

$$d^n(A, E) = \inf_{X^n} \sup_{f \in A \cap X^n} \| f \|, \qquad (10.4)$$

where the X^n are subspaces of E of codimension n, is the Gelfand width. It is used in the problem of best coding methods and the recovery of functions and their derivatives.

Trivially,

$$d_n(A, E) \leqslant \lambda_n(A, E), \qquad d_n(A, E) \leqslant \varepsilon_n(A, E), \qquad (10.5)$$

in a Hilbert space E, $d_n(A) = \lambda_n(A) = d^n(A)$. One might expect these widths to be, at least asymptotically, all equal, and equal to the degree of approximation of A, let us say, by polynomials of degree $n - 1$. But this is true only in simple cases.

An important and far from trivial result is one by Gohberg and Krein (1957), namely, that if $X_n \subset X_{n+1} \subset E$, and if U_{n+1} is the unit ball of X_{n+1}, then $d_n(U_{n+1}, E) = 1$; this helps many investigations. A very important tool is the computation of widths of some elementary sets (ellipsoids, octahedra) in finite-dimensional spaces. Many of the more difficult results have been obtained in this way. In 1936, Kolmogorov expressed the widths $d_n(W_2^r(1))_2$ of the unit ball of the space W_2^r in L_2 by means of the eigenvalues of certain differential operators. Later, in 1960–1966 Tihomirov ([0-T$_1$], [10-T]) and others computed the widths of several function sets. For instance, they have the weak equivalence

$$d_n\big(W_p^r(1)\big)_q \approx \begin{cases} n^{-r} & \text{for} \quad p \geqslant q, \\ n^{-r+(1/q-1/p)} & \text{for} \quad 1 \leqslant p < q \leqslant 2. \end{cases} \qquad (10.6)$$

(The case when $p=1$, $q=2$ is an early result of Rudin [10-R].) For the periodic case Tihomirov has the exact formula

$$d_{2n-1}\big(W_\infty^{*r}(1)\big)_\infty = d_{2n} = \lambda_{2n-1} = \lambda_{2n} = \mathcal{K}_r n^{-r}, \qquad (10.7)$$

with the Favard constant (5.3). Here, and in most of what follows, the extreme spaces are either the spaces $\mathcal{T}_{2n-1}$ or spaces of splines with equidistant knots. Sometimes even a linear operator that achieves extremal degree of approximation is known. In such cases, the estimation of d_n from below becomes the most difficult part of the theorem. For instance, for (10.7), certain subspaces of C^* with an interpolation property and ideas related to Favard's theorem of §5 are used. Exact determinations of widths are possible mostly only for periodic functions. We mention a few further theorems:

$$d_{2n}\big(W_1^{*r}\big) = d_{2n-1}\big(W_1^{*r}\big) = d_{2n-1}\big(W^{*r-1}V\big) = \mathcal{K}_r n^{-r}, \quad r = 1, 2, \ldots$$

in the norm of the space L_1^* (Makovoz [10-M$_1$]), and for $H^\omega[0,1]$,

$$d_n\big(H^\omega, L_p\big) = \frac{1}{2}\left(n\int_0^{1/n} \omega(t)^p\, dt\right)^{1/p}, \qquad 1 \leqslant p \leqslant 3, \quad n = 1, 2, \ldots,$$

where ω is a concave modulus of continuity (Korneichuk [10-K$_3$]).

    In the following papers, exact values of widths are given, which are not immediately numerical. Micchelli and Pinkus [10-M$_2$, 10-M$_3$] give the n-widths of a set spanned by functions $k_1(t), \ldots, k_n(t)$ and $\int_0^1 K(x,t)h(t)\, dt$, where $h \in L_\infty$ or $h \in L_1$ with $\|h\| \leqslant 1$; some properties of total positivity of the $k_i(t)$, $K(x,t)$ must be imposed. Fisher and Micchelli [10-F$_1$] show that the n-width of the restriction of the ball of the space H_1 to a compact subset of the disk is equal to the infimum of norms of certain Blaschke products. Blaschke products also appear in another paper of the last two authors about optimization problems for holomorphic functions [10-F$_2$].

    In the paper [10-K$_3$], (see also [10-K$_2$]) Kashin settled the problem of n-widths for the smoothness spaces W_p^r in arbitrary L_q. For $rp > 1$,

$$d_n\big(W_p^r, L_q\big) \approx \begin{cases} n^{-r} & \text{if} \quad p \geqslant q \text{ or } 2 < p < q; \\ n^{-r-1/2+1/p} & \text{if} \quad p \leqslant 2 < q \\ n^{-r-1/q+1/p} & \text{if} \quad 1 \leqslant p < q \leqslant 2. \end{cases} \qquad (10.8)$$

The cases $p \geqslant q$ and $1 \leqslant p < q \leqslant 2$ are trivial in the sense that the width coincides with the degree of approximation of W_p^r in L_q by trigonometric polynomials. Kashin obtains the other cases of (10.8) by a probabilistic method; first he reduces the problem to the existence of a large $m \times n$ matrix with elements $+1$, -1 satisfying certain conditions. Kashin also announces interesting theorems about general orthogonal series $\Sigma a_n \phi_n(x)$ in L_2, which he apparently also obtains with probabilistic methods. Kashin's results are probably the most important recent achievements in the theory of widths.

For a compact set A in a metric space E there exists, for each $\varepsilon > 0$, the smallest number $N_\varepsilon(A)$ of balls of radius $\leqslant \varepsilon$ that cover A, and the largest number $M_\varepsilon(A)$ of points y_i in a finite set $\Sigma \subset A$ with $\rho(y_i, y_j) > \varepsilon$, $i \neq j$. Then $H_\varepsilon(A) = \log N_\varepsilon(A)$ is the entropy of A in E, and $C_\varepsilon(A) = \log M_\varepsilon(A)$ is the capacity of A. (The first definition, generally accepted now, is a slight simplification of Kolmogorov's original definition.) The inequality $C_{2\varepsilon}(A) \leqslant H_\varepsilon(A)$ is often used to estimate $H_\varepsilon(A)$ from below. For functions of s variables, Kolmogorov found the entropy of the unit ball of $W^r H^\alpha$ in the uniform norm, and Vituškin did the same for some sets of analytic functions.

For a Banach space, let the sequence of linearly independent elements $\Phi = \{\phi_k\}_1^\infty$ span E and let $\Delta = \{\delta_k\}_0^\infty$ satisfy $\delta_k \downarrow 0$. The full approximation set $A := A(\Phi, \Delta)$ consists of all $f \in E$ that satisfy $E_{X_n}(f) \leqslant \delta_n$, $n = 0, 1, \ldots$, where X_n is spanned by $\phi_1, \ldots, \phi_n$. It is possible to determine the entropy of A in a way that does not depend on the metric of E or of the selected Φ, but is expressed solely in terms of the δ_n (Lorentz [10-L]). This uses some geometric properties of finite-dimensional Banach spaces (properties of ellipsoids, e.g.). Normally, a set $B \subset E$ is not a full approximation set, but it can often be embedded between two such sets,

$$A(\Phi, C_1 \Delta) \subset B \subset A(\Phi, C_2 \Delta), \qquad 0 < C_1 < C_2 < +\infty. \qquad (10.9)$$

In this way one gets the results of Kolmogorov and Vituškin, in the uniform and the L_p norms.

The later paper of Birman and Solomjak [10-B] does not fit into this scheme because of the presence of two different norms. They found

$$H_\varepsilon\left(W_p^r(1)\right)_q \approx \left(\frac{1}{\varepsilon}\right)^{r/s} \qquad (10.10)$$

(where s is the dimension). This expression is independent of p, q. One should consult the interesting paper of Höllig [10-H], where the quantities d_n, λ_n, ε_n are computed for balls in Besov spaces, and another proof of (10.10) is given.

REFERENCES

[10-B]   M. S. Birman and M. Z. Solomjak, Piecewise polynomial approximation of functions of classes W_p^α, Mat. Sb. 73 (1967), 295–317.

[10-F$_1$] S. D. Fisher and C. A. Micchelli, The n-widths of sets of analytic functions, *Duke Math. J.* **47** (1980), 789–801.

[10-F$_2$] S. D. Fisher and C. A. Micchelli, Optimal sampling of holomorphic functions, to appear in *Amer. J. Math.*

[10-H] K. Höllig, Diameters of classes of smooth functions. In *Quantitative Approximation* (R. A. DeVore and K. Scherer, eds.), pp. 163–176, Academic Press, New York, 1980.

[10-K$_1$] B. S. Kashin, The width of certain finite dimensional sets and classes of smooth functions, *Izv. Akad. Nauk SSSR* **41** (1977), 334–351.

[10-K$_2$] B. S. Kashin, On estimates of diameters. In *Quantitative Approximation*, (R. A. DeVore and K. Scherer, eds.), pp. 177–184, Academic Press, New York, 1980.

[10-K$_3$] N. P. Korneichuk, Diameters and optimal recovery. In *Approximation and Function Spaces* (Z. Ciesielski, ed.), pp. 355–370, North Holland, New York, 1981.

[10-L] G. G. Lorentz, Metric entropy and approximation, *Bull. Amer. Math. Soc.* **72** (1966), 903–937.

[10-M$_1$] Yu. I. Makovoz, On a method of estimation from below of widths of sets in Banach spaces, *Mat. Sb.* **87** (1972), 136–142.

[10-M$_2$] C. A. Micchelli and A. Pinkus, Total positivity and the exact n-widths of certain sets in L^1, *Pacific J. Math.* **71** (1977), 499–515.

[10-M$_3$] C. A. Micchelli and A. Pinkus, On n-widths in L^∞, *Trans. Amer. Math. Soc.* **24** (1978), 51–77.

[10-R] W. Rudin, L_2-approximation by partial sums of orthogonal developments, *Duke Math. J.* **19** (1952), 1–4.

[10-T] V. M. Tihomirov, Widths of sets in functional spaces and the theory of best approximation, *Uspehi Mat. Nauk* **15** (1960), 81–120.

§11. MULTIVARIATE APPROXIMATION

11.1. Polynomial Approximation and Interpolation

We approximate here with polynomials P of *total degree* r, $r = 0, 1, \ldots,$ in the variable $\mathbf{x} \in \mathbb{R}^n$,

$$P(\mathbf{x}) = \sum_{|\mathbf{k}| \leqslant n} a_{\mathbf{k}} \mathbf{x}^{\mathbf{k}}, \qquad \mathbf{k} = (k_1, \ldots, k_n), \qquad |\mathbf{k}| = k_1 + \cdots + k_n$$

(11.1)

with $\mathbf{x}^{\mathbf{k}} = x_1^{k_1} \cdots x_n^{k_n}$. It has $\binom{n+r}{n}$ free coefficients.

One of the basic statements is here (Sobolev [11-S])

$$\|f - P\|_p \leqslant \text{const.} \sum_{|\mathbf{k}| = r} \|D^{\mathbf{k}} f\|_p \delta, \qquad f \in L_p(\Omega),$$

(11.2)

where the derivatives are taken in the distributional sense and Ω is a cube of side length δ. Brudnyi (1970, 1977) proved a similar estimate with $\omega_r(f, \delta)_p$ on the right and for polynomials P of *coordinate degree* $\mathbf{r} = (r_1, \ldots, r_n)$, where

$k_i \leqslant r_i$, $i = 1, \ldots, n$, in (11.1) Dahmen, DeVore, and Scherer [11-D] treat all intermediate cases between the two definitions of the degree; in particular, they settle the question which derivatives then appear in (11.2).

Interpolation by polynomials of several variables has been neglected for a long time. The reason for this is obvious. Since there are no Chebyshev systems in the space $\mathbb{R}^n$, $n \geqslant 2$, the Lagrange interpolation problem

$$P(\mathbf{x}_j) = f(\mathbf{x}_j), \qquad j = 1, \ldots, \binom{n+r}{n}, \tag{11.3}$$

must be unsolvable for some f and some selection of the knots $X = \{\mathbf{x}_j\}$. For such X, the Lagrange interpolation problem is not regular (or not poised, or not correct). In Kergin interpolation (briefly discussed in §2.5 of this book; see also Micchelli [11-M]), this is overcome. In addition to (11.3), some derivatives are matched with those of f in some not completely specified points. On the contrary, Hakopian [11-H] assumes that the $\mathbf{x}_j$ are in general position, in other words, that a certain determinant does not vanish. Much clarification is still required here.

11.2. Multivariate Splines

At present, multivariate splines constitute the most intriguing and fastest-growing chapter of approximation theory. The development began only a few years ago (with Kergin interpolation as one of the catalysts), and led to a large literature, at this time mostly unpublished (but often available as reports from Math. Research Center (MRC), University of Wisconsin (Madison), IBM Research Institute, Yorktown Heights, N.Y., Center for Approximation Theory, Texas A&M University, College Station, TX). The main concern seems to be the definition and properties of B-splines. Several groups are at work at present: de Boor, Höllig, DeVore, and others; Micchelli, Dahmen, and others; Chui, Schumaker, and Wang; and Goodman at the University of Dundee. I have decided not to review this field, but to refer the reader to a masterly (and at this time fairly complete) exposition of it in de Boor [11-B].

REFERENCES

[11-B]   C. de Boor, *Topics in Multivariate Approximation Theory*, MRC Report 2379, 1982.

[11-D]   W. Dahmen, R. DeVore, and K. Scherer, Multidimensional spline approximation, *SIAM J. Numer. Anal.* **17** (1980), 380–402.

[11-H]   H. Hakopian, Multivariate divided differences and multivariate interpolation of Lagrange and Hermite type, *J. Approx. Theory* **34** (1982), 286–305.

[11-M]   C. A. Micchelli, A constructive approach to Kergin interpolation in $\mathbb{R}^k$: Multivariate B-splines and Lagrange interpolation, *Rocky Mountain J. Math.* **10** (1980), 485–497.

[11-S]   S. L. Sobolev, *Applications of Functional Analysis in Mathematical Physics* (Transl. Math. Monogr. No. 7), Amer. Math. Soc., Providence, RI, 1963.

§12. APPROXIMATION IN BANACH SPACES

Let M be a subset of a Banach space E. If $x \notin M$, we denote by $P_M(x)$ the set of all $y \in M$ with $\|x - y\| = \rho(x, M)$. This is the metric projection onto M; $P_M(x)$ is a subset that may be empty. If $P_M(x)$ is never empty, M is called proximinal; if $P_M(x)$ always consists of a single point, then M is called a Chebyshev subset of E. The problem is to describe all proximinal sets and all Chebyshev sets. Another problem is to decide when $P_M(x)$ is continuous.

These problems are, in general, very difficult. As a guiding example we can take the following result. In an n-dimensional Banach space E_n, the class of all Chebyshev sets coincides with that of all closed convex sets if and only if E_n is strictly convex and smooth.

To facilitate the problem, one introduces the following definition. A subset M of a Banach space is a *sun* if it is proximinal and if for each $x \notin M$ there is a $y \in P_M x$ with the property that all points of the ray from y through x in E have y as one of their best approximants in M. Suns can also be described by means of a condition similar to Kolmogorov's criterion for an element y of best approximation (see §1). One seeks relations among the notions: (a) M is convex; (b) M is a sun; (c) M is a Chebyshev subset of E. We list some of the most striking ones: (a) $\Rightarrow$ (b); (b) $\Rightarrow$ (a) if M is proximinal and E is smooth. Also, (b) $\Rightarrow$ (c) if E is strictly convex; (c) does not always imply (b). Finally, (c) $\Rightarrow$ (a) if E is smooth and M is boundedly compact. But many problems remain open, for example, the question whether each Chebyshev set in a Hilbert space is necessarily convex.

Concerning the continuity of P_M we have: If E is uniformly convex, or more generally if it belongs to the class of spaces of Fan and Glicksberg [12-F$_1$], then the metric projection $P_M x$ on a closed and convex set M is single valued for each $x \notin M$ and continuous.

Several mathematicians—Efimov, Klee, Stečkin, Vlasov—contributed to this theory, and later Amir, Blatter, Brosowski, Deutsch, Dunham, Oshman, Wulbert, and Ziegler, too. See the reviews of Vlasov [12-V] and, for the continuity of P_M, of Deutsch [12-D$_1$].

If P_M is not a point-to-point map, one can consider a continuous (or even a linear continuous) selection of P_M. See §1 for an example. For some general theorems, see Deutsch and Kenderov [12-D$_2$], also [12-D$_3$].

If M is a linear subspace of E and P_M is not linear, one can try to approximate it by a linear projection Π_M of E onto M. Here we have the Lebesgue inequality

$$\|x - \Pi_M x\| \leqslant (1 + \|\Pi_M\|) E_M(x). \tag{12.1}$$

Therefore, it is natural to consider a minimal projection, one with smallest norm $\|\Pi_M\|$. If M is n-dimensional, there always exists a projection of norm $\leqslant \sqrt{n}$ (Kadec and Snobar; see Chalmers [12-C$_2$]).

To find a minimal projection is in general very difficult. It has been known for a long time that for $E = C^*$ and $M = \mathfrak{I}_n$, a minimal projection is given by the Fourier partial sum $s_n(f)$ of $f \in C^*$ (see [0-L, p. 98]). That this is unique has been shown by five authors (Cheney et al., [12-C$_1$]). There is a similar result by Fisher, Morris, and Wulbert [12-F$_2$].

An interesting new problem of approximation theory is that of *optimal recovery*, which is formulated by Micchelli and Rivlin [12-M] as follows. Let X, Y, Z be linear normed spaces, let $K \subset X$, and let U be a given linear operator from X to Z. We wish to estimate Ux for $x \in K$ on the basis of the following information. There is another linear operator V from X to Y, and the value of Vx, $x \in K$, is known with an error $\leqslant \varepsilon$, so that for each $x \in K$ an element $y \in Y$ is known so that $\|Vx - y\|_Y \leqslant \varepsilon$. To determine Ux, we take some operator A from Y to Z, called an algorithm, and get Ux with an error not exceeding

$$E_A = \sup_{x \in K} \langle \|Ux - Ay\| : x \in K, \|Ix - y\| \leqslant \varepsilon \rangle.$$

The problem is to find the smallest possible E_A, and the corresponding A—the optimal algorithm. Apparently there is no general theory, but many interesting examples; see also [12-N].

REFERENCES

[12-C$_1$] E. W. Cheney, C. R. Hobby, P. D. Morris, F. Schurer, and D. E. Wulbert, On the minimal property of the Fourier projection, *Trans. Amer. Math. Soc.* **143** (1969), 249–258.

[12-C$_2$] B. L. Chalmers, A natural simple projection with norm $\leqslant \sqrt{n}$, *J. Approx. Theory* **32** (1981), 226–232.

[12-D$_1$] F. Deutsch, Existence of best approximation, *J. Approx. Theory* **28** (1980), 132–154.

[12-D$_2$] F. Deutsch and P. Kenderov, When does the metric projection admit a linear selection? In *Approximation Theory* Vol. III (E. W. Cheney, ed.) pp. 327–333, Academic Press, New York, 1980.

[12-D$_3$] F. Deutsch, Linear selections for the metric projection, *J. Funct. Analysis*, in print.

[12-F$_1$] K. Fan and I. Glicksberg, Some geometric properties of a sphere in a normed linear space, *Duke Math. J.* **25** (1958), 553–568.

[12-F$_2$] S. D. Fischer, P. D. Morris, and D. E. Wulbert, Unique minimality of Fourier projections, *Trans. Amer. Math. Soc.*, in press.

[12-M] C. A. Micchelli and T. J. Rivlin, A survey of optimal recovery in *Optimal Estimations in Approximation Theory*, pp. 1–54, Plenum Press, New York, 1977.

[12-N] D. J. Newman and T. J. Rivlin, Optimal recovery among polynomials. In *Approximation Theory and Applications* (Z. Ziegler, ed.), pp. 291–302, Academic Press, 1981.

[12-V] L. P. Vlasov, Approximate properties of sets in normed linear spaces, *Russian Math. Surveys* **28** (1973), 1–66.

Sept. 1982   G. G. Lorentz

Chapter 1

Basic Definitions and Properties

§1.1. INTRODUCTION

In the problem of interpolation of given data $c_{i,k}$ by a polynomial P_n of degree at most n,

$$P_n^{(k)}(x_i) = c_{i,k} \qquad (n+1 \text{ equations}), \qquad (1.1.1)$$

two cases should be considered. We have *Hermite interpolation* if for each i, the orders k of the derivatives in (1.1.1) form an unbroken sequence, $k = 0, 1, \ldots, k_i$. (The special case of Lagrange interpolation obtains when all k_i are 0.) Here the interpolation polynomial always exists, is unique, and can be given by an explicit formula.

If some of the sequences are broken, we have the *lacunary* or *Birkhoff interpolation*. The two cases are as different as, let us say, the theory of linear and nonlinear differential equations. It is, however, convenient to consider Hermite interpolation to be a special case of lacunary interpolation.

Pairs (i, k) which appear in (1.1.1) are best described by means of the *interpolation matrix* $E = [e_{i,k}]_{i=1, k=0}^{m, \ n}$ of the problem. We put $e_{i,k} = 1$ if the pair (i, k) appears in (1.1.1), and $e_{i,k} = 0$ if it does not. In 1966 Schoenberg (see [151]) posed the problem of determining all those E for which the problem (1.1.1) is always (i.e., for all choices of $x_i, c_{i,k}$) solvable. We call such matrices E *regular* and the remaining matrices *singular*. Since Schoenberg's paper, regularity and singularity have been the emphasis of the research. But problem (1.1.1) poses interesting challenges for fixed knots x_i, also. G. D. Birkhoff himself (at the age of 18, in the only paper [7] that he wrote on the subject) discussed the remainder of the interpolation, assuming a priori that problem (1.1.1) is solvable. In this way he obtained his theorem about the sign of the kernel of the interpolation formula; its generalization proved to be important for the general theory. Since 1955, Turán, his students, and his followers have studied cases in which regularity is enforced by a special choice of the knots x_i; they are usually related to zeros of certain orthogonal polynomials. Chapters 11 and 12 contain examples of these investigations. Except perhaps some of the results in Chapter 11, no general theorems have been obtained here.

First papers of the general theory by Atkinson and Sharma and D. Ferguson gave the basic sufficient conditions for regularity. Surprisingly, no significant improvements of their conditions have been found since, in spite of the invention of new methods: coalescence of rows (due to Karlin and Karon and to Ferguson) and independent knots. Interest shifted to singularity criteria: several were given by Lorentz, by Karlin and Karon, and by others. In Chapter 8, with DeVore, Meir and Sharma, and others, we study special three-row matrices. What can be done here is a good illustration of the general case: Only sometimes can regularity or singularity be decided in simple terms; in general we are thrown back to a study of zeros of polynomials of one or several variables. Chapters 9 and 10 give applications of lacunary interpolation to approximation by polynomials, and to quadrature formulas. Finally, in Chapters 13 and 14, we discuss interpolation by splines rather than polynomials. Here our new set of definitions and some new proofs will help, we hope, to promote better understanding of this subject.

§1.2. BASIC DEFINITIONS

Let $\mathcal{G} = \{g_0, g_1, \ldots, g_N\}$ be a system of linearly independent, n times continuously differentiable real-valued functions on a set A that is either an interval $[a, b]$ or the circle $\mathbf{T}$. A linear combination $P = \sum_{j=0}^{N} a_j g_j$ with real a_j will be called a *polynomial* in the system $\mathcal{G}$.

We can also consider complex-valued functions g_j on a subset A of the complex plane, and polynomials P with complex coefficients.

A matrix

$$E = [e_{i,k}]_{i=1, k=0}^{m \quad n}, \qquad m \geqslant 1, \quad n \geqslant 0, \qquad (1.2.1)$$

is an *interpolation matrix* if its elements $e_{i,k}$ are 0 or 1 and if the number of 1's in E is equal to $N + 1$, $|E| = \sum e_{i,k} = N + 1$. In general we do not allow empty rows, that is, an i for which $e_{i,k} = 0$, $k = 0, \ldots, n$. A *set of knots* $X = \{x_1, \ldots, x_m\}$ consists of m distinct points of the set A. The elements E,

X, $\mathcal{G}$, and the data $c_{i,k}$ (defined for $e_{i,k} = 1$) determine a *Birkhoff interpola-tion problem* which consists in finding a polynomial P satisfying

$$P^{(k)}(x_i) = c_{i,k}, \qquad e_{i,k} = 1. \qquad (1.2.2)$$

The system (1.2.2) consists of $N+1$ linear equations with $N+1$ unknowns a_j. The pair E, X is called *regular* if equations (1.2.2) have a unique solution for each given set of $c_{i,k}$; otherwise, the pair E, X is *singular*. A pair E, X is regular if and only if the determinant of the system

$$D(E, X) = \det\left[g_0^{(k)}(x_i), \ldots, g_N^{(k)}(x_i); e_{i,k} = 1 \right] \qquad (1.2.3)$$

is different from 0. Formula (1.2.3) displays only one row of the determi-nant, namely, the row corresponding to a pair (i, k) with $e_{i,k} = 1$. We order the pairs in (1.2.3) *lexicographically*: the pair (i, k) precedes (i', k') if and only if $i < i'$, or $i = i'$ and $k < k'$. By $A(E, X)$ we denote the $(N+1) \times (N+1)$ matrix that appears in (1.2.3).

A basic notion, due to I. J. Schoenberg [151], is that of a *poised* or *regular* interpolation matrix. A matrix E has this property if the pair E, X is regular for each set of knots X of a given class. Several types of regularity can be considered. In particular, E is *order regular* if X is allowed to be an arbitrary ordered subset $x_1 < x_2 < \cdots < x_m$ of $[a, b]$ or $\mathbf{T}$. (For the circle $\mathbf{T}$ these inequalities mean that points $x_1, x_2, \ldots, x_m, x_1$ are encountered in this order as we move along $\mathbf{T}$ counterclockwise.) The matrix E is *real regular* (or *complex regular*) if E, X is regular for any set of distinct real (or complex) knots X. For the real knots, we distinguish [26] between *strong singularity*, when $D(E, X)$ takes values of different sign, and *weak singularity*, when $D(E, X)$ vanishes without change of sign.

A matrix is singular if and only if some nontrivial polynomial P is *annihilated by* E, X for some admissible X; this means that P satisfies the homogeneous equations $P^{(k)}(x_i) = 0$ for $e_{i,k} = 1$. For a singular matrix we consider $r(E)$, the lowest possible rank of the matrix $A(E, X)$. Then

$$d = d(E) = N + 1 - r(E), \qquad (1.2.4)$$

the defect of E, is the largest possible dimension of the subspace of polynomials P annihilated by E, X for some X.

Originally, the problem under discussion has been to find all order regular interpolation matrices E in a simple way by the properties of their entries $e_{i,k}$. This problem proved to be exceedingly difficult, and only for some related problems (complex regularity, Theorem 4.10; conditional regularity, Theorem 1.3) have simple answers been found.

Examples. A *Lagrange interpolation matrix* has $m = N + 1$, $n = 0$; it is a column of 1's. Equations (1.2.2) become here $P(x_i) = c_i$, $i = 1, \ldots, N + 1$. Obviously, we have regularity for a system of functions $\mathcal{G}$ if and only if each nontrivial polynomial P has at most N distinct zeros in A. By a theorem of Haar [I, p. 27], this happens if and only if each continuous function on A has a unique polynomial of best approximation. A *Taylor matrix* E is a $1 \times (n+1)$ row of 1's; here the equations (1.2.2) are $P^{(k)}(x_1) = c_k$, $k = 0, \ldots, n$. An *Abel matrix* is an $(n+1) \times (n+1)$ matrix with $N = n$ and exactly one 1 in every row and every column.

We call a row i of the matrix E *Hermitian* if for some r_i, $e_{i,k} = 1$ for $k < r_i$, and $e_{i,k} = 0$ for $k \geqslant r_i$. A *Hermitian matrix* E has only Hermitian rows. A Hermitian matrix is regular, at least for interpolation by algebraic polynomials.

For a *quasi-Hermitian matrix*, only the interior rows, $1 < i < m$, are required to be Hermitian.

The following lemma is useful in deciding the strong singularity of E.

Lemma 1.1. *Let $\Phi(X)$ be a continuous function of $X = \{x_1, \ldots, x_m\}$ on the simplex $T: a \leqslant x_1 < x_2 < \cdots < x_m \leqslant b$, which changes sign and vanishes only on a nowhere-dense set in T. Then Φ changes sign in one of the variables x_i.*

Proof. Let $\Phi(X^0) < 0 < \Phi(Y^0)$ for $X^0 = \{x_1^0, \ldots, x_m^0\}$, $Y^0 = \{y_1^0, \ldots, y_m^0\}$. First, let $x_i^0 < y_i^0$, $i = 1, \ldots, m$. Then the same inequalities hold for X^0, Y^0 replaced by $X \in U$, $Y \in V$ where U and V are neighborhoods of X^0, Y^0, respectively. The functions Φ_r, $r = 0, \ldots, m$, defined on $U \times V$ by $\Phi_r(X, Y) = \Phi(\{x_1, \ldots, x_r, y_{r+1}, \ldots, y_m\})$ are continuous and vanish only on a nowhere-dense set of $U \times V$; hence the same is true for $\Psi = \prod_{r=0}^{m} \Phi_r$. Thus, there are points $X \in U$, $Y \in V$ such that $\Phi_r(X, Y) \neq 0$, $r = 0, \ldots, m$. Since Φ_0 and Φ_m have opposite signs, there is a first r with $\Phi_r \Phi_{r+1} < 0$, and Φ changes sign in x_{r+1}.

In the general case, let $z_i = \min(x_i^0, y_i^0) - \varepsilon$, $\varepsilon > 0$ small, $Z = \{z_1, \ldots, z_m\}$. By changing X^0 or Y^0 slightly, we can assume that $Z \in T$ and $\Phi(Z) \neq 0$. Then the preceding argument can be applied to one of the pairs $\Phi(X), \Phi(Z)$, or $\Phi(Z), \Phi(Y)$. $\square$

§1.3. THE ALGEBRAIC CASE

In particular, let $A = [a, b]$ and let $\mathcal{G}$ be the powers

$$g_0(x) = 1, \quad g_1(x) = \frac{x}{1!}, \quad \cdots, \quad g_N(x) = \frac{x^N}{N!}; \qquad (1.3.1)$$

then P is an algebraic polynomial of degree $\leqslant N$. Since $P^{(k)} \equiv 0$ for $k > N$, we can assume that the interpolation matrix (1.2.1) satisfies $n \leqslant N$, and by adding zero columns if necessary, we can make $n = N$. We shall call an interpolation matrix *normal* if it has as many 1's as columns.

For a normal matrix and the system (1.3.1) with $N = n$, the determinant (1.2.3) is equal to

$$D(E, X) = \det\left[\frac{x_i^{-k}}{(-k)!}, \ldots, \frac{x_i^{n-k}}{(n-k)!}\,; e_{i, k} = 1\right] \qquad (1.3.2)$$

if we agree to replace $1/r!$ by 0 if $r < 0$. Thus, the (i, k)th row of $D(E, X)$ contains 0's in the first k positions. The determinant is a polynomial in the variables x_i. If $D(E, X)$ is not identically 0, the set of points $X = \{x_1, \ldots, x_m\}$ in $\mathbf{R}^m$ where it vanishes is a closed nowhere-dense set of measure 0. Lemma 1.1 is applicable; if E is strongly (order) singular, then $D(E, X)$ changes sign for one of the variables x_i if the others are properly fixed.

For arbitrary a, α we write $aX = \{ax_1, \ldots, ax_m\}$; $X + \alpha = \{x_1 + \alpha, \ldots, x_m + \alpha\}$, where $X = \{x_1, \ldots, x_m\}$.

Proposition 1.2. *The determinant $D(E, X)$ is a homogeneous polynomial in $x_1, \ldots, x_m$ of total degree*

$$1 + 2 + \cdots + n - \sum_{e_{i, k} = 1} k = \rho \qquad (1.3.3)$$

and satisfies, for real α and $a \geq 0$,

$$D(E, aX) = a^\rho D(E, X), \qquad D(E, X + \alpha) = D(E, X). \qquad (1.3.4)$$

[If $\rho < 0$, $D(E, X)$ is identically 0.]

Proof. The first formula is obtained by taking out the factor a^{-k} from the row (i, k) of $D(E, aX)$, and then the factor a^r from the rth column of the resulting determinant.

For the second formula, we note that $D(E, X + \alpha)$ is a polynomial in α. Let us compute its derivative. The derivative of each of the columns of this determinant is the preceding column, and the derivative of the first column is 0. Hence $D(E, X + \alpha)$ is independent of α. $\qquad\square$

The degree of $D(E, X)$ in the separate variables x_i will be found in Chapter 3.

As a corollary we obtain that under linear transformations, the determinant is multiplied by a nonzero constant. Hence, the properties of regularity and strong or weak singularity of E do not depend on the choice of the interval $[a, b]$. Also, we can always assume, in dealing with regularity questions, that x_1, x_m of X are prescribed numbers. This applies also to complex knots X. In particular: A two-row interpolation matrix is regular if and only if it is real regular.

The algebraic complements of the elements of the last column of the determinant (1.3.2) are denoted by $D_{i, k}(X)$. They differ only by a sign from $D(E_{i, k}, X)$, where the matrix $E_{i, k}$ is derived from E by replacing $e_{i, k} = 1$ by

0 and by omitting the last column of E. Applying formula (1.3.4) to minors, we obtain

$$D(E, X) = \det\left[\frac{(x_i + \alpha)^{-k}}{(-k)!}, \ldots, \frac{(x_i + \alpha)^{n-k-1}}{(n-k-1)!}, \frac{x_i^{n-k}}{(n-k)!}; e_{i,k} = 1\right].$$

$$(1.3.5)$$

By induction in n we obtain a more general identity

$$D(E, X) = \det\left[\frac{(x_i + \alpha_0)^{-k}}{(-k)!}, \ldots, \frac{(x_i + \alpha_{n-1})^{n-k-1}}{(n-k-1)!},\right.$$

$$\left.\frac{(x_i + \alpha_n)^{n-k}}{(n-k)!}; e_{i,k} = 1\right], \qquad (1.3.6)$$

where $\alpha_0, \ldots, \alpha_n$ are arbitrary constants.

Examples. The normal $(n+1) \times (n+1)$ matrix E of Lagrange interpolation has 1's in positions $e_{i,0} = 1$, $i = 1, \ldots, n+1$. The determinant (1.3.2) is the Vandermonde determinant

$$\det\left[1, \frac{x_i}{1!}, \ldots, \frac{x_i^n}{n!}; i = 1, \ldots, n+1\right] = \text{const} \prod_{i > j} (x_i - x_j) \neq 0.$$

The $1 \times (n+1)$ Taylor matrix is also regular, with the determinant (1.3.2) $D(E, X) = 1$. For an arbitrary system $\mathcal{G}$, and $X = \{x\}$, the determinant for the Taylor matrix is the Wronskian

$$D(E, X) = W(x) = \det\left[g_0^{(k)}(x), \ldots, g_n^{(k)}(x); k = 0, \ldots, n\right].$$

By a direct computation of $D(E, X)$, with $X = \{0, x, 1\}$ and $0 < x < 1$, we see that in

$$E_1 = \begin{bmatrix} 1 & 0 & 0 \\ 0 & 1 & 0 \\ 1 & 0 & 0 \end{bmatrix}, \qquad E_2 = \begin{bmatrix} 1 & 1 & 0 & 0 & 0 \\ 0 & 1 & 0 & 1 & 0 \\ 1 & 0 & 0 & 0 & 0 \end{bmatrix},$$

$$E_3 = \begin{bmatrix} 1 & 1 & 0 & 0 & 0 & 0 \\ 0 & 1 & 0 & 0 & 1 & 0 \\ 1 & 1 & 0 & 0 & 0 & 0 \end{bmatrix}, \qquad (1.3.7)$$

the matrix E_1 is strongly singular, the matrix E_2 is weakly singular, and the matrix E_3 is regular in spite of the two odd supported sequences (see §1.5) that it contains. We have, for example,

$$D(E_3, X) = \text{const } x(x-1)(15x^2 - 15x + 4) \neq 0 \quad \text{if} \quad 0 < x < 1.$$

§1.4. PÓLYA AND BIRKHOFF MATRICES

Let E be an $m \times (n+1)$ interpolation matrix. Then $m_k = \Sigma_i e_{i,k}$ is the number of 1's in column k, and

$$M_r = \sum_{k=0}^{r} m_k = \sum_{k=0}^{r} \sum_{i=1}^{m} e_{i,k} \qquad (1.4.1)$$

is the number of 1's in columns of E numbered $0, 1, \ldots, r$. For normal matrices, the condition

$$M_r \geqslant r+1, \qquad r = 0, 1, \ldots, n, \qquad (1.4.2)$$

is called the *Pólya condition*. Then, automatically, $M_n = n+1$. Subtracting from this the inequality (1.4.2), we see that for normal matrices, (1.4.2) is equivalent to

$$\sum_{k=r+1}^{n} m_k \leqslant n - r, \qquad r = 0, 1, \ldots, n. \qquad (1.4.3)$$

This inequality we call the *upper Pólya condition*; it is useful for matrices with less than $n+1$ ones; condition (1.4.2) is not adequate in this case. We have then that a submatrix E_1 of E satisfies the condition (1.4.3) if E does. (A submatrix is obtained from E by omitting some 1's and some empty rows.)

Exercise. Let l_k, $k = 1, \ldots, n$, be the total number of 1's in all sequences of 1's of E, which begin in the kth column. Then $L_k = l_0 + \cdots + l_k$ is the number of 1's that are in the columns $0, \ldots, k$ or belong to a sequence beginning in one of these columns. Prove that the condition $L_r \geqslant r+1$, $r = 0, \ldots, n$, is equivalent to (1.4.2).

We show that the Pólya condition is *necessary* [151] for (real, order or complex) regularity of a pair E, X with respect to the powers (1.3.1). Indeed, if (1.4.2) is not satisfied, then $M_r \leqslant r$ for some $r \leqslant n$. Let P_r be a polynomial of degree $\leqslant r$; then automatically $P_r^{(k)}(x_i) = 0$ for $k > r$. On the other hand, there are at most r equations $P_r^{(k)}(x_i) = 0$, $e_{i,k} = 1$ with $k \leqslant r$. Since any r homogeneous linear equations with $r+1$ unknowns have a nontrivial solution, there exists a nontrivial polynomial of degree r, $r \leqslant n$, annihilated by E, X.

Another proof of our statement is based on the remark that a row (i, k) of the determinant (1.3.2) begins with k zeros. If $M_r \leqslant r$, there are at most r rows (i, k) with $k \leqslant r$. Consequently, all but at most r rows begin with $r+1$ zeros. The first $r+1$ columns of $D(E, X)$ are linearly dependent, and we have $D(E, X) = 0$.

It may be difficult to decide when a matrix E is regular, that is, when $D(E, X) \neq 0$ *for all X*. A simpler question is when E is *conditionally regular*.

This means that $D(E, X)$ is not identically 0, or $D(E, X) \neq 0$ *for some X* (for algebraic interpolation this means that $D(E, X) \neq 0$ except for a set of measure 0 in $\mathbf{R}^m$). We have just proved part (i) of this theorem:

Theorem 1.3 (Ferguson [37]; Nemeth [128]).   (i) *A normal interpolation matrix that is conditionally regular for algebraic interpolation must satisfy the Pólya condition.*

(ii) *If the Wronskian $W(x)$ of the system $\mathcal{G}$ is not identically zero, and E satisfies the Pólya condition, then E is conditionally regular. In particular: A normal matrix is conditionally regular for algebraic interpolation if and only if it satisfies the Pólya condition.*

An attempt to prove (ii) is already in G. D. Birkhoff [7], but his proof is incorrect. D. Ferguson and Nemeth proved (ii) for algebraic interpolation. For a simple proof of (ii) see Chapter 4.

A normal matrix satisfies the *Birkhoff condition* if

$$M_r \geqslant r + 2, \qquad r = 0, 1, \ldots, n - 1; \qquad (1.4.4)$$

this condition appears explicitly in [7, p. 113]. We assume that this condition is trivially satisfied if $n = 0$. If $n \geqslant 1$, the condition implies that column 0 of E has at least two 1's, and that $M_{n-1} = M_n = n + 1$, hence that the last column consists of 0's.

A normal matrix that satisfies the Pólya or the Birkhoff condition will be called simply a *Pólya* or a *Birkhoff matrix*, respectively. From a Birkhoff matrix E an $m \times n$ Pólya matrix results if a 1 in E is replaced by 0 and the last column is omitted.

A normal interpolation matrix E is *decomposable* [162], $E = E_1 \oplus E_2$, if it can be split vertically into two normal matrices E_1, E_2. Thus, if E_1 has $r + 1$ columns, then they must be the first $r + 1$ columns of E and contain exactly $r + 1$ ones, while E_2 consists of the last $n - r$ columns of E with $n - r$ ones. This means that $M_r = r + 1$. Hence: *A Pólya matrix E is indecomposable if and only if it satisfies the Birkhoff condition.*

For $E = E_1 \oplus E_2$, let us consider the matrix $A(E, X)$ for the system of powers. We rearrange the rows of this matrix to have as the first $r + 1$ rows those corresponding to $e_{i,k} = 1$ from E_1, in their lexicographic order, then rows corresponding to 1's from E_2. This transforms $A(E, X)$ into

$$\left[\begin{array}{c|c} A(E_1, X) & * \\ \hline 0 & A(E_2, X) \end{array} \right] \begin{array}{l} \} \ r+1 \\ \} \ n-r \end{array} \qquad (1.4.5)$$
$$\underbrace{}_{r+1} \ \underbrace{}_{n-r}$$

Let X_1, X_2 be the subsets of X significant for E_1, E_2. It follows that

$$D(E, X) = \pm D(E_1, X_1) D(E_2, X_2). \qquad (1.4.6)$$

As a corollary we obtain, for order, real, or complex regularity or singularity:

Theorem 1.4 (Atkinson-Sharma [2]; Ferguson [37]). *A decomposable matrix $E = E_1 \oplus E_2$ is regular for algebraic interpolation if and only if each of the components E_1, E_2 is regular. It is strongly (weakly) singular if one of the matrices E_1, E_2 is regular and the other is strongly (weakly) singular.*

Another corollary of (1.4.5) is

$$\operatorname{rank} A(E, X) \geqslant \operatorname{rank} A(E_1, X_1) + \operatorname{rank} A(E_2, X_2). \qquad (1.4.7)$$

Of importance is the *canonical decomposition* of a Pólya matrix E:

$$E = E_1 \oplus \cdots \oplus E_\mu. \qquad (1.4.8)$$

This is the finest possible decomposition of E; it is obviously unique.

Using (1.4.7) and the definition of the defect (1.2.4) we have, for each set X,

$$\operatorname{rank} A(E, X) \geqslant \sum_{\lambda=1}^{\mu} \operatorname{rank} A(E_\lambda, X_\lambda) \geqslant (n+1) - \sum_{\lambda=1}^{\mu} d(E_\lambda).$$

Hence for the defects of the matrices E and E_λ we obtain

$$d(E) \leqslant \sum_{\lambda=1}^{\mu} d(E_\lambda). \qquad (1.4.9)$$

Examples

1. The Abel $(n+1) \times (n+1)$ interpolation matrix E has only a single 1 in each column; its canonical decomposition consists of one-column normal matrices; and it is regular by Theorem 1.4.

2. For polynomials P of degree $n = mp - 1$, we prescribe the derivatives $P^{(sp)}(x_i)$, $s = 0, \ldots, m-1$, at p distinct points (Poritsky [139]). The interpolation matrix is regular, since the components of its canonical decomposition are $p \times p$ Lagrange matrices.

§1.5. REGULARITY THEOREMS

To establish order regularity of a matrix E, we need a new condition. Let

$$1 \quad 1 \quad 1 \quad 0 \quad 0 \quad 1 \quad 1 \quad 0 \quad 1 \quad 0 \quad 0$$

be one of the rows of E. It contains three *sequences* (*of 1's*). Two of them are *odd* (have an odd number of elements); one is even. The first of them is a *Hermite sequence*, since it begins in column 0. A sequence of the ith row of E is *supported* if there are 1's in E to both the northwest and the southwest of its first element. More precisely, if (i, k) is the position of the first 1 of the sequence, there must be 1's $e_{i_1, k_1} = e_{i_2, k_2} = 1$ with $i_1 < i < i_2$, $k_1 < k$, $k_2 < k$. Thus, no sequence in the first or the last row of E, nor any sequence with its first element in the 0th column can be supported. Even

Birkhoff knew the significance of *odd supported sequences*, calling a matrix *conservative* if it did not contain them.

The following fundamental theorem was first proved by K. Atkinson and A. Sharma [2]. The paper of Ferguson [37] also contains ideas that are sufficient to give this theorem. It has since been found that the theorem also follows from Birkhoff's results about his kernels (see Chapter 7).

Theorem 1.5. *A normal interpolation matrix is order regular for algebraic interpolation if it satisfies the Pólya condition and contains no odd supported sequences.*

We shall see that this theorem is nothing but a sophisticated form of Rolle's theorem. The following lemmas are formulated for an analytic function f defined on $[a, b]$.

Lemma 1.6. *Between two adjacent zeros of f there is (counting multiplicities) an odd number of zeros of f'.*

Proof. Let $\alpha < \beta$ be the zeros. On (α, β) the function f is of constant sign; let, for example, $f(x) > 0$. Then necessarily $f'(x) > 0$ for x close to α, $x > \alpha$; and $f'(x) < 0$ for x close to β, $x < \beta$. Thus, f' changes sign on (α, β). At its zeros of odd or of even order, f' does or does not change sign, respectively. Hence f' has an odd number of zeros of odd order on (α, β). $\square$

A function f is *annihilated by* an $m \times (n+1)$ interpolation matrix E and a set of knots X if f satisfies the homogeneous equations

$$f^{(k)}(x_i) = 0, \qquad e_{i,k} = 1. \qquad (1.5.1)$$

Lemma 1.7. *Let f be annihilated by E, X with $|E| \geqslant n + 1$ and $X \subset [a, b]$. If E has no odd supported sequences, then either*

$$f^{(n)}(\xi) = 0 \qquad for\ some \quad x_1 < \xi < x_m, \qquad (1.5.2)$$

or there is a $\mu \times n$ interpolation matrix E' with at least n 1's and no odd supported sequences and a set $X' = \{x'_1, \dots, x'_\mu\}$, $x_1 \leqslant x'_1$, $x'_\mu \leqslant x_m$, in $[a, b]$ that annihilates the derivative f'.

Proof. Assume that (1.5.2) is not true. If column 0 of E has at least two 1's, we denote by α, β the extreme zeros of f in $[a, b]$ belonging to X. We can assume that all zeros of f in $[\alpha, \beta]$ belong to X. Otherwise we could add them to this set, and introduce in E correspondingly new rows of the type $(1, 0, \dots, 0)$. This will not destroy our assumptions on E: If a sequence is supported in the new matrix, it was also supported in the old one.

We omit in E the 0th column. If this column contains less than two 1's, then E' is the remaining matrix. Otherwise we can assume that $\alpha < \beta$ exist. Then E' will be this remaining matrix with some additional 1's. Let $\gamma < \delta$ be two adjacent zeros of f in $[\alpha, \beta]$. If ξ, $\gamma < \xi < \delta$, is a zero of f' stipulated by (1.5.1), then it comes from a 1 that begins a supported sequence in a column. This sequence is even and ends with a 1 in position $< n$, for otherwise (1.5.2) would hold.

The total number of zeros of f' in (γ, δ), stipulated by (1.5.1), is even. But according to Lemma 1.6, f' has an odd number of zeros in this interval. There is therefore an additional zero of f', which may appear as a zero not contained in X or as an additional multiplicity of a zero already known. We introduce in E' an additional 1 corresponding to each such zero, either in a new row or as an additional 1 in an old row, just after the sequence in question. The new set X' is obtained by adding the new zeros of f' to X, and by discarding those x_i that are not zeros of a derivative $f^{(r)}$, $r = 1, 2, \ldots, n$. Clearly, E' has at least n ones, and each sequence supported in E' is also supported in E, so that E' has no odd supported sequences.   □

Applying Lemma 1.7 several times, we obtain

Proposition 1.8.   *For E, f as in Lemma 1.7, there exists a ξ, $x_1 \leqslant \xi \leqslant x_m$, for which $f^{(n)}(\xi) = 0$. (If E has no 1's in the last column, we can take $x_1 < \xi < x_m$.)*

Proof of Theorem 1.5.   Let E be an $m \times (n + 1)$ interpolation matrix satisfying the hypotheses of the theorem; let the corresponding set X be arbitrary; and let P be a polynomial annihilated by E, X. We apply Proposition 1.8 to P and the matrix E_r, which is formed by the first $r + 1$ columns of E. The matrix E_r satisfies either all conditions of Lemma 1.7 or $e_{i,r} = 1$ for some i. Therefore for each $r = 0, \ldots, n$, there are points $x_1 \leqslant \xi_r \leqslant x_m$, with the property that $P^{(r)}(\xi_r) = 0$. This equation for $r = n$ means that the coefficient of x^n in P is 0, then for $r = n - 1$ we see that also x^{n-1} has coefficient 0, and so on. Thus, P is identically 0 and E is regular.   □

Remark.   Even without the assumption that E is normal in Theorem 1.5, it follows that a polynomial P, annihilated by E, X, is identically 0. We call E, X *complete for $\mathcal{P}_n$* if this is the case.

Examples

1. A Lagrange, Taylor, Hermite, or even quasi-Hermitian matrix E has no supported sequences and is therefore regular for polynomials of proper degree.

2. A two-row Pólya matrix has no supported sequences and therefore is real and even complex regular. We see that *the Pólya condition is necessary and sufficient for the regularity of a two-row matrix* (Pólya [136], Whittaker [O]).

3. The matrix E_3 of Examples (1.3.7) shows that the conditions of Theorem 1.5 are not necessary. There exist normal regular matrices that contain odd supported sequences.

We can combine Theorems 1.5 and 1.4, generalizing the former. For this purpose, let $E = E_1 \oplus \cdots \oplus E_\mu$ be the canonical decomposition of a

Pólya matrix E. A sequence of 1's in a row of E will be called an *essentially supported odd sequence* if it is entirely contained in one of the matrices E_λ and is an odd supported sequence in E_λ.

We have

Theorem 1.9. *A Pólya matrix is order regular for algebraic interpolation if it contains no essentially supported odd sequences.*

For complex regularity in the algebraic case we have a complete description obtained for Birkhoff matrices by Ferguson [37]:

Theorem 1.10. *A Pólya matrix is complex regular if and only if its canonical decomposition consists of matrices that are either Hermitian or have at most two nonzero rows.*

For the proof see Chapter 4.

§1.6. THE INTERPOLATING POLYNOMIAL

An explicit formula for the polynomial that solves a regular interpolation problem (1.2.2) can be easily obtained. We shall do it for the case when the data $c_{i,k}$ are given as the values of the derivatives of a known function: $c_{i,k} = f^{(k)}(x_i)$, for $e_{i,k} = 1$, where $f \in C^n[a, b]$. To the equations (1.2.2) we add the relation

$$a_0 g_0(t) + \cdots + a_N g_N(t) = P(t).$$

Elimination of the coefficients a_k yields

$$P(t) = \sum_{j=0}^{N} \frac{D(E, X, \mathcal{G}_j)}{D(E, X)} g_j(t) \tag{1.6.1}$$

where $\mathcal{G}_j$ is the set of functions obtained from $\mathcal{G} = \{g_0, \ldots, g_N\}$ by replacing g_j by f.

There is a simple formula, due to Mühlbach [127], that reduces the problem (1.2.2) with a Birkhoff matrix E to an interpolation problem with a smaller matrix if $\mathcal{G} = \{1, x, \ldots, x^n/n!\}$. We assume that the pair E, X is regular and that some submatrix $E_{i_0, k_0} = E^*$ of E is regular where E^* is obtained from E by replacing $e_{i_0, k_0} = 1$ by $e_{i_0, k_0} = 0$ and omitting the last column. We let $g_n(t) = t^n/n!$ and denote by $P_n(f, E, X; t) = P_n(t)$ the polynomial that interpolates a function $f \in C^n$. Then

$$P_n(f, E, X; t) = P_{n-1}(f, E^*, X; t) + c[g_n(t) - P_{n-1}(g_n, E^*, X; t)] \tag{1.6.2}$$

where c is the constant

$$c = \frac{f^{(k_0)}(x_{i_0}) - P_{n-1}^{(k_0)}(f, E^*, X; x_{i_0})}{g_n^{(k_0)}(x_{i_0}) - P_{n-1}^{(k_0)}(g_n, E^*, X; x_{i_0})}. \tag{1.6.3}$$

Indeed, let $Q(t)$ be the right-hand side of (1.6.2). If $e_{i,k}^* = 1$, we have $Q^{(k)}(x_i) = f^{(k)}(x_i)$ by the definition of P_{n-1}. Also $Q^{(k_0)}(x_{i_0}) = f^{(k_0)}(x_{i_0})$ by the definition of c. It remains to check that c is well defined. If the denominator of (1.6.3) were zero, then the $(n-1)$-degree polynomial $P_{n-1}(g_n, E^*, X; t)$ would also interpolate $g_n(t) = t^n/n!$ for the pair E, X. Since E, X is regular, then we would have $g_n(t) \equiv P_{n-1}(g_n, E^*, X; t)$, a contradiction.

Chapter 2

Further Elementary Theorems

§2.1. ESTIMATION OF THE DEFECT; THEOREM OF BUDAN–FOURIER

Let E be a normal interpolation matrix. Its defect is defined by

$$d = d(E) = (n+1) - \min_X \operatorname{rank} A(E, X). \qquad (2.1.1)$$

We can apply Theorem 1.5 to obtain the following estimate, which is better than those stated in [18] and [93]. We use the method of [18].

Theorem 2.1. *For a Pólya matrix E with p supported odd sequences,*

$$d(E) \leqslant \left[\frac{p+1}{2} \right]; \qquad (2.1.2)$$

and if p_λ is the number of supported odd sequences of the component E_λ of the canonical decomposition of E, $\lambda = 1, \ldots, \mu$, then

$$d(E) \leqslant \sum_{\lambda=1}^{\mu} \left[\frac{p_\lambda + 1}{2} \right], \qquad (2.1.3)$$

where $[y]$ denotes the integer part of y.

Proof. Inequality (1.4.9) shows that it is sufficient to prove (2.1.2). Let X be a given set of knots. First let $p \geqslant 2$. We single out in E two odd supported sequences, the first with its initial 1 in lowest possible column:

$$I_1: \quad e_{i_1, k_1} = \cdots = e_{i_1, k_1'} = 1; \qquad I_2: \quad e_{i_2, k_2} = \cdots = e_{i_2, k_2'} = 1.$$

We replace E by a new matrix E', obtained from E by shifting the 1 in position i_2, k_2' into the new position $i_1, k_1 - 1$. This operation will make the sequence I_2 even, and will transform I_1 into an even or an unsupported sequence. The first possibility will be realized if $e_{i_1, k_1 - 2} \neq 1$, and also if $e_{i_1, k_1 - 2} = 1$ and the sequence ending in this position is even. We will have the second possibility if $e_{i_1, k_1 - 2} = 1$ and this sequence is odd. It is easy to see that sequences supported in E' are supported already in E. Thus, E' will have $p - 2$ odd supported sequences.

After $[p/2]$ operations we will reduce E to a matrix with either none or just one odd supported sequence. In the latter case we move the last 1 of this sequence to the end of any Hermite sequence in E. This will give a conservative Pólya matrix, and E, X will become a regular pair $\tilde{E}, \tilde{X}$.

The number of operations performed is $\frac{1}{2}p$ if p is even, $\frac{1}{2}(p - 1) + 1 = \frac{1}{2}(p + 1)$ if p is odd, $[(p + 1)/2]$ in any case. All rows of the matrix $A(E, X)$, except $[(p + 1)/2]$ rows, are also contained in $A(\tilde{E}, \tilde{X})$. Hence

$$\text{rank } A(E, X) \geq (n + 1) - \left[\frac{p + 1}{2} \right]. \qquad \square$$

Examples

1. The inequality (2.1.2) cannot, in general, be improved. For positive integers r, p let the matrix

$$E = \begin{bmatrix} 1 & \cdot & \cdot & \cdot & 1 & 0 & \cdots & 0 \\ 0 & 1 & 0 & 1 & \cdot & \cdot & \cdots & \\ 1 & \cdot & \cdot & \cdot & 1 & 0 & \cdots & 0 \end{bmatrix} \qquad (2.1.4)$$

have p groups 01 in the second row, and contain $n + 1 = 2r + p$ ones. For $X = \{-1, 0, 1\}$, all polynomials $(x^2 - 1)^r x^{2k}$, $k = 0, 1, \ldots$, are annihilated by E, X, and exactly $[(p + 1)/2]$ of them have degree $\leq n$. Hence $d(E) \geq [(p + 1)/2]$, and we have equality in (2.1.2).

2. If $p = 1$ and E is a Birkhoff matrix, then $d(E) = 1$. This follows from the fundamental Theorem 6.2 of Chapter 6. For further examples see [18].

To state the classical theorem of Budan–Fourier about zeros of polynomials we need some simple notions. Let $\{c_k\}_{k=0}^n$ be a sequence of real numbers. Two elements c_k, c_l, $k < l$, of the sequence produce a change of sign (a constancy of sign) if $c_k c_l < 0$ (or $c_k c_l > 0$) and if $c_j = 0$ for $k < j < l$. We define $\mathfrak{S}^-\{c_k\}$ (or $\mathfrak{C}^-\{c_k\}$) to be the number of all changes (or constancies) of sign of the sequence c_k. In this definition, 0's of the sequence c_k are disregarded. In contrast, let $\mathfrak{S}^+\{c_k\}$ or ($\mathfrak{C}^+\{c_k\}$) be the number of changes of sign (or constancies) of the sequence if to 0's $c_j = 0$ we assign signs ($+$ or $-$) so as to maximize this number. It is easy to describe $\mathfrak{C}^+\{c_k\}$: To obtain this number, we consider all maximal subsequences of 0's $c_j = 0$, $k < j < l$,

$c_k, c_l \neq 0$ (or $k = -1$ and $c_l \neq 0$, or $l = n+1$ and $c_k \neq 0$) and assign to all c_j the sign of c_k or of c_l. If $q = l - k - 1$ is the number of 0's, this procedure produces exactly q new constancies of sign (compared with their number for the pair c_k, c_l). Consequently,

$$\mathfrak{C}^+\{c_k\} = \mathfrak{C}^-\{c_k\} + Q = \mathfrak{S}^+\{(-1)^k c_k\}, \qquad (2.1.5)$$

where Q is the total number of 0's. We also have $\mathfrak{C}^-\{(-1)^k c_k\} = \mathfrak{S}^-\{c_k\}$. Adding this and (2.1.5), we obtain

$$\mathfrak{S}^-\{c_k\} + \mathfrak{S}^+\{(-1)^k c_k\} = n. \qquad (2.1.6)$$

Theorem 2.2 (Budan–Fourier). *The number Z_0 of zeros in (a, b) of an algebraic polynomial P of degree n, counting the multiplicities, satisfies*

$$Z_0 \leqslant \mathfrak{S}^-\{P^{(k)}(a)\}_{k=0}^n - \mathfrak{S}^+\{P^{(k)}(b)\}_{k=0}^n. \qquad (2.1.7)$$

We shall see that this is a consequence of a generalized form of the Atkinson–Sharma theorem.

Let f be an n times continuously differentiable function on $[a, b]$. We need the notion of zeros of f in the *wide sense* in this interval. For $a < \alpha < b$, α is a zero of f if and only if $f(\alpha) = 0$. On the other hand, a is a zero of f under the new definition if either $f(a) = 0$ or else $f(a)f^{(k)}(a) > 0$, where k is the smallest integer $k = 1, \ldots, n$ with $f^{(k)}(a) \neq 0$. Similarly, b is a zero if $f(b) = 0$ or if $f(b)(-1)^k f^{(k)}(b) > 0$, with the similarly defined k.

It is essential that we have the following Rolle's lemma for an n times continuously differentiable function f on $[a, b]$:

Lemma 2.3. *Between any two adjacent zeros $\alpha < \beta$ in the wide sense of f on $[a, b]$ there is an odd number of zeros of f' (with multiplicities counted).*

Proof. Let, for example, $f(x) > 0$ on (α, β). It is sufficient to show that $f'(x) > 0$ for $x > \alpha$, x close to α, and that $f'(x) < 0$ for $x < \beta$, x close to β. This is obvious for ordinary zeros α, β, but it is also true if one or both of α, β are generalized zeros. Let, for example, $\beta = b$, $f(b) \neq 0$. Then $f(b) > 0$; hence $(-1)^k f^{(k)}(b) > 0$, and from Taylor's expansion,

$$f'(x) = \frac{(x-b)^{k-1}}{(k-1)!} f^{(k)}(b) + \cdots < 0 \qquad \text{for} \quad x < b$$

if x is close to b. $\square$

We shall say that a matrix E and a set of knots $X: a \leqslant x_0 < \cdots < x_m \leqslant b$ annihilate the function f in the *wide sense* if for each pair i, k with $e_{i,k} = 1$, x_i is a zero of $f^{(k)}$ in the wide sense.

Using Lemma 2.3 in the same way in which Lemma 1.6 was used to prove Proposition 1.8 we have

Theorem 2.4. *Let f be an n times continuously differentiable function on $[a, b]$. If f is annihilated in the wide sense by a conservative matrix E with*

at least $n + 1$ ones, then for some ξ,

$$f^{(n)}(\xi) = 0, \qquad a \leqslant \xi \leqslant b. \tag{2.1.8}$$

This gives the generalized Budan–Fourier theorem [97]:

Theorem 2.5. *Let P be a polynomial of degree (exactly) n that is annihilated in the wide sense by a conservative matrix E and a set of knots $X = \{a = x_1 < \cdots < x_m = b\}$. Let Z be the number of zeros of P and its derivatives in (a, b) described by E, X (i.e., Z is the number of 1's in the interior rows of E), and let the zeros of P at a and b (in the wide sense) be given exactly by the 1's in rows 1 and m. Then*

$$Z \leqslant \mathfrak{S}^- \{P^{(k)}(a)\}_{k=0}^n - \mathfrak{S}^+ \{P^{(k)}(b)\}_{k=0}^n. \tag{2.1.9}$$

Proof. The number $\mathfrak{S}_a$ of zeros of P and of its derivatives in the wide sense at $x = x_1 = a$ is $\mathfrak{S}_a = \mathfrak{C} + Q$, where Q is the number of ordinary zeros and $\mathfrak{C}$ is the number of constancies of sign in the sequence $P(a), P'(a), \ldots, P^{(n)}(a)$ (disregarding zero terms). By (2.1.5) and (2.1.6),

$$\mathfrak{S}_a = \mathfrak{S}^+ \{(-1)^k P^{(k)}(a)\} = n - \mathfrak{S}^- \{P^{(k)}(a)\}.$$

Similarly, $\mathfrak{S}_b = \mathfrak{S}^+ \{P^{(k)}(b)\}$. The total number of 1's in the matrix E is therefore

$$n - \mathfrak{S}^- \{P^{(k)}(a)\} + \mathfrak{S}^+ \{P^{(k)}(b)\} + Z.$$

If (2.1.9) were wrong, this number would be $\geqslant n + 1$. Theorem 2.4 would then yield $P^{(n)}(\xi) = 0$, a contradiction.   □

Clearly, Theorem 2.2 is a special case of this.

§2.2. COUNT OF PÓLYA AND BIRKHOFF MATRICES

How many normal $m \times n$ matrices of 0's and 1's are there? How many of them are Pólya or Birkhoff matrices? For the sake of convenience, we allow here rows consisting of 0's. Let $M(m, n)$, $P(m, n)$, and $B(m, n)$ denote the number of normal $m \times n$ matrices, Pólya matrices, and Birkhoff matrices, respectively. (We shall use the same notation for the classes of these matrices.) Clearly,

$$B(m, n) \leqslant P(m, n) \leqslant M(m, n).$$

We shall see that $P(m, n) = o(M(m, n))$ for large n, but that $B(m, n) \geqslant cP(m, n)$ for some constant $c > 0$. It is trivial that $M(m, n) = \binom{mn}{n}$; by Stirling's formula, we have the strong equivalence for $n \to \infty$,

$$M(m, n) = \binom{mn}{n} \approx \frac{1}{\sqrt{2\pi n}} m^n \left(1 + \frac{1}{m-1}\right)^{(m-1)n + 1/2}$$

$$\leqslant \text{const } n^{-1/2} m^n e^n. \tag{2.2.1}$$

To study the $m \times n$ Birkhoff matrices E, we have to consider arbitrary $m \times (n-1)$ matrices E' with n ones; denote this class of matrices by $\bar{M}(m, n-1)$. To each matrix $E' \in \bar{M}(m, n-1)$ there correspond its Pólya functions m_l, $l = 0, 1, \ldots, n-2$, where m_l is the number of 1's in the lth column of E'. Obviously, $\sum_{l=0}^{n-2} m_l = n$. We shall also consider the vector $\bar{m} = (m_0, \ldots, m_{n-2})$ of the Pólya functions. Since an $m \times n$ Birkhoff matrix E has an $(n-1)$st column of 0's, it produces a matrix in $\bar{M}(m, n-1)$ when the $(n-1)$st column is omitted. We shall study cyclic permutations of columns of matrices E'.

The cyclic permutation σ will replace column 0 of E' by column 1, column 1 by column 2, ..., and finally column $n-2$ by column 0. Along with E', the $(n-1)$-tuple $\bar{m}$ will also undergo cyclic permutations. The kth iteration of σ, σ^k, will produce

$$\sigma^k \bar{m} = (m_k, \ldots, m_{n-2}, m_0, \ldots, m_{k-1}), \qquad k = 0, \ldots, n-2.$$

We say that two matrices of the class $\bar{M}(m, n-1)$ are *equivalent* if one of them can be obtained from the other by cyclic permutations of the columns. This is an equivalence relation for matrices E' of $\bar{M}(m, n-1)$.

We shall need the functional

$$A(E') = A(\bar{m}) = (n-2)m_0 + (n-3)m_1 + \cdots + 0 m_{n-2}. \quad (2.2.2)$$

[The functional $A(\bar{m})$ can be interpreted as the area under a certain graph connected with $\bar{m}$, but we will not use this.]

Theorem 2.6 (Lorentz and Riemenschneider [100]). *Each equivalence class of matrices E' in $\bar{M}(m, n-1)$ consists of $n-1$ different matrices. Exactly one matrix in each equivalence class produces a Birkhoff matrix when an $(n-1)$st column of 0's is added. All $m \times n$ Birkhoff matrices are obtained in this way. In particular,*

$$B(m, n) = \frac{1}{n-1}\binom{m(n-1)}{n}, \qquad n \geq 2. \quad (2.2.3)$$

Proof. We have

$$A(\sigma^k \bar{m}) - A(\sigma^{k+1} \bar{m}) = (n-2)m_k - m_{k+1} - \cdots - m_{n-2} - m_0 - \cdots - m_{k-1}$$
$$= (n-1)m_k - n,$$

so that

$$A(\sigma^k \bar{m}) - A(\sigma^l \bar{m}) = (n-1)\sum_{j=k}^{l-1} m_j - (l-k)n, \qquad 0 \leq k \leq l \leq n-2.$$

$$(2.2.4)$$

To prove that all matrices of an equivalence class are different, it suffices to show that all numbers $A(\sigma^k \bar{m})$, $k = 0, \ldots, n$, are different. If the difference (2.2.4) is 0, then $(n-1)\sum_k^{l-1} m_j = (l-k)n$, and consequently $\sum_k^{l-1} m_j$ is divisible by n. There are only two possible cases: Either this sum is n or the sum is 0. The first case is impossible, since it yields $l - k = n - 1$. In the second case, $l = k$, as required.

It remains to show that exactly one of the cyclic permutations of E' produces an $m \times n$ Birkhoff matrix [if an $(n-1)$st column of 0's is added]. This is the permutation with the largest value of $A(\sigma^k \bar{m})$. Let this largest value correspond to $k = 0$, for example. According to (2.2.4), this is equivalent to

$$A(\bar{m}) - A(\sigma^k \bar{m}) = (n-1) \sum_{j=0}^{k-1} m_j - kn > 0, \qquad k = 1, \ldots, n-2,$$

(2.2.5)

or to $\sum_{j=0}^{k-1} m_j > k + k/(n-1)$, and this is identical to the Birkhoff condition $\sum_{j=0}^{k-1} m_j \geqslant k + 1, k = 1, \ldots, n-2$. That this is the only matrix producing a Birkhoff matrix also follows from (2.2.5) by taking $A(\sigma^k \bar{m})$ to be the maximum and reversing the inequality. □

In a similar way, all $m \times n$ Pólya matrices E can be counted. This time we consider arbitrary $m \times (n+1)$ matrices E' with exactly n ones and the $(n+1)$-tuples of their Pólya functions $\bar{m} = (m_0, \ldots, m_n)$. Some of the matrices E' produce $m \times n$ matrices with n ones if their nth column is omitted, namely, those with $m_n = 0$. As before, we consider the equivalence classes of such matrices under cyclic permutation of their columns. Exactly one matrix in each equivalence class produces an $m \times n$ Pólya matrix when the nth column is omitted—the one maximizing the functional

$$A(\bar{m}) = nm_0 + (n-1)m_1 + \cdots + 0m_n. \qquad (2.2.6)$$

In this way, we obtain

Theorem 2.7 [100]. *The number of $m \times n$ Pólya matrices is exactly*

$$P(m, n) = \frac{1}{n+1} \binom{m(n+1)}{n}. \qquad (2.2.7)$$

Examples. By means of (2.2.3), (2.2.7), and Stirling's formula, we obtain, as in (2.2.1), the strong equivalences

$$M(n, n) \approx \frac{1}{\sqrt{2\pi n}} n^n \left(1 + \frac{1}{n-1}\right)^{(n-1)n} \approx \frac{1}{\sqrt{2\pi e}} n^{n-1/2} e^n,$$

$$P(n, n) \approx \frac{1}{\sqrt{2\pi n}} \frac{1}{n} n^n \left(1 + \frac{1}{n}\right)^{n^2 + n} \approx \sqrt{\frac{e}{2\pi}} n^{n-3/2} e^n,$$

$$B(n, n) \approx \frac{1}{\sqrt{2\pi}} e^{-3/2} n^{n-3/2} e^n.$$

Proposition 2.8. *For $m \geq 3$, $n \geq 2$, there are positive constants c_1, c_2* such that

$$\frac{B(m,n)}{P(m,n)} \geq c_1, \qquad \frac{P(m,n)}{M(m,n)} \leq \frac{c_2}{n}. \tag{2.2.8}$$

Proof. We prove, for instance, the second relation,

$$\frac{P(m,n)}{M(m,n)} = \frac{1}{n+1} \frac{\binom{mn+m}{n}}{\binom{mn}{n}} \leq \frac{1}{n} \prod_{k=0}^{n-1} \frac{mn+m-k}{mn-k}$$

$$= \frac{1}{n} \prod_{k=0}^{n-1} \left(1 + \frac{1}{n - \frac{k}{m}} \right) \leq \frac{1}{n} \left(1 + \frac{3/2}{n} \right)^n \leq \frac{e^{3/2}}{n}. \qquad \square$$

Since all regular matrices must satisfy the Pólya condition, the second inequality (2.2.8) shows that most of the $M(m,n)$ normal $m \times n$ matrices are singular. It is more difficult to establish the singularity of almost all Pólya and almost all Birkhoff matrices. We shall treat this question in Chapter 6.

§2.3. SYMMETRY

Using the notion of symmetry, we obtain here new necessary conditions or sufficient conditions for the regularity of a matrix for algebraic interpolation. Symmetric matrices were studied by Turán (see Chapter 11), Windauer [190], and Lorentz [95].

A matrix $E = [e_{i,k}]_{i=-m}^{m} {}_{k=0}^{n}$ is *symmetric* if $e_{i,k} = e_{-i,k}$ for all i, k. For the purpose of generality, we allow E to have an empty row $i = 0$. A matching set of knots $X: x_{-m} < \cdots < x_0 < \cdots < x_m$ is *symmetric* if $x_i = -x_{-i}$, $x_0 = 0$. A symmetric matrix E is *symmetrically regular* if each pair E, X with symmetric X is regular. We show that this notion can be reduced to the order regularity of some matrices with fewer rows.

Let E^e—the even part of E—consist of rows $1, \ldots, m$ of E and of row 0 of E with all elements of this row in odd positions replaced by 1's. Let E^o be similarly defined with all elements in even positions of row 0 replaced by 1's.

Let Z^e and Z^o denote the number of 0's in row 0 in even and odd positions, respectively, and let $\overline{M}$ be the number of 1's in rows $1, \ldots, m$ of E, while $Z_k^e, Z_k^o, \overline{M}_k$ refer to the same numbers restricted to columns $0, \ldots, k$ of the matrix E. Since the Pólya function of E is given by

$$M_k(E) = 2\overline{M}_k + (k+1) - Z_k^e - Z_k^o,$$

the Pólya condition for E is

$$Z_k^e + Z_k^o \leq 2\overline{M}_k, \qquad k = 0, \ldots, n. \tag{2.3.1}$$

Likewise, from $M_k(E^e)=\overline{M}_k+(k+1)-Z_k^e$ and $M_k(E^o)=\overline{M}_k+(k+1)-Z_k^o$, we see that the Pólya conditions for E^e and E^o are

$$Z_k^e \leqslant \overline{M}_k \quad \text{and} \quad Z_k^o \leqslant \overline{M}_k, \qquad k=0,\ldots,n. \tag{2.3.2}$$

Proposition 2.9 [95]. *A symmetric Pólya matrix E is symmetrically regular if and only if both E^e and E^o are regular Pólya matrices.*

Proof. (a) Let E be a symmetrically regular Pólya matrix. Since $|E^e|+|E^o|=n+1+|E|$, one of the numbers $|E^e|$ or $|E^o|$ does not exceed $n+1$, say $|E^e| \leqslant n+1$. If E^e is not normal or if it is not regular, then for some set of knots $X_0 = \{0 = x_0 < \cdots < x_m\}$ there is a nontrivial polynomial P of degree $\leqslant n$ annihilated by E^e, X_0. This implies that P is even; hence P is also annihilated by E, $X = \{-x_m,\ldots, -x_1, x_0,\ldots,x_m\}$. We obtain a contradiction, $P \equiv 0$. Therefore E^e is normal and regular. Similarly, E^o is normal and regular.

(b) Let E^e, E^o be regular Pólya matrices. For a symmetric set of knots X let P be a polynomial annihilated by E, X. If P_e and P_o are the even and the odd parts of P, then P_e is annihilated by E^e, X_0, so $P_e \equiv 0$. Similarly $P_o \equiv 0$. $\qquad\square$

Corollary 2.10. *Necessary conditions for the regularity of a symmetric matrix are* (2.3.2) *and*

$$Z^e = Z^o = \overline{M}; \tag{2.3.3}$$

the conditions are sufficient for the symmetric regularity of a three-row symmetric matrix.

Indeed, in the second case, E^e and E^o are two-row Pólya matrices.

Examples
   1. Of the matrices

$$E' = \begin{bmatrix} 1 & 1 & 0 & 0 & 0 & 0 & 0 \\ 0 & 1 & 0 & 1 & 0 & 1 & 0 \\ 1 & 1 & 0 & 0 & 0 & 0 & 0 \end{bmatrix},$$

$$E'' = \begin{bmatrix} 1 & 1 & 0 & 0 & 0 & 1 & 0 & 0 & 0 & 0 \\ 0 & 1 & 0 & 1 & 0 & 0 & 1 & 0 & 1 & 0 \\ 1 & 1 & 0 & 0 & 0 & 1 & 0 & 0 & 0 & 0 \end{bmatrix},$$

$$E''' = \begin{bmatrix} 1 & 1 & 0 & 0 & 0 & 0 \\ 1 & 0 & 0 & 1 & 0 & 0 \\ 1 & 1 & 0 & 0 & 0 & 0 \end{bmatrix},$$

the first is symmetrically singular by (2.3.3). The second satisfies (2.3.3) but is symmetrically singular since it does not satisfy (2.3.2), $Z_4^e = 3$, $\overline{M}_4 = 2$. The third is symmetrically regular by the corollary yet is singular by Theorem 6.2.

2. Turán's matrices of the $(0, 2)$ interpolation are normal matrices E_p, $p = 3, 4, \ldots$, with p rows of the form

$$1 \ 0 \ 1 \ 0 \ 0 \ 0 \ \cdots.$$

They are all symmetrically singular. For matrices with odd p, condition (2.3.3) is violated. For even p (when an additional central row of 0's should be inserted), this follows from Theorem 6.2, because E^e has rows with exactly one odd supported sequence.

Symmetry ideas can be useful for nonsymmetric matrices, as is shown by the following.

Exercise [95]. In the matrix E consider coupled rows $i = s_j, -s_j'$, with $0 < s_1 < \cdots < s_r$ and $0 < s_1' < \cdots < s_r'$. Let S_j denote the number of k's for which $e_{-s_j', k} = e_{s_j, k}$, in other words, the number of 1's in identical positions of the coupled rows. Then a necessary condition for the regularity of E is the inequality

$$S_1 + \cdots + S_r \leqslant \min(Z^e, Z^o). \tag{2.3.4}$$

§2.4. TRIGONOMETRIC INTERPOLATION

For the system

$$\mathcal{G} = \{1, \cos x, \sin x, \ldots, \cos Nx, \sin Nx\} \tag{2.4.1}$$

or the equivalent system $\mathcal{G}^* = \{e^{ikx}, \ k = -N, \ldots, N\}$ on the circle $\mathbf{T}$, the linear combinations are the trigonometric polynomials

$$T_N(x) = a_0 + \sum_{k=1}^{N} (a_k \cos kx + b_k \sin kx). \tag{2.4.2}$$

Each differentiable function on $\mathbf{T}$ has an *even* number of zeros (counting multiplicities); in particular, a nontrivial polynomial T_N has at most $2N$ zeros. An interpolation matrix $E = [e_{i,k}]_{i=1, k=0}^{m, \ n}$ has $2N + 1$ ones; there is no a priori reason to assume that it is normal, that is, satisfies $n = 2N$. For a fixed set of knots X on $\mathbf{T}$, the determinant (1.2.3) is translation invariant:

$$D(E, X + \alpha) = D(E, X). \tag{2.4.3}$$

Indeed,

$$\frac{d}{d\alpha} D(E, X + \alpha) = 0,$$

since the derivative of an odd-numbered column of $D(E, X + \alpha)$ is proportional to the next even-numbered column and vice versa.

The condition

$$M_0 \geqslant 1 \tag{2.4.4}$$

takes the place of Pólya's condition. If it is not satisfied, $D(E, X) \equiv 0$.

Theorem 2.11. (i) *An interpolation matrix is conditionally regular with respect to trigonometric interpolation only if it satisfies* (2.4.4).

(ii) *For* (*normal*) *Pólya matrices, this condition is also sufficient for conditional regularity.*

For the proof of statement (ii), which is not necessarily true for arbitrary interpolation matrices, see Chapter 4.

A trigonometric form of Theorem 1.5 is easy to establish. First we must interpret, in this new situation, the notion of supported sequences. Together with functions on **T**, the interpolation matrix E should be considered periodic: Its last row m should precede the first row. For instance, in a two-row matrix

$$\begin{bmatrix} 0 & 1 & 1 & 0 & 0 & \cdots \\ 1 & 1 & 0 & 1 & 0 & \cdots \end{bmatrix},$$

the first row should be understood to contain a supported sequence. Thus, if $m \geq 2$ and (2.4.4) holds, a supported sequence is simply a sequence that does not begin in column 0. We have ([62], [92])

Theorem 2.12. *An interpolation matrix E with $m \geq 2$ is regular for trigonometric interpolation if it satisfies* (2.4.4) *and if all its odd sequences begin in column* 0.

The *proof* is the same as that of Theorem 1.5. Rolle's theorem on the circle guarantees p zeros for f' if the function f itself has p zeros on **T**, $p \geq 1$. Let T_N be a polynomial (2.4.2) annihilated by E. Lemma 1.7 takes the following form here: If T_N is annihilated by an $m \times (n+1)$ matrix E with at least $2N+1$ ones that does not contain odd sequences beginning in a column different from the 0th, then T_N' is annihilated by an $m' \times n$ matrix E' with $\geq 2N+1$ ones and without such sequences.

After n applications of the lemma, we see that $T_N^{(n)}$ has at least $2N+1$ zeros. Hence $T_N^{(n)} \equiv 0$, $a_k = b_k = 0$, $k = 1, \ldots, N$, and condition (2.4.4) gives $a_0 = 0$. □

Symmetry considerations yield further necessary conditions for regularity. See [95] for examples.

A complete answer to the question of regularity for trigonometric Birkhoff interpolation is known in only one simple situation.

Theorem 2.13 (Johnson [62]). *A one-row matrix E that satisfies* (2.4.4) *is regular if and only if it has $N+1$ ones in even columns and N ones in odd columns.*

Proof. By translation we can assume that $x_1 = 0$. The equations $T_N^{(k)}(0) = c_k$ are here, with $0^0 = 1$,

$$a_0 0^k + (-1)^{k/2} \sum_{j=1}^{N} j^k a_j = c_k, \qquad k \text{ even}, \tag{2.4.5}$$

$$(-1)^{k-1/2} \sum_{j=1}^{N} j^k b_j = c_k, \qquad k \text{ odd.} \tag{2.4.6}$$

A necessary condition for their unique solvability is that the number of equations be equal to the number of unknowns in each of these systems. This is also sufficient because the determinants of the systems are not 0. This follows from the fact that the powers $x^{k_j}, j = 0, \ldots, N$, $0 \leqslant k_0 < \cdots < k_N$, form a Chebyshev system on any interval $(0, A]$ (see also §4.6). □

Exercise. Show that the condition of Theorem 2.13 is necessary and sufficient for a two-row matrix E and the knot set $X = \{0, \pi\}$ to be regular. It is also necessary that any two rows of a regular matrix have at most $N + 1$ (or N) 1's in even (odd) columns (Johnson [62]). [*Hint*: For necessity use the argument of Theorem 2.13; for sufficiency expand $D(E, X)$ by means of Laplace's theorem.]

Examples. The matrices

$$E_1 = \begin{bmatrix} 1 & 1 & 0 \\ 0 & 1 & 0 \end{bmatrix}, \qquad E_2 = \begin{bmatrix} 1 & 0 & 1 & 0 & 1 \\ 1 & 1 & 0 & 0 & 0 \end{bmatrix}$$

are both regular for algebraic polynomials, but singular for trigonometric interpolation. If $X = \{0, \pi\}$, then $1 - \cos x$ is annihilated by E_1, X and $2 \sin x + \sin 2x$ is annihilated by E_2, X. This shows that the different definitions of supported sequences are needed in Theorems 1.5 and 2.12.

§2.5. NOTES

Not much is known about interpolation of functions of several variables. There does not even exist a generally accepted definition of Hermite interpolation. For some possible definitions and problems see Glaeser [196] and Turán [179, p. 82].

    An interesting method of interpolation was given by Kergin ([197]; see also [200]). We use the definition of Cavaretta, Micchelli, and Sharma [194, 195]. Let E be a regular Birkhoff interpolation matrix; let $\mathbf{X}: \mathbf{x}_1, \ldots, \mathbf{x}_m$ be some points of $\mathbf{R}^k$. We seek a bounded linear operator $\mathbf{P} = \mathbf{P}_{\mathbf{X}}$, defined on the space of functions $\mathbf{f} \in C^n(\mathbf{R}^k)$, for which $\mathbf{P}_{\mathbf{X}}(\mathbf{f})(\mathbf{t})$ is a polynomial in $\mathbf{t} \in \mathbf{R}^k$ of the total degree n and which, for "ridge functions" $\mathbf{f} = \mathbf{f}_\lambda$, for each $\lambda \in \mathbf{R}^k$, reduces to the ordinary Birkhoff interpolation polynomial $P(f, E; \lambda \cdot \mathbf{X}, \lambda \cdot \mathbf{t})$. Here $\mathbf{f}_\lambda(\mathbf{t}) = f(\lambda \cdot \mathbf{t})$, with a function of one variable f. For example, for $k = 3$, $n = 1$, and E a 2×2 Hermite matrix, the Kergin.

interpolation polynomial at $\mathbf{x}_1 = (0,0,0)$ and $\mathbf{x}_2 = (a, b, c)$ is

$$\mathbf{P}(\mathbf{f}; x, y, z) = \mathbf{f}(0,0,0) + \int_0^1 \left[x \frac{\partial \mathbf{f}}{\partial x}(\mathbf{q}_u) + y \frac{\partial \mathbf{f}}{\partial y}(\mathbf{q}_u) + z \frac{\partial \mathbf{f}}{\partial z}(\mathbf{q}_u) \right] du,$$

where $\mathbf{q}_u$ is the point (au, bu, cu). There are many examples in the papers quoted. For non-Hermitian E, the result is essentially negative [194, 195]: the operator $\mathbf{P}$ exists if and only if the matrix E is complex regular (see Theorem 1.10).

Chapter 3

Coalescence of Rows

§3.1. INTRODUCTION

Coalescence of two rows of a matrix has been implicitly used by Ferguson [37, p. 14]. It has been formally defined by Karlin and Karon [70], who studied its influence on the regularity of the matrix and gave the Taylor expansion (3.5.2). Lorentz and Zeller [108] proved that the leading coefficient in this formula is different from 0. Lorentz [94] studied the coalescence of several rows. There are many applications of this method, for example, in [70, 91, 94, 103]. We discuss them in Chapter 4.

Let $E = [e_{i,k}]$ be an $m \times (n+1)$ matrix, not necessarily normal, satisfying the Pólya condition (1.4.3). We interpret E as a vertical grid of boxes. If $e_{i,k} = 1$, then a ball occupies the ith box in the kth column. We place a tray of $n+1$ boxes under the column of the grid. Then the balls are permitted to fall from the grid into the boxes of the tray in such a way that if the box immediately below is occupied, the ball rolls to the first available box on the right. The condition (1.4.3) assures us that no ball will roll out of the tray. The distribution of balls in the tray constitutes the one-row matrix obtained by coalescence of the m rows of E. It is to be expected that the final arrangement of the balls in the tray is independent of the manner in which the balls were allowed to fall. Figure 3.1 is an example of coalescence of a two-row matrix.

1st row	1	0	1	1	1	0	0	1	0	0	0	$\cdots$
2nd row	0	1	1	0	0	0	0	1	1	0	0	$\cdots$
coalesced row	1	1	1	1	1	1	0	1	1	1	0	$\cdots$
precoalesced 1st row	1	0	0	1	1	1	0	0	0	1	0	$\cdots$

FIGURE 3.1

§3.2. LEVEL FUNCTIONS AND COALESCENCE

Let $m(k)$, $k = 0,\ldots,n$, be a nonnegative function with integral values. [We put $m(-1) = 0$, $m(k) = 0$ for $k > n$.] We use m_k interchangeably with $m(k)$, and denote by $M(k) = M_k$ the *sum* of m, given by $M(k) = \sum_{l=0}^{k} m(l)$. Functions m, M mutually determine each other.

The *level function* $m^0(k) = m_k^0$, $k = 0,1,\ldots$, of $m(k)$ takes only values 0 and 1 and is defined by induction. Let $m_{-1}^0 = 0$. We put $m_0^0 = 1$ if and only if $m_0 \geq 1$; if m_l^0 and hence its sum M_l^0 are known for $l = 0,\ldots,k$, we define $m_{k+1}^0 = 1$ if $M_{k+1} - M_k^0 \geq 1$, and $m_{k+1}^0 = 0$ otherwise. By induction it follows that $M_k^0 \leq M_k$, $k = 0,1,\ldots,n$.

An equivalent definition is as follows. Let $\mu_{-1} = 0$ and $\mu_k = (\mu_{k-1} - 1)_+ + m_k$; that is, let

$$\mu_k = \left(\cdots\left((m_0 - 1)_+ + m_1 - 1\right)_+ \cdots + m_{k-1} - 1\right)_+ + m_k, \qquad k = 0,1,\ldots. \tag{3.2.1}$$

(For a real a we put $a_+ = a$ if $a \geq 0$, $a_+ = 0$ otherwise.) Then $m_k^0 = 1$ if and only if $\mu_k \geq 1$. Indeed,

$$\mu_{k+1} = M_{k+1} - M_k^0, \qquad k = 0,1,\ldots. \tag{3.2.2}$$

This is true if $k = 0$, and for $k \geq 0$ follows by induction:

$$\mu_{k+1} = (\mu_k - 1)_+ + m_{k+1} = \left(M_k - M_{k-1}^0 - 1\right)_+ + m_{k+1}$$
$$= M_k - M_{k-1}^0 - m_k^0 + m_{k+1} = M_{k+1} - M_k^0.$$

From (3.2.1) and the formula $\mu_l = (\cdots(\mu_k - 1)_+ + \cdots)_+ + m_l$ we derive simple corollaries. (a) If $m_l \geq 1$, then $m_l^0 \geq 1$. (b) If $m_k \geq 2$, $m_j \geq 1$, $j - k + 1,\ldots,l - 1$, then $m_l^0 \geq 1$. (c) Let $M(n_1 - 1) = M^0(n_1 - 1)$ for some n_1, $0 < n_1 < n$. From (3.2.2), $\mu_{n_1} = m_{n_1}$ and therefore

$$\mu_k = \left(\cdots\left(m_{n_1} - 1\right)_+ + \cdots + m_{k-1} - 1\right)_+ + m_k, \qquad k \geq n_1.$$

This means that in this case, μ_k and m_k^0 for $k \geq n_1$ depend only on the values of m_k in the interval $n_1 \leq k \leq n$. (d) If $M \leq M_1$, then $M^0 \leq M_1^0$.

Lemma 3.1. *Let G be a function with integral values and with $0 \leq g(l) \leq 1$. Then $G(l) \leq M_1(l) + M_2(l)$ implies that $G(l) \leq M_1^0(l) + M_2(l)$.*

Proof. The inequality in question is certainly true for $l = 0$. Assume that it holds for $l = 0,\ldots,k$. If $m_1^0(k+1) = 1$, then $G(k+1) \leq G(k) + 1 \leq$

$M_1^0(k) + M_2(k) + 1 \leqslant M_1^0(k+1) + M_2(k+1)$. If, on the other hand, $m_1^0(k + 1) = 0$, then $M_1(k+1) = M_1^0(k)$ and

$$G(k+1) \leqslant M_1(k+1) + M_2(k+1) \leqslant M_1^0(k) + M_2(k+1). \qquad \square$$

Proposition 3.2.   *The level function M^0 of M is the largest function G satisfying $0 \leqslant g(l) \leqslant 1$, $G(l) \leqslant M(l)$, $l = 0, 1, \ldots$.*

This follows from Lemma 3.1 by taking $M_1 = M$, $M_2 = 0$.

The *coefficient of collision* $\alpha(M)$ of a function M measures the distance from the function M to M^0, and is defined by

$$\alpha(M) = \sum_{k=0}^{\infty} \left(M(k) - M^0(k) \right). \tag{3.2.3}$$

(Only finitely many terms of this sum are different from zero.)

The basic properties of level functions are as follows.

Theorem 3.3.   *We have*

$$\left(M_1 + M_2 \right)^0 = \left(M_1^0 + M_2 \right)^0 = \left(M_1^0 + M_2^0 \right)^0. \tag{3.2.4}$$

$$\left((M_1 + M_2)^0 + M_3 \right)^0 = \left(M_1 + (M_2 + M_3)^0 \right)^0$$
$$= (M_1 + M_2 + M_3)^0. \tag{3.2.5}$$

$$\alpha(M_1 + M_2) = \alpha(M_1) + \alpha\left(M_1^0 + M_2 \right)$$
$$= \alpha(M_1) + \alpha(M_2) + \alpha\left(M_1^0 + M_2^0 \right). \tag{3.2.6}$$

$$\alpha(M) \leqslant \alpha(M') \qquad \text{if} \quad M \leqslant M'. \tag{3.2.7}$$

Proof.   The first two statements follow from Proposition 3.2, and the last statement follows from the third. To prove (3.2.6) we observe that by (3.2.3)

$$\alpha(M_1 + M_2) = \sum_{k=0}^{\infty} \left[M_1(k) + M_2(k) - (M_1 + M_2)^0(k) \right]$$
$$= \sum_{k=0}^{\infty} \left[M_1(k) - M_1^0(k) \right]$$
$$+ \sum_{k=0}^{\infty} \left[M_1^0(k) + M_2(k) - \left(M_1^0 + M_2 \right)^0(k) \right]. \qquad \square$$

Exercise.   Prove that for any three functions M, M_1, M_2

$$\alpha(M + M_1) + \alpha(M + M_2) \leqslant \alpha(M) + \alpha(M + M_1 + M_2).$$

An important property that a function M can have is the upper Pólya condition

$$\sum_{l=k}^{n} m_l \leqslant n - k + 1, \qquad k = 0, 1, \ldots, n. \tag{3.2.8}$$

(In particular, this implies $m_n \leqslant 1$.) We note some consequences of this condition.

Proposition 3.4. (i) *The sum of two functions* $m + m'$ *satisfies the upper Pólya condition* (3.2.8) *if and only if* $m^0 + m'$ *does.*
(ii) *If* m *satisfies* (3.2.8), *then* $m_k^0 = 0$, $k > n$, *and* $M_n^0 = M_n$.

Proof. We consider the following operation, which shifts a unit from position k to $k + 1$, and is allowed if $m_k \geqslant 2$. [From (3.2.8) it follows that necessarily $k < n$.] The operation transforms m into $\bar{m}$, where $\bar{m}_k = m_k - 1$, $\bar{m}_{k+1} = m_{k+1} + 1$, $\bar{m}_l = m_l$ for all other l. In other words, $\bar{M}_k = M_k - 1$, $\bar{M}_l = M_l$ for $l \neq k$. Thus, $\bar{M} \leqslant M$.

We note that $\bar{M}^0 = M^0$. Indeed, the two sets of inequalities $G_l \leqslant M_l$ and $G_l \leqslant \bar{M}_l$ are equivalent. We have only to show that from the first inequality there follows $G_k \leqslant \bar{M}_k$. We see this from

$$G_k \leqslant G_{k-1} + 1 \leqslant M_{k-1} + 1 = M_k - m_k + 1 \leqslant M_k - 1.$$

Sums of type (3.2.8) for the function $m + m'$ do not exceed the corresponding sums for $\bar{m} + m'$. Hence $m + m'$ satisfies the upper Pólya condition if $\bar{m} + m'$ does. To prove the inverse, we assume that $m + m'$ satisfies (3.2.8) and have only to check the relation $\sum_{k+1}^n (\bar{m}_l + m'_l) \leqslant n - k$. But this follows from $m_k \geqslant 2$ and $\sum_k^n (m_l + m'_l) \leqslant n - k + 1$.

After finitely many operations of this type, we obtain from m a function $\bar{m}$ with $\bar{m}_l \leqslant 1$, $l = 0, \dots, n$, and then $\bar{m} = m^0$.

For each shift we have $k < n$, the shift does not change $m_k = 0$ for $k > n$ nor M_n. Hence we must have (ii).   □

§3.3. COALESCENCE IN A MATRIX

Let E be an $m \times (n + 1)$ matrix with Pólya functions m_k, M_k, $k = 0, \dots, n$, which satisfy the Pólya condition (3.2.8). Then also any submatrix E' of E (with functions $m'_k \leqslant m_k$) satisfies this condition.

In this section we interpret E as an unordered set of its rows. We write $E = E_1 \cup E_2$ if E is the disjoint union of two sets of rows E_1, E_2. Let M_k^0 be the level function of M_k. Since $m_k^0 = 0$ or 1, $k = 0, \dots, n$, this function defines a single row E^0, which we call *the coalescence of* E *to a one-row matrix*; $\alpha(E) = \alpha(M)$ *is the coefficient of collision of* E. If $\alpha(E) = 0$, then $M = M^0$; in this case E^0 can be obtained by simple addition of rows of E. If E^0 has a 0 in position k, then the whole kth column of E consists of 0's. If $\bar{E}$ is a submatrix of E, then the 1's of $\bar{E}^0$ form a subset of the 1's in E^0. Theorem 3.3. takes the following form:

$$\left(E_1 \cup E_2\right)^0 = \left(E_1^0 \cup E_2\right)^0 = \left(E_1^0 \cup E_2^0\right)^0. \tag{3.3.1}$$

$$\left((E_1 \cup E_2)^0 \cup E_3\right)^0 = \left(E_1 \cup (E_2 \cup E_3)^0\right)^0 = (E_1 \cup E_2 \cup E_3)^0. \tag{3.3.2}$$

$$\alpha(E_1 \cup E_2) = \alpha(E_1) + \alpha(E_1^0 \cup E_2) = \alpha(E_1) + \alpha(E_2) + \alpha(E_1^0 \cup E_2^0). \tag{3.3.3}$$

$$\alpha(E') \leqslant \alpha(E) \quad \text{if } E' \text{ is a submatrix of } E. \tag{3.3.4}$$

We shall call $E_1^0 \cup E_2$ the *coalescence* of E_1 in E to one row; by Proposition 3.4, $E_1^0 \cup E_2$ satisfies the upper Pólya condition (3.2.8) whenever E does.

Examples

1. For a Hermitian matrix E with rows F_i and numbers of 1's p_i, we obtain by induction:

$$\alpha(E) = \sum_{i<j} p_i p_j. \tag{3.3.5}$$

In particular,

$$\alpha(F_1 \cup F_2) = p_1 p_2. \tag{3.3.6}$$

2. If rows F_1, F_2 are Hermitian with p_1, p_2 ones except for an additional 1 in position $p_2 + s$ in row F_2, then $(F_1 \cup F_2)^0$ has $p_1 + p_2$ Hermitian 1's and a 1 in position $\max(p_1 + p_2, p_2 + s)$; and

$$\alpha(F_1 \cup F_2) = p_1 p_2 + (p_1 - s)_+. \tag{3.3.7}$$

Let $E = E_1 \oplus E_2$ be a vertical decomposition of E into two matrices satisfying (3.2.8). By Proposition 3.4, the level function m_1^0 of E_1 is 0 for $k > n_1$ (where n_1 is the number of the last column of E_1). Functions M_1^0 and M_2^0 have disjoint support, hence

$$\alpha(E_1 \oplus E_2) = \alpha(E_1) + \alpha(E_2). \tag{3.3.8}$$

As an important special case, we shall discuss the coalescence of two rows F_i, F_j in the matrix E, which satisfies (3.2.8). For the Pólya function m of $F_i \cup F_j$, we define

$$\rho_k = m_k - m_k^0, \quad k = 0, \ldots, n, \quad \rho_k = 0 \text{ for all other } k.$$

The possible values of ρ_k are $1, -1, 0$; we have $\rho_k = 1$ precisely when $m_k = 2$, $m_k^0 = 1$, that is, when

$$e_{i,k} = e_{j,k} = 1. \tag{3.3.9}$$

We have $\rho_k = -1$ exactly when $m_k = 0$, $m_k^0 = 1$, that is, when

$$e_{i,k} = e_{j,k} = 0, \tag{3.3.10}$$

and in addition $m_k^0 = 1$. For all other k, $\rho_k = 0$. Since $\sum_0^n \rho_k = M_n - M_n^0 = 0$, we have as many k with $\rho_k = 1$ as with $\rho_k = -1$. Let $k_1 < k_2 < \cdots < k_q$ and $k_1' < k_2' < \cdots < k_q'$ be the respective positions of k's of these two kinds. Since $\sum_k^n \rho_s = M_{k-1}^0 - M_{k-1} \leqslant 0$ for each k, there are at least as many $k_s' \geqslant k$

as there are $k_s \geq k$. This proves that $k_s < k'_s$, $s = 1, \ldots, q$. The k'_s may also be described in the following way:

(*) k'_1 is the smallest $k > k_1$ that satisfies (3.3.10); k'_s is the smallest $k > k_s$, $k > k'_{s-1}$ that satisfies (3.3.10).

The k'_s defined here must exist, for otherwise we would have $m(k_s) = 2$, $m(k) \geq 1$, $k > k_s$, in contradiction to (3.2.8). This being established, the inequalities $m(k) \geq 1$, $k_s < k < k'_s$, and (b) of Section 3.2 yield $m^0(k'_s) = 1$. Hence the k'_s defined by (*) are the same as the old k'_s.

It is sometimes more convenient to add to the k_s and the k'_s the positions of all $e_{i,k} = 1$ for which $e_{j,k} = 0$. In this way we obtain positions $l_1 < \cdots < l_p$ and $l'_1 < \cdots < l'_p$, respectively (with $q \leq p \leq n + 1$). The l_s give positions of all 1's in row F_i. It follows from (*) that the l'_s can be defined by:

(**) l'_1 is the first $k \geq l_1$ with $e_{j,k} = 0$; l'_s is the first $k \geq l_s$, $k > l'_{s-1}$ for which $e_{j,k} = 0$.

The row $\tilde{F}_i$ with 1's in positions $l'_1 < \cdots < l'_p$ will be called *the precoalescence of F_i with respect to F_j* (see Figure 3.1 for an example).

Proposition 3.5. *Row $(F_i \cup F_j)^0$ can be obtained by adding row $\tilde{F}_i$ to row F_j.*

This is the Karlin–Karon [70] definition of coalescence.

The coefficient of collision $\alpha_{i,j} = \alpha(F_i \cup F_j)$ can be computed by means of the formula

$$\alpha_{i,j} = \sum_{k=0}^{n} \left(M_k - M_k^0 \right) = \sum_{k=0}^{n} (n-k)\rho_k = \sum_{s=1}^{q} (k'_s - k_s) = \sum_{s=1}^{p} (l'_s - l_s).$$

$$(3.3.11)$$

Rows F_i, F_j are in collision if this coefficient is not 0.

We shall also consider the coalescence of $F_i \cup F'$, where $F' = E'^0$ and E' is the union of some rows of E not containing F_i. What was said earlier remains valid, with "positions with $e_{j,k} = 0$" in (**) replaced by "positions of 0's of row F'."

For example, if $F_i \cup E'$ is a Pólya matrix, then $F_i \cup E'^0$ is also one. There are then as many 1's in F_i as there are 0's in E'^0. This means that in this case, l'_s, $s = 1, \ldots, p$, are simply positions of 0's of E'^0. In particular, for a Birkhoff matrix, $l'_p = n$.

As another example, and also for further use, we consider what happens to the coefficient of collision if a new 1 in position $k \neq l_t$, $t = 1, \ldots, p$, is added to row F_i, while $F' = E'^0$ remains unchanged.

Lemma 3.6. *The addition of a 1 in position k to the row F_i increases $\alpha(F_i \cup F')$ by the amount $k' - k \geqslant 0$, where k' is the position of the first 0 of row F', $k' \geqslant k$, which is different from all l_t'.*

Proof. Let s be the largest integer for which $l_s < k$, s_1 the smallest integer with $k' < l_{s_1+1}'$. We have $s_1 \geqslant s$. The numbers l_t of the new row F_i in increasing order will be $l_1 < \cdots < l_s < k < l_{s+1} < \cdots < l_p$. From $(**)$ we find the corresponding numbers l_t' for the new row. To $l_1 < \cdots < l_s$ will correspond the old $l_1' < \cdots < l_s'$; to $k < l_{s+1} < \cdots < l_{s_1}$ will correspond $l_{s+1}' < \cdots < l_{s_1}' < k'$; and to $l_{s_1+1} < \cdots < l_p$ again the old $l_{s_1+1}' < \cdots < l_p'$. The total change of α, computed by (3.3.11), will be $k' - k$. $\square$

If E' is a submatrix of E'', then the 1's of E'^0 form a subset of those of E''^0. In this case, the numbers l_s', l_s'' of precoalescences of a row F_i with these rows will satisfy $l_s'' \geqslant l_s'$, $s = 1, \ldots, p$. In particular, $\alpha(E \cup E'^0) \leqslant \alpha(E \cup E''^0)$.

The largest coefficient of collision corresponds to the matrix $E' = E \setminus F_i$. This coalescence of F_i with $E \setminus F_i$ will be called the *maximal coalescence of the row F_i in E.* It has the coefficient

$$\gamma_i = \alpha\big(F_i \cup (E \setminus F_i)^0\big) = \sum_{s=1}^p (l_s^* - l_s), \qquad (3.3.12)$$

where l_s^* are the positions of 1's in the precoalescence of F_i with respect to $E \setminus F_i$.

§3.4. SHIFTS AND DIFFERENTIATION OF DETERMINANTS

As before, we discuss the coalescence of a row F_i of E with a row F, with the coefficient of collision $\alpha = \alpha(F_i \cup F)$. Here, F is either a row F_j, $j \neq i$, of E, or a coalescence of several rows. In order to transform F_i into the precoalesced row $\tilde{F}_i$, we define shifts of rows F_i. The *shift* $\Lambda: k \to k+1$ is applicable if $e_{i,k} = 1$, $e_{i,k+1} = 0$; it moves the 1 from position k to position $k+1$, producing the new row ΛF_i, and from E the matrix ΛE. In other words, Λ is applicable to F_i if $k = l_s$ for some s and $l_{s+1} > l_s + 1$; in the new row this l_s will be replaced by $l_s + 1$, all other l_t remaining unchanged.

A shift Λ is *reducing* with respect to a row $F' = E'^0$ if it diminishes the coefficient of collision $\alpha = \alpha(F_i \cup F')$. This happens if

$$k = l_s, \qquad l_s' > l_s. \qquad (3.4.1)$$

In this case, the l_t' will not change if row F_i is replaced by ΛF_i. By (3.3.11), α will be replaced by $\alpha - 1$.

On the other hand, if $l_s' = l_s = k$, the differences $l_t' - l_t$ will not change for $t < s$, while for $t \geqslant s$ their sum will not decrease. This follows from Lemma 3.6 applied to the segment $k+1 \leqslant l \leqslant n$ of F_i. Thus, $\Lambda: k \to k+1$ is reducing if and only if we have (3.4.1).

If $\alpha > 0$, there are reducing shifts. Indeed, there must be a largest s with $l_s < l'_s$. Then for $k = l_s$ we have $k = l_s < l'_s < l_{s+1}$. Another remark is that each shift $\Lambda: k \to k+1$ that transforms a Pólya matrix E into another Pólya matrix ΛE is reducing with respect to $\alpha(E)$, for the shift does not change the quantities $M^0(l) = l+1$, $l = 0, \ldots, n$, and $M(l)$, $l \neq k$, while $M(k)$ is reduced by one.

As an example of applications of shifts we examine the coalescence of two rows in a Birkhoff matrix. In an interpolation matrix E, let $E_1 = F_i \cup F_{i+1}$, $E_2 = E \setminus E_1$, $E' = E_1^0 \cup E_2$.

Proposition 3.7. *If E is a Birkhoff matrix, then E' is also one, except when E has only two 1's in column 0, $e_{i,0} = e_{i+1,0} = 1$. In this exceptional case, if E has more than two rows and if μ is the first column containing a 1 outside of E_1, then E' has the canonical decomposition*

$$E' = E'_0 \oplus \cdots \oplus E'_\mu \tag{3.4.2}$$

where E'_λ, $\lambda < \mu$, are one-column matrices, while E'_μ is a Birkhoff matrix.

Proof. (a) In the general case, let M_1, M_2, M be Pólya functions of matrices E_1, E_2, E. We want to establish the Birkhoff inequality

$$M_1^0(l) + M_2(l) \geqslant l+2, \qquad l = 0, \ldots, n-1. \tag{3.4.3}$$

Since M_1^0 depends only on the final column positions of the 1's in E_1^0, we may perform reducing shifts $k \to k+1$ for *either* row F_i or row F_{i+1}. If $M_1^0 \neq M_1$, then for the largest k with $m_1(k) = 2$, the shift $k \to k+1$ is applicable for one of the rows and is reducing. This shift in E_1 replaces M_1 by $\overline{M}_1$, and we have $\overline{M}_1(k) = M_1(k) - 1$, $\overline{M}_1(l) = M_1(l)$ $(l \neq k)$. To show that the function $\overline{M}_1(l) + M_2(l)$ satisfies (3.4.3), it suffices to consider $l = k$. If $k > 0$, we have

$$\overline{M}_1(k) + M_2(k) \geqslant M_1(k) - 1 + M_2(k-1) = M(k-1) + 1 \geqslant k+2.$$

If $k = 0$, then $m_2(0) \geqslant 1$ and $\overline{M}_1(0) + M_2(0) \geqslant 2$. By repeated shifts (there can be only one shift $0 \to 1$), we obtain (3.4.3).

(b) In the exceptional case, the foregoing argument still applies for shifts with $k > 0$. After finitely many such shifts, we have $m_1(k) = 2$ only for $k = 0$. Applying the shift $0 \to 1$, we reduce E to $E'_1 \oplus \overline{E}$. Repeating the argument, we reduce E to $E'_1 \oplus \cdots \oplus E'_{\mu-1} \oplus E_\mu$ where E_μ is a Birkhoff matrix that is not exceptional, and case (a) applies. $\qquad\square$

A *multiple shift* $\tilde{\Lambda}_\beta = \Lambda_\beta \cdots \Lambda_2 \Lambda_1$ of row F_i of order β is a product of β simple shifts; it is applicable to F_i if Λ_1 is applicable to F_i, Λ_2 is applicable to $\Lambda_1 F_i$, and so on. Since a simple shift can reduce α by at most 1, we have, for each $F' = E'^0$,

$$\alpha\big(\tilde{\Lambda}_\beta F_i \cup F'\big) \geqslant \alpha - \beta. \tag{3.4.4}$$

If $\tilde{\Lambda}_\alpha F_i$ and F' are without collision, each simple shift in $\tilde{\Lambda}_\alpha = \Lambda_\alpha \cdots \Lambda_1$ must be reducing. We have the following useful statements about the coefficient of collision of row F_i. Here F' is either any row F_j, $j \neq i$, or a coalescence of such rows.

Proposition 3.8. (i) *For each multiple shift $\tilde{\Lambda}_\beta$ of order $\beta < \alpha$, rows $\tilde{\Lambda}_\beta F_i$ and F' are in collision.*

(ii) *If $\tilde{\Lambda}_\alpha F_i$ and F' have no collision, the shift $\tilde{\Lambda}_\alpha$ carries 1's of F_i in positions $l_1 < \cdots < l_p$ into 1's in positions $l'_1 < \cdots < l'_p$ of row $\tilde{F}_i$.*

(iii) *If the shift $\tilde{\Lambda}_\beta$ takes $l_1 < \cdots < l_p$ of row F_i into $\bar{l}_1 < \cdots < \bar{l}_p$ and if $\tilde{\Lambda}_\beta E$ is a Pólya matrix, then $\bar{l}_s \leqslant l_s^*$, $s = 1, \ldots, p$. In particular, for $\beta > \gamma$, $\tilde{\Lambda}_\beta E$ is not a Pólya matrix.*

(iv) *There is only one shift $\tilde{\Lambda}_\gamma$ of row F_i of order γ that takes a Pólya matrix E into another Pólya matrix, namely, the one that shifts the 1's from positions l_s into positions l_s^*, $s = 1, \ldots, p$.*

Proof. Only (iii) has to be established. For a fixed k, let $\bar{l}$ (or l^*) be the number of $\bar{l}_s \leqslant k$ (or of $l_s^* \leqslant k$). Then the row $\tilde{\Lambda}_\beta F_i$ has $\bar{l}$ ones in positions $\leqslant k$, row $F' = (E \setminus F_i)^0$ has $(k+1) - l^*$ such ones. The Pólya inequality for the matrix $\tilde{\Lambda}_\beta F_i \cup F'$ gives $\bar{l} + (k+1) - l^* \geqslant k+1$, or $\bar{l} \geqslant l^*$. Hence $\bar{l}_s \leqslant l_s^*$, $s = 1, \ldots, p$.   □

The multiple shift $\tilde{\Lambda}_\alpha$ of (ii) has in general several representations $\tilde{\Lambda}_\alpha = \Lambda_\alpha \cdots \Lambda_1$ as a product of simple shifts. The *number of different representations* is an important combinatorial quantity

$$C = C(l_1, \ldots, l_p; l'_1, \ldots, l'_p). \qquad (3.4.5)$$

We call it the *coalescence constant*. In Corollary 4.17 of Chapter 4, we give an explicit representation for C. As an example, $C(0;3) = 1$; $C(0,1;3,4) = 5$.

Shifts appear in the differentiation of determinants. From now on, we have to interpret a matrix E as an *ordered* set of its rows, $F_1, \ldots, F_m$. Let

$$D = D(E, X) = \det\left[g_0^{(k)}(x_i), \ldots, g_N^{(k)}(x_i); e_{i,k} = 1 \right] \qquad (3.4.6)$$

be the determinant (1.2.3) of the interpolation problem that corresponds to an $m \times (n+1)$ interpolation matrix E, a set of knots X, and a system of functions $\mathcal{G} = \{g_0, \ldots, g_N\}$ that are analytic on $[a, b]$. We fix an i, $1 \leqslant i \leqslant m$, and study the determinant as a function $D(x_i)$ of the variable x_i. The rows in (3.4.6) that contain x_i correspond to 1's $e_{i,k} = 1$ of the ith row. The derivative of $D(x_i)$ with respect to x_i is equal to the sum of determinants obtained by differentiating exactly one of the rows (i, k) with $e_{i,k} = 1$. If $e_{i,k+1} = 1$, the determinant will have two equal rows and is 0. Otherwise the differentiation has the effect of replacing F_i by ΛF_i for the shift Λ: $k \to k+1$. We have therefore

$$\frac{\partial}{\partial x_i} D = \sum_\Lambda D(\Lambda E, X), \qquad (3.4.7)$$

for all shifts Λ of row F_i. Similarly,

$$\frac{\partial^\beta D}{\partial x_i^\beta} = \sum_{\tilde{\Lambda}_\beta} D\left(\tilde{\Lambda}_\beta E, X\right) \tag{3.4.8}$$

for all multiple shifts of order β of F_i. In particular, we can find the values of these derivatives for $x_i = x_j$. We have $D(\tilde{\Lambda}_\beta E, X)|_{x_i = x_j} = 0$ if there is collision of rows $\tilde{\Lambda}_\beta F_i$ and F_j. By taking $F' = F_j$ in Proposition 3.8, we have

$$\left.\frac{\partial^\beta D}{\partial x_i^\beta}\right|_{x_i = x_j} = 0 \qquad \text{for} \quad \beta < \alpha, \tag{3.4.9}$$

$$\left.\frac{\partial^\alpha D}{\partial x_i^\alpha}\right|_{x_i = x_j} = CD(\bar{E}, X)|_{x_i = x_j}, \tag{3.4.10}$$

where $\bar{E}$ is obtained from E by replacing F_i by $\tilde{F}_i = \Lambda_\alpha F_i$, and C is the constant (3.4.5).

§3.5. PERMUTATION NUMBERS AND TAYLOR EXPANSIONS

In a matrix E with *ordered* rows, we define the (directed!) *coalescence of row* F_i *to row* F_j. This is the $(m-1) \times (n+1)$ matrix E' obtained from E by omitting row F_i and by replacing F_j by $(F_i \cup F_j)^0$. Let also X' be the set of knots X with x_i omitted. The determinant $D(E', X')$ differs only in sign $(-1)^\sigma$ from the determinant in the right-hand side of (3.4.10). Here $\sigma = \sigma_{i,j}$ (which is important only mod 2) is the number of permutations of rows that are needed to bring the rows of the determinant in (3.4.10) to the lexicographical order induced by E'. Thus

$$\sigma_{i,j} = \tau p + \sigma', \tag{3.5.1}$$

where τ is the number of 1's in rows of E contained between rows i and j, and σ' is the number of interchanges that are required to bring the sequence of integers

$$l'_1, \ldots, l'_p, \bar{l}_1, \ldots, \bar{l}_q$$

to its natural order (here, $\bar{l}_1 < \cdots < \bar{l}_q$ are the positions of 1's in row F_j). We have then

Theorem 3.9. *For an interpolation matrix E, the determinant $D(E, X)$ has for $x_i \to x_j$ the Taylor expansion*

$$D(E, X) = C\frac{(x_i - x_j)^\alpha}{\alpha!}(-1)^\sigma D(E', X') + \cdots, \tag{3.5.2}$$

with $\alpha = \alpha_{i,j} = \alpha(F_i \cup F_j)$, $\sigma = \sigma_{i,j}$, and $C \neq 0$.

This is a result of Karlin and Karon [70], except that they did not use shifts and did not show that $C \neq 0$. That C is the constant (3.4.5) appears in [108].

If E is a Pólya matrix, then E' is also a Pólya matrix and $D(E', X')$ $\neq 0$ for almost all X' in $\mathbf{R}^{m-1}$. For these X', $D(x_i)$ has a root at x_j of multiplicity exactly equal to $\alpha_{i, j}$.

For the functions $\mathcal{G} = \{1, x, \ldots, x^n/n!\}$ on $[a, b]$, maximal coalescence becomes important. By (3.4.8), (iii) of Proposition 3.8, and Theorem 1.3, the derivative $\partial^\beta D/\partial x_i^\beta$ is identically 0 for $\beta > \gamma$. We see that $D(x_i)$ is a polynomial of degree $\leq \gamma = \gamma_i$ in x_i. For $\beta = \gamma$, we get from (iv) of Proposition 3.8 that $\partial^\gamma D/\partial x_i^\gamma = C^* D(\bar{E}, X)$, where $C^* = C(l_1, \ldots, l_p; l_1^*, \ldots, l_p^*)$ and $\bar{E}$ is the Pólya matrix $\tilde{\Lambda}_\gamma E$. If E^* is obtained by addition of row $\tilde{\Lambda}_\gamma F_i$ to F_j, and if X^* is the set of knots X with x_i omitted, the last determinant is $(-1)^{\sigma^*} D(E^*, X^*)$, with the permutation number σ^* defined similarly to $\sigma_{i, j}$ of (3.5.1). We have [91]

Theorem 3.10. *For a Pólya matrix E and the system $\mathcal{G} = \{1, \ldots, x^n/n!\}$, we have the expansion*

$$D(E, X) = (-1)^\sigma \frac{C}{\alpha!} (x_i - x_j)^\alpha D(E', X')$$

$$+ \cdots + (-1)^{\sigma^*} \frac{C^*}{\gamma!} (x_i - x_j)^\gamma D(E^*, X^*). \quad (3.5.3)$$

For almost all X, $D(x_i)$ is of exact degree γ and has a zero of exact multiplicity α at x_j.

Differentiability Requirements for Coalescence

For simplicity, we have assumed that the functions $g_i \in \mathcal{G}$ are analytic. If this is not the case, we must assume their differentiability of very high order to ensure the applicability of the method. Indeed, a row of one of the determinants in (3.4.8) with $\beta = \alpha$ could be $(g_0^{(k+\alpha)}(x_i), \ldots, g_n^{(k+\alpha)}(x_i))$, and $k + \alpha$ may be much larger than n. For example, if n is odd and the matrix consists of just two Hermitian rows of lengths $(n+1)/2$ each, then by (3.3.6), $\alpha = (n+1)^2/4$, and k may be $(n-1)/2$. Hence the order of differentiability required is $(n^2 + 4n - 1)/4$.

Exercises

1. Except for its sign, the determinant $D(E^*, X^*)$ is independent of the choice of the row F_j.

2. The largest possible value of the coefficient of collision of two rows in an $m \times (n+1)$ normal matrix is attained when all 1's are contained in two rows and the rows are Hermitian and as equal as possible in length.

Accordingly, the maximal α is

$$\alpha = \begin{cases} \dfrac{n^2 + 4n - 1}{4} & \text{if } n \text{ is odd,} \\[2ex] \dfrac{n^2 + 2n}{4} & \text{if } n \text{ is even.} \end{cases}$$

3. In order to justify the coalescence method of proof for Theorem 3.9, it is sufficient to assume that the functions of the system are p times continuously differentiable, where $p \geqslant (n^2 + 4n - 1)/4$.

4. Prove that the determinant $D(E, X)$ is divisible by the product $\prod_{i<j}(x_i - x_j)^{\alpha_{i,j}}$. Prove that it is equal to a constant multiple of this product if the canonical decomposition of the matrix E consists only of Hermitian and two-row matrices.

Chapter 4

Applications of Coalescence

§4.1 CONDITIONALLY REGULAR AND STRONGLY SINGULAR MATRICES

As a first application of the theorems of Chapter 3, we discuss the determinant $D(E, X)$ as a function of the variables $x_1, \ldots, x_m$. The matrix E is called (*order*) *conditionally regular* if the determinant is not identically 0, $D(E, X) \not\equiv 0$, $a \leqslant x_1 < \cdots < x_m \leqslant b$. If the functions $g_k(x)$ are analytic on $[a, b]$, and the determinant is identically 0, then it will also vanish for any other order of the knots x_i. On the other hand if $D(E, X) \not\equiv 0$, $x_1 < \cdots < x_m$, then this determinant can vanish only on a nowhere-dense set of measure 0 in $\mathbf{R}^m$, again for any order of the x_i. Thus there is no difference between *real* and *order* conditional regularity.

In the expansion (3.5.2), the first term predominates for $x_i \to x_j$, and we see that $D(E, X)$ is surely not identically 0 if $D(E', X')$ is not. From (3.5.3) for $x_i \to \infty$ (and the polynomial case) we obtain the same statement with respect to $D(E^*, X^*)$.

After $m - 1$ row coalescences, an interpolation matrix E will be reduced to a one-row matrix E^0 with $N + 1$ ones. If $0 \leqslant k_0 < \cdots < k_N$ are the positions of 1's in E^0, its determinant $D(E^0, x)$ with respect to the system $\mathcal{G}$ will be

$$D(E^0, x) = \det\left[g_0^{(k_i)}(x), \ldots, g_N^{(k_i)}(x); i = 0, \ldots, N \right]. \qquad (4.1.1)$$

We obtain the following general theorem, with several interesting special cases:

Theorem 4.1. *An $m \times (n+1)$ matrix E with $N+1$ ones is conditionally regular with respect to the system $\mathcal{G} = \{g_0, \ldots, g_N\}$ of analytic functions if the determinant (4.1.1) is not identically 0.*

Corollary 4.2 [91]. *A Pólya matrix E is conditionally regular with respect to $\mathcal{G}$ if its Wronskian is not identically 0:*

$$W(x) = \det\left[g_0^{(k)}(x), \ldots, g_n^{(k)}(x); k = 0, \ldots, n \right] \not\equiv 0.$$

Theorem 1.3 is a special case of this for $\mathcal{G} = \{1, x, \ldots, x^n/n!\}$.

Corollary 4.3. *Any interpolation matrix [with the additional property $M(0) > 0$ if $a_0 = 0$] is conditionally regular with respect to the system $\mathcal{G} = \{e^{a_0 x}, \ldots, e^{a_N x}\}$ on $[a, b]$, $-\infty < a < b < +\infty$, if $0 \leqslant a_0 < a_1 < \cdots < a_N$.*

Indeed, $D(E^0, x)$ of (4.1.1) is here a positive multiple of the determinant

$$\det\left[a_0^{k_i}, a_1^{k_i}, \ldots, a_N^{k_j}; i = 0, \ldots, N \right],$$

which is not 0 because the functions $x^{k_0}, \ldots, x^{k_N}$ form a Chebyshev system on each interval $[\alpha, \beta]$, $\alpha \geqslant 0$; if $\alpha = 0$, we must assume that $k_0 = 0$, that is, that $M(0) > 0$. □

For the trigonometric system on $\mathbf{T}$,

$$\mathcal{G} = \{1, \cos x, \sin x, \ldots, \cos Nx, \sin Nx\},$$

and an interpolation matrix E with $2N+1$ ones, the determinant $D(E^0, x)$ is translation invariant. The matrix E^0 is conditionally regular if and only if it is regular.

From Theorem 2.13 of Johnson we obtain

Corollary 4.4. *An interpolation matrix E is conditionally regular with respect to the trigonometric system if its coalescence to one row E^0 has $N+1$ ones in even positions and N ones in odd positions and if $M(0) > 0$.*

In particular, a Pólya matrix E is conditionally regular as stated in Theorem 2.11.

A matrix E is *strongly singular* (with respect to the system $\mathcal{G}$) if the determinant $D(E, X)$ changes sign, that is, if there are two sets of knots X_1, X_2 for which $D(E, X_1) < 0 < D(E, X_2)$. We must distinguish the *real* case and the *ordered* case when $x_1 < \cdots < x_m$. In the latter case, we must restrict coalescences to adjacent rows.

Let X'_1, X'_2 be two sets of knots x_l, $l \neq i$; let X_1, X_2 be obtained from these by adding an x_i. If for the coalescence E' of E of Theorem 3.9, $D(E', X'_1) < 0 < D(E', X'_2)$, then formula (3.5.2) reveals that for x_i close to but different from x_j, $D(E, X_1)$ and $D(E, X_2)$ are also of opposite sign. For the matrix E^* of Theorem 3.10 and polynomial interpolation, we derive similar conclusions from (3.5.3); we make $x_i \to -\infty$ or $x_i \to +\infty$ (in the ordered case, we must assume $i = 1$ or $i = m$, respectively). Hence [70, 91]:

Theorem 4.5. *Let E' be obtained from the matrix E by coalescence of two rows F_i and F_j, let E^* be obtained by maximal coalescence of F_i in E. Then E is (order or real) strongly singular if E' or E^* has this property. For order singularity, for E', F_i and F_j are adjacent rows; for E^*, $i = 1$ or $i = m$.*

Let us examine what cases are possible for a matrix E and the coalesced matrix E' in the case of polynomial interpolation. We distinguish three classes of matrices: regular (r) matrices, weakly singular (ws) matrices, and strongly singular (ss) matrices. A priori, there are nine possibilities for E, E'. The cases $E' \in$ (ss) and $E \in$ (r) or $\in$ (ws) are ruled out by Theorem 4.5. Many examples for E', $E \in$ (ss) are given by Theorem 6.2, according to which a Birkhoff matrix is strongly singular if one of its rows contains precisely one odd supported sequence. Cases $E' \in$ (r) and $E \in$ (ss), (ws), or (r) are realized if we coalesce matrices of (1.3.7) to two rows. The three remaining examples are due to Kimchi and Richter-Dyn ([76] and oral communication). Let

$$E' = \begin{bmatrix} 0 & 1 & 1 & 0 & 0 & 0 \\ 0 & 0 & 1 & 0 & 1 & 0 \\ 1 & 1 & 0 & 0 & 0 & 0 \end{bmatrix}, \quad E_4 = \begin{bmatrix} 0 & 1 & 1 & 0 & 0 & 0 \\ 0 & 0 & 1 & 0 & 0 & 0 \\ 0 & 0 & 0 & 0 & 1 & 0 \\ 1 & 1 & 0 & 0 & 0 & 0 \end{bmatrix},$$

$$E_5 = \begin{bmatrix} 0 & 1 & 1 & 0 & 0 & 0 \\ 0 & 0 & 1 & 0 & 1 & 0 \\ 1 & 0 & 0 & 0 & 0 & 0 \\ 0 & 1 & 0 & 0 & 0 & 0 \end{bmatrix}, \tag{4.1.2}$$

with E' obtainable by coalescence from either E_4 or E_5. Here, the first and the third matrix are (ws), by Theorem 1.4 applied to matrix E_2 in (1.3.7). The second matrix is (ss) by Theorems 1.4 and 6.1.

Finally let

$$E = \begin{bmatrix} 0 & 1 & 1 & 0 & 0 & 0 \\ 0 & 0 & 1 & 0 & 1 & 0 \\ 1 & 0 & 0 & 0 & 0 & 0 \\ 1 & 0 & 0 & 0 & 0 & 0 \end{bmatrix}. \tag{4.1.3}$$

The coalescence E' [see (4.1.2)] of E has been seen to be (ws), but the matrix E itself belongs to the class (r). This can be established by computing the determinant, but also in the following way. Let P_5 be a polynomial annihilated by E and the set of knots X: $x_1 < x_2 = 0 < x_3 < x_4$. Then the polynomial P_5' is even, has a double root at $x = x_1$, hence is of the form $P_5' = c(x^2 - x_1^2)^2$. Since it also has an odd number of roots between x_3 and x_4, we must have $P_5' \equiv 0$.

§4.2. THE FUNCTION $\delta(E)$

In a Pólya matrix E, we single out a leading row, for example the first row, and denote by $\alpha_2,\ldots,\alpha_m,\gamma$ the coefficients of collision of row 1 with rows $2,\ldots,m$, and the maximal coefficient of collision of row 1 in E, respectively. We put

$$\delta = \delta(E) = \gamma - \sum_{i=2}^{m} \alpha_i. \qquad (4.2.1)$$

For polynomial interpolation, Theorems 3.9 and 3.10 imply the following. For properly chosen knots $x_2,\ldots,x_m$, the determinant $D(E, X)$ is a polynomial of degree γ in x_1, which has roots of orders α_i at x_i, $i = 2,\ldots,m$. This shows that always $\delta \geqslant 0$. We shall prove this and more (see Theorem 4.9) in a different way.

If E has the canonical decomposition $E = E_1 \oplus \cdots \oplus E_\mu$, then from the formula (3.3.8) we get

$$\delta(E) = \sum_{\lambda=1}^{\mu} \delta(E_\lambda). \qquad (4.2.2)$$

We mention some further properties of the function $\delta(E)$. First, $\delta(E)$ can be reduced to δ's of matrices with fewer rows. If $E = F_1 \cup \cdots \cup F_m, m \geqslant 3$, we put

$$\bar{E} = F_1 \cup F_2 \cup F', \qquad \bar{\bar{E}} = F_1 \cup F_3 \cup \cdots \cup F_m, \qquad F' = (F_3 \cup \cdots \cup F_m)^0.$$

From (4.2.1) we have then

$$\delta(E) = \alpha\left(F_1 \cup (F_2 \cup F')^0\right) - \alpha(F_1 \cup F_2) - \alpha(F_1 \cup F')$$

$$+ \alpha\left(F_1 \cup (F_3 \cup \cdots \cup F_m)^0\right) - \alpha(F_1 \cup F_3) - \cdots - \alpha(F_1 \cup F_m)$$

$$= \delta(\bar{E}) + \delta(\bar{\bar{E}}). \qquad (4.2.3)$$

The value of $\delta(E)$ may depend on the choice of the leading row in E, but not if $m = 3$. In this case, if δ_1 and δ_2 correspond to F_1 and F_2 as leading rows,

$$\delta_1(E) = \delta_2(E). \qquad (4.2.4)$$

Indeed, this relation is equivalent to

$$\alpha\left(F_1 \cup (F_2 \cup F_3)^0\right) + \alpha(F_2 \cup F_3) = \alpha\left(F_2 \cup (F_1 \cup F_3)^0\right) + \alpha(F_1 \cup F_3),$$

and this is true, for by (3.3.3) both sides are equal to $\alpha(F_1 \cup F_2 \cup F_3)$.

Examples

1. For a two-row matrix, $\gamma = \alpha_2$, hence $\delta = 0$.
2. Also, for a Hermitian matrix E with rows F_i of lengths $p_i > 0$, $i = 1,\ldots,m$, we have $\delta = 0$. By (3.3.6), $\alpha_i = \alpha(F_1 \cup F_i) = p_1 p_i$, $i = 2,\ldots,m$, and by the same formula, $\gamma = p_1(p_2 + \cdots + p_m)$.

3. Let E be a Birkhoff matrix of the following type. The (nonzero) rows F_i of E have Hermitian sequences of lengths $p_i \geqslant 0$, $i = 1, \ldots, m$, $m \geqslant 3$. In addition, F_1 has a sequence of 1's for $k = n - r, \ldots, n - 1$, with $n - r > p_1$, and F_2 has a single 1 in position $p_2 + s$; at least one of the numbers $r, s \geqslant 0$ is not 0. Then $\delta \geqslant 1$.

Since E is a Birkhoff matrix, we have

$$p_2 + s < n - r, \qquad p_1 + \cdots + p_m + r = n. \tag{4.2.5}$$

Using Examples 1, 2 of Section 3.3, we have

$$\alpha_i = p_1 p_i, \qquad i = 3, \ldots, m,$$

$$\alpha_2 = p_1 p_2 + (p_1 - s)_+ = p_1 p_2 - s + \max(p_1, s).$$

The coalesced row $F' = (F_2 \cup \cdots \cup F_m)^0$ will have a Hermitian sequence of length $p_2 + \cdots + p_m$, and in addition a 1 at $k_0 = \max(p_2 + \cdots + p_m, p_2 + s)$. In coalescence of F' with F_1, the coefficient of collision of the first $p_2 + \cdots + p_m$ ones in F' with F_1 will be $p_1(p_2 + \cdots + p_m)$, that of the 1 at k_0 with F_1 will be $n - k_0$, hence

$$\gamma = p_1(p_2 + \cdots + p_m) + n - k_0,$$

$$\delta = \gamma - \alpha_2 - \cdots - \alpha_m = n + s - \max(p_1, s) - \max(p_2 + \cdots + p_m, p_2 + s).$$

Relations (4.2.5) yield $\delta \geqslant 1$ in all four possible cases.

Here is a first application of the function δ:

Theorem 4.6.   *If $\delta(E)$ is odd, then E is strongly real singular.*

Proof.   Indeed, $(x_1 - x_2)^{-\alpha_2} \ldots (x_1 - x_m)^{-\alpha_m} D(E, X)$ is a polynomial in x_1 of odd degree, and not vanishing (for proper choice of knots) at $x_2, \ldots, x_m$. Hence it has a real root x_1 different from $x_2, \ldots, x_m$.   □

A matrix E is *complex regular* if its determinant $D(E, X) \neq 0$ for any choice of distinct *complex* knots $x_1, \ldots, x_m$. The following simple result connects this notion with the function $\delta(E)$:

Proposition 4.7.   *A Pólya matrix E is complex regular if and only if $\delta(E) = 0$.*

Proof.   For the polynomial $D(x_1) = D(E, X)$, with fixed complex knots $x_2, \ldots, x_m$ and variable x_1, we denote by g its degree, by $\beta_2, \ldots, \beta_m$ the multiplicities of its roots $x_2, \ldots, x_m$. By Theorems 3.9 and 3.10, $g \leqslant \gamma$, $\beta_i \geqslant \alpha_i$, $i = 2, \ldots, m$. Hence if $\delta = 0$, then $g \leqslant \beta_2 + \cdots + \beta_m$, and we must have $g = \beta_2 + \cdots + \beta_m$. This means that, for each choice of $x_2, \ldots, x_m$, $D(x_1)$ has no other zeros than $x_2, \ldots, x_m$ or that E is complex regular. Conversely, if $\delta > 0$, then for almost all choices of knots, $g = \gamma$, $\beta_i = \alpha_i$, $i = 2, \ldots, m$. For knots $x_2, \ldots, x_m$ of this type, $D(x_1)$ will have a zero different from them, and E will be complex singular.   □

For example, a two-row Pólya matrix is complex regular, since it is order regular (see Section 1.3). Each Hermitian matrix E is complex regular.

We can prove this by computing the determinant $D(E, X)$ (see Karlin and Studden [H, pp. 7 and 9]), or by writing the fundamental polynomials of Hermite interpolation (see Goncharov [F, p. 64]). For us, the most natural approach is by means of Proposition 4.7 and Example 2.

§4.3. THE THEOREM OF FERGUSON

The proof of Ferguson's theorem given here is based on a simplifying operation T for E introduced in [103]. After finitely many applications of T, the problem is reduced to one of two simple cases: (a) E is as in Example 3 of the last section; (b) E is a two-row matrix.

For a Birkhoff matrix E with leading row F_1, we define an operation $T(E) = E_1$ as follows. In the rows $F_2,\ldots,F_m$ we select $e_{i,k} = 1$, with largest possible k. If there are several k of this kind, we select a row with the maximal number of 1's in it. Without loss of generality, let this be $e_{2,k} = 1$. We replace this 1 by 0, and omit the last column of E (which consists of 0's), obtaining an $m \times n$ matrix $E' = T'(E)$. This E' is a Pólya (but not necessarily a Birkhoff) matrix. However, the Birkhoff condition is satisfied for all columns $l < k$ of E', and also for column k, if there is more than a single 1 in this column. It follows that all 1's of rows $2,\ldots,m$ of E' are contained in the first matrix E_1 of the canonical decomposition $E' = E_1 \oplus \cdots \oplus E_\mu$ into Birkhoff matrices. Thus, all matrices E_λ, $\lambda \geq 2$ (if they exist), are one-column matrices with a 1 in row 1, and with $\delta(E_\lambda) = 0$. From (4.2.2), $\delta(E_1) = \delta(E')$. We define $T(E) = E_1$.

For simplicity, we assume here and in Theorem 4.9 that the leading row F_1 has been so selected that $e_{1,0} = 1$.

Lemma 4.8. *For a Birkhoff matrix E, and each selection of the leading row,*

$$\delta(TE) \leq \delta(E). \tag{4.3.1}$$

Moreover, we have strict inequality in (4.3.1) if after application of T, the matrix becomes a two-row matrix.

Proof. We show that

$$\delta(E') \leq \delta(E). \tag{4.3.2}$$

It is clear that the numbers $\alpha_i(E)$, $i = 3,\ldots,m$, are not affected by the operation T.

Let $l_1 < \cdots < l_p$ be the positions of 1's in row 1, $l_1' < \cdots < l_p'$ the positions of 1's in precoalescence of row 1 with respect to row 2; let $l_1^* < \cdots < l_p^*$ be the positions of 1's in the maximal precoalescence of row 1. None of the l_j', l_j^* is equal to k.

Since k is the highest nonzero column in $F_2 \cup \cdots \cup F_m$, only those l_j^* which satisfy $l_j^* > k$ will change if $e_{2,k} = 1$ is removed. The row

$(F_2 \cup \cdots \cup F_m)^0$ has a sequence of 1's, in positions $k, \ldots, k_1$, $k_1 \geqslant k$, followed by 0's; hence the $l_j^* > k$ are precisely the integers $k_1 + 1, \ldots, n$. If $e_{2,k} = 1$ is removed, these l_j^* decrease by exactly one.

Similarly only $l_j' > k$ can change if $e_{2,k} = 1$ is removed; they will decrease by one or remain constant. Since $l_j' > k$ implies $l_j^* > k$, the sum $\gamma - \alpha_2 = \sum_{j=1}^{p} (l_j^* - l_j')$ can only decrease.

Suppose that TE is a two-row matrix. Then E is a three-row Birkhoff matrix where row F_2 contains only a single nonzero entry, $e_{2,k} = 1$, and in row F_3, $e_{3,0} = 1$ and we can assume that $e_{3,l} = 0$ for $l \geqslant k$. By Example 1, $\delta(TE) = 0$. In Lemma 3.6, we take $F_i = F_3$ and $F' = F_1$. Since $F_1 \cup F_3$ satisfies the Pólya condition for $k = 0, \ldots, n-1$, the number k' of Lemma 3.6 is n. Therefore, by this lemma, $\gamma - \alpha_3 = n - k$, since the addition of a 1, $e_{3,k} = 1$, is equivalent to coalescing rows 2 and 3. On the other hand, $\alpha_2 = k_1 - k$ where k_1 is the position of the first 0 in row 1 with $k_1 \geqslant k$. Since E is a Birkhoff matrix and $e_{2,k} = e_{3,0} = 1$, we have $k_1 \leqslant n-1$. Hence,

$$\delta(E) = \gamma - \alpha_3 - \alpha_2 = n - k_1 \geqslant 1. \qquad \Box$$

Theorem 4.9. *For a Pólya matrix E, (i) $\delta(E) \geqslant 0$. (ii) We have $\delta(E) > 0$ if and only if one of the matrices of the canonical decomposition of E is non-Hermitian, with at least three nonzero rows.*

Proof. Because of (4.2.2), we may assume that E is a Birkhoff matrix. We apply the operation T of Lemma 4.8 several times. After finitely many steps, a Hermitian matrix results. Then $\delta = 0$. This gives (i).

To prove (ii), we assume that E is a Birkhoff matrix with $m \geqslant 3$ that has gaps. A gap is a maximal sequence of 0's followed by a 1 in a row. At each application of T, at most two gaps in E are destroyed, possibly in the row where $e_{i,k} = 1$ has been replaced by 0, and in row 1, after the canonical decomposition.

We can assume that at each step E has a nonzero first row and at least three rows. Then our process will stop at the moment when the gaps disappear. At the last application of T, the matrix E must be of the type of Example 3 (§4.2). For this matrix, $\delta \geqslant 1$, and we obtain (ii) again by Lemma 4.8. $\qquad \Box$

Combining Proposition 4.7 and Theorem 4.9, we have

Theorem 4.10. *A Pólya matrix E is complex regular if and only if its canonical decomposition consists of matrices that are Hermitian, or have at most two nonzero rows.*

For Birkhoff matrices E, this has been given by D. Ferguson [37]; the general case appears in [92].

Exercise. Use property (4.2.4) in order to avoid reduction to two rows in the proof of Theorem 4.9, thereby eliminating the need for the special argument in Lemma 4.8.

§4.4. SINGULARITY AND PERMUTATION NUMBERS

Let E be a Pólya matrix, $\mathcal{G}$ a system of analytic functions $\{g_1, \ldots, g_m\}$ on $[a, b]$, and X: $x_1 < \cdots < x_m$ an increasing set of knots. There is yet another way to obtain new singularity criteria for the matrix E which are based on the expansions (3.5.2) and (3.5.3) of the determinant $D(E, X)$. The permutation numbers of Section 3.5 will be used in this approach.

For two adjacent rows i and $j = i - 1$ of E, we consider the coalescence of F_i to F_{i-1}, given by $x_i \to x_{i-1}$; the permutation number $\sigma = \sigma_i$ (which depends only on E and i) appears in the formula (3.5.2). From this formula we derive the following. Let X' be a set of knots $x_j, j \neq i$, for which $D(E', X') \neq 0$ (this is true for almost all $X' \in \mathbf{R}^{m-1}$). If X is obtained by adding x_i to X', and if x_i is sufficiently close to x_{i-1}, then

$$\operatorname{sign} D(E, X) = (-1)^\sigma \operatorname{sign} D(E', X'). \tag{4.4.1}$$

A similar statement (for the polynomial case) can be made for the determinant $D(E^*, X^*)$ in (3.5.3) and for the permutation number σ^* of the maximal coalescence in E of F_m.

Let $F_1, \ldots, F_p, p > 2$, be adjacent rows of E (listed in order from top to bottom); let their coalescence be $F' = (F_1 \cup \cdots \cup F_p)^0$. There are at least two different ways to obtain F' by coalescing rows F_j one by one. For example, when $p = 3$, there are exactly two ways if we proceed in decreasing order, namely:

$$\left(\left(F_3 \cup (F_2 \cup F_1)^0 \right)^0 \right)^0, \qquad \left((F_3 \cup F_2)^0 \cup F_1 \right)^0. \tag{4.4.2}$$

Let $\sigma_1, \ldots, \sigma_{p-1}$ and $\sigma'_1, \ldots, \sigma'_{p-1}$ be the two sets of permutation numbers corresponding to any two such ways.

Theorem 4.11. *A Pólya matrix E is strongly order singular if*

$$\sigma_1 + \cdots + \sigma_{p-1} \not\equiv \sigma'_1 + \cdots + \sigma'_{p-1} \pmod{2}. \tag{4.4.3}$$

Proof. Let E' be the matrix E with rows $F_1, \ldots, F_p$ coalesced to F'. There is a set of knots X' for which $D(E', X') \neq 0$. If ε is the sign of this determinant, we can, using (4.4.1) several times, find sets of knots X, Y for which

$$\operatorname{sign} D(E, X) = (-1)^{\Sigma \sigma_i} \varepsilon, \qquad \operatorname{sign} D(E, Y) = (-1)^{\Sigma \sigma'_i} \varepsilon. \qquad \square$$

Remarks. 1. We have real strong singularity of E if (4.4.3) holds for any set of rows $F_1, \ldots, F_p$, not necessarily adjacent.

2. Some of the σ_j, σ'_j in (4.4.3) may correspond to maximal coalescences.

3. If $F_1, \ldots, F_p$ are not listed in order, then a sign may be introduced by the collision numbers α of the coalescences. (See §4.5, where this is used.)

Theorem 4.11 appears in Lorentz [94]; several other authors [70, 164, 130] have obtained, by different methods, variations or special cases of this, for three-row matrices E. See §4.5.

As an example, we discuss in more detail the case of three rows F_1, F_2, F_3. Of course, this case is of interest also if E has more than three rows.

We occasionally write a row F of 0's and 1's as an increasing sequence of integers, $F = (\lambda_1, \ldots, \lambda_t)$. Here λ_j is the position of the jth 1 in F. Precoalescences of F with F_1, with $(F_1 \cup F_2)^0$, and so on, will be denoted by $(F)_1 = (\lambda'_1, \ldots, \lambda'_t)$, $(F)_{1,2}$, and so on. These operations transform the increasing sequence of integers $\lambda_1, \ldots, \lambda_t$ into another such sequence. It is useful to extend these operations to any rearrangement $\lambda_1^*, \ldots, \lambda_t^*$ of the λ_j. For example, if $(\lambda_1, \ldots, \lambda_t)_1 = \mu_1, \ldots, \mu_t$, with $\mu_j = F(\lambda_j)$, we put $(\lambda_1^*, \ldots, \lambda_t^*)_1 = F(\lambda_1^*), \ldots, F(\lambda_t^*)$. Thus, if $F = (1010100100) = (0,2,4,7)$, $F_1 = (1001100110) = (0,3,4,7,8)$, then $(0,2,4,7)_1 = (1,2,5,9)$, and $(4,2,7,0)_1 = (5,2,9,1)$.

With these notations, we can describe the permutation numbers of the two coalescences (4.4.2). The original determinant $D(E, X)$ has rows of the type

$$\{g_0^{(l)}(x), \ldots, g_N^{(l)}(x)\}. \tag{4.4.4}$$

To the row F_1 of E and the knot x_i correspond p rows (4.4.4) with $l = l_j^1$, $j = 1, \ldots, p$, and $x = x_i$; similarly, to the row F_2 will correspond q rows (4.4.4), $l = l_j^2$ and $x = x_{i+1}$. After coalescence of F_2 to F_1, the last set of q rows will be replaced by rows of type (4.4.4) with orders of derivatives taken from the sequence $(l_1^2, \ldots, l_q^2)_1$, and with $x = x_i$. This describes all $p + q$ rows of the determinant $D(\bar{E}, X')$ of §3.4 that contain x_i. They have derivatives of orders

$$l_1^1, \ldots, l_p^1, \left(l_1^2, \ldots, l_q^2\right)_1, \tag{4.4.5}$$

following as in this sequence. Hence σ_1 is the number of interchanges (defined modulo 2) that bring the sequence of integers (4.4.5) to its natural order $k_1 < \cdots < k_{p+q}$.

After this, the coalescence of F_3 to $(F_1 \cup F_2)^0$ will produce a determinant $D(\bar{E}, X')$ with rows containing x_i having derivatives of orders $k_1, \ldots, k_{p+q}, (l_1^3, \ldots, l_r^3)_{1,2}$, and listed in this sequence. Then σ_2 will be the number of interchanges that makes this sequence monotone. It follows from this that there are $\sigma_1 + \sigma_2$ permutations that bring the sequence of $p + q + r$ integers

$$l_1^1, \ldots, l_p^1, \left(l_1^2, \ldots, l_q^2\right)_1 \left(l_1^3, \ldots, l_r^3\right)_{1,2} \tag{4.4.6}$$

to its natural order.

Likewise, the sum of the permutation numbers $\sigma'_1 + \sigma'_2$ of the second coalescence (4.4.2) is the number of interchanges that make monotone the sequence

$$l_1^1, \ldots, l_p^1, \left(l_1^2, \ldots, l_q^2, \left(l_1^3, \ldots, l_r^3\right)_2\right)_1. \tag{4.4.7}$$

We can omit the first p elements in (4.4.6) and (4.4.7) and obtain

Theorem 4.12 [94]. *The Pólya matrix E is strongly order singular if it contains three adjacent rows for which the two sequences*

$$S_1 = \left(l_1^2, \ldots, l_q^2\right)_1 \left(l_1^3, \ldots, l_r^3\right)_{1,2},$$

$$S_2 = \left(l_1^2, \ldots, l_q^2, \left(l_1^3, \ldots, l_r^3\right)_2\right)_1 \tag{4.4.8}$$

belong to different permutation classes.

An interesting corollary of this theorem is that three rows of a matrix E can be "so bad" that E will be strongly singular no matter what the nature of its other rows.

§4.5.   APPLICATIONS

The results of §4.4 have some elementary but useful applications. We begin with some very simple examples.

Examples

1. Let F_1, F_2, F_3 consist of a single 1 in positions k, l, k. If $l = k + 1$, then

$$(F_2)_1 = k + 1, \qquad (F_3)_{1,2} = k + 2, \qquad S_1 = (k + 1, k + 2);$$
$$(F_3)_2 = k, \qquad S_2 = \left(F_2(F_3)_2\right)_1 = (k + 1, k)_1 = (k + 2, k + 1).$$

The matrix is singular, which also follows from Theorem 6.1. If $l > k + 1$, application of (4.4.8) is not conclusive.

2. Let F_1, F_3 be Hermitian rows of lengths p, q, while F_2 has 1's in positions $k_1 < \cdots < k_r$, $k_r \leqslant p, q$. Then S_1 is monotone increasing, while S_2 needs $k_1 + \cdots + k_r - \frac{1}{2}r(r - 1)$ permutations to achieve this. Therefore E is singular if $k_1 + \cdots + k_r \not\equiv \frac{1}{2}r(r - 1) \pmod 2$.

It is interesting to compare Theorem 4.12 with criteria of singularity of three-row matrices E found in the literature [70, 164, 130].

For a sequence of distinct integers $S = (l_1, \ldots, l_p)$ we denote by $\sigma(S)$ the class of permutations that bring S to the natural order: We put $\sigma(S) = 0$ if an even number of interchanges are required, $\sigma(S) = 1$ otherwise. Let F_1, F_2 be two adjacent rows of E in locations $i, i + 1$. With the notations of §4.4, $\sigma_1 = \sigma(F_1, (F_2)_1) = \sigma(l_1^1, \ldots, l_p^1, (l_1^2, \ldots, l_q^2)_1)$, $\sigma_2 = \sigma((F_1)_2, F_2) = \sigma((l_1^1, \ldots, l_p^1)_2, l_1^2, \ldots, l_q^2)$ we have

$$\sigma(F_1, (F_2)_1) \equiv \sigma((F_1)_2, F_2) + \alpha_{1,2}. \tag{4.5.1}$$

We use matrices $\bar{E}, E'$ of §3.4. In §4.4 we saw that for x_{i+1} close to x_i, the determinant $D(E, X)$ has the same sign as $D(\bar{E}, X')$, where $\bar{E}$ is obtained by coalescing F_2 to F_1, and the same sign as $(-1)^{\sigma_2} D(E', X')$. Similarly, coalescing F_1 to F_2, we see under the same conditions that $D(E, X)$ has the sign of $(x_i - x_{i+1})^{\alpha_{1,2}} (-1)^{\sigma_1} D(E', X')$. Comparing signs, we obtain (4.5.1).

Proposition 4.13 (Sharma and Tzimbalario). *A three-row matrix that has rows $F_1 = (l_1^1, \ldots, l_p^1)$ and $F_3 = (l_1^3, \ldots, l_r^3)$ and has 0's in positions $k_1 < \cdots < k_{p+r}$ of row F_2 is strongly order singular if $l_j^1 \leq k_j$, $j = 1, \ldots, p$, $l_j^3 \leq k_j$, $j = 1, \ldots, r$, and if*

$$\sum_{j=1}^{p} (k_{r+j} - k_j) + pr \equiv 1 \quad (\bmod 2). \tag{4.5.2}$$

Proof. The singularity criteria of Theorem 4.12 can be written $\sigma(S_1) - \sigma(S_2) \equiv 1$, where S_1 and S_2 are the sequences (4.4.6) and (4.4.7), respectively. By means of (4.5.1) we compute

$$\sigma(S_1) - \sigma(S_2) \equiv \sigma((F_1)_2, F_2, (F_3)_{1,2}) + \alpha_{1,2}$$

$$+ \sigma((F_1)_{2,3}, F_2, (F_3)_2) + \alpha_{1,(2,3)}$$

$$\equiv \sigma((F_1)_2, (F_3)_{1,2}) + \sigma((F_1)_{2,3}, (F_3)_2) + \alpha_{1,2} + \alpha_{1,(2,3)}$$

$$\equiv \sigma(k_1, \ldots, k_{p+r}) + \sigma(k_{r+1}, \ldots, k_{r+p}, k_1, \ldots, k_r)$$

$$+ \sum_{j=1}^{p} (k_j - l_j^1) + \sum_{j=1}^{p} (k_{r+j} - l_j^1) \equiv rp + \sum_{j=1}^{p} (k_{r+j} - k_j). \quad \square$$

Karlin and Karon [70] were the first to obtain, for three-row matrices, a condition of singularity of the form $\Omega \equiv 1$, where Ω depends in a simple way on E. We do not reproduce their expression of Ω; it takes some work (carried out in [102]) to show that the condition is equivalent to that of Theorem 4.12 for three-row matrices. Karlin and Karon used properties of totally positive matrices and not coalescence.

Exercises

1. [130]. If E of Proposition 4.13 has Hermitian first and last rows with $p = 1$, then E is strongly order singular if it has an odd number of odd supported sequences.

2. One can replace coalescences (4.4.2) by different ones, for instance, the coalescence of F_2 to F_1 by that of F_1 to F_2, and so on. By means of (4.5.1) and (3.3.3), prove that all criteria obtained in this way are equivalent to that of Theorem 4.12.

The fact that the matrix E_1 of (1.3.7) is singular has the following generalization for matrices of arbitrary length (see [99]).

Proposition 4.14. *If matrices E_1 and E_2 are identical except for columns k and $k+1$ from three adjacent rows F_1, F_2, F_3 which are A for E_1 and B for E_2:*

$$A = \begin{bmatrix} 1 & 0 \\ 1 & 0 \\ 1 & 0 \end{bmatrix}, \qquad B = \begin{bmatrix} 1 & 0 \\ 0 & 1 \\ 1 & 0 \end{bmatrix}, \tag{4.5.3}$$

then at least one of the matrices E_1, E_2 is strongly order singular.

Proof. We compare the sequences (4.4.8) for the matrices E_1 and E_2. There is a 1 in position k of row F_1. It follows that the 1 in columns k or $k+1$ of row F_2 in precoalescence with row F_1 goes to the same position for both E_1 and E_2. Hence sequences S_1 coincide for these matrices.

On the other hand, when row F_3 is precoalesced with row F_2, the 1 in column $l \leqslant k$ of row F_3 that goes to position k for E_2 must go to position $k+1$ for E_1, all remaining positions are unchanged. This means that the sequences S_2 are

$$\left(l_1^2, \ldots, l_t^2, k, \ldots, l_q^2, l_1^{3\prime}, \ldots, l_{s-1}^{3\prime}, k+1, \ldots, l_r^{3\prime} \right)_1$$

and

$$\left(l_1^2, \ldots, l_t^2, k+1, \ldots, l_q^2, l_1^{3\prime}, \ldots, l_{s-1}^{3\prime}, k, \ldots, l_r^{3\prime} \right)_1$$

for E_1 and E_2, respectively (here l_j^3 goes to $l_j^{3\prime}$ in precoalescence of row F_3 with row F_2). Since the relative positions are unchanged by the operation $(\ )_1$, we see that the sequences are of different permutation classes.   □

§4.6. THE COALESCENCE CONSTANT; SIGNS OF DETERMINANTS

From Theorem 3.10 we can derive signs and values of useful determinants.

The constant $C = C(l_1, \ldots, l_p; l_1', \ldots, l_p')$, which appears in (3.4.5), was introduced in a paper [108] by Lorentz and Zeller. It is determined by the coalescence of a row $F_2 = (l_1, \ldots, l_p)$, $0 \leqslant l_1 < \cdots < l_p \leqslant n$, with another row F_1, the 1's at l_j going under precoalescence with F_1 to positions $l_j' \geqslant l_j$, $j = 1, \ldots, p$. At the l_j' we have 0's of F_1, and if $E = F_1 \cup F_2$ is a Pólya matrix, at the remaining positions $k = k_j$, $j = 1, \ldots, q$, $p + q = n+1$ are the 1's of F_1.

Proposition 4.15. *Let E be a $2 \times (n+1)$ Pólya matrix with rows $F_1 = (k_1, \ldots, k_q)$ and $F_2 = (l_1, \ldots, l_p)$. If $\alpha = \alpha(F_1 \cup F_2)$ and $X = \{0, x\}$, then*

$$D(E, X) = \frac{C}{\alpha!}(-1)^{q(q-1)/2 + k_1 + \cdots + k_q} x^\alpha. \tag{4.6.1}$$

Proof. By Theorem 3.10, $D(E, X)$ is a polynomial in x of degree $\leqslant \alpha$ with a zero at $x = 0$ of order $\geqslant \alpha$. Relation (4.6.1) will follow from (3.5.3) if we show that

$$\sigma \equiv \frac{q(q-1)}{2} + k_1 + \cdots + k_q \pmod{2}. \tag{4.6.2}$$

From the discussion of §4.4 we see that σ is the number of interchanges required to bring $k_1, \ldots, k_q, l_1', \ldots, l_p'$ into the natural order. Since $q = n + 1 - p$, k_q moves past $p - (n - k_q) = k_q - (q-1)$ of the l_j', k_{q-1} moves past $k_{q-1} - (q-2)$ of them, and so on. Hence we obtain (4.6.2).   □

The first q rows of the determinant $D(E, X)$ contain only single 1's, with the 1 in row j, $j = 1, \ldots, q$, being in position k_j.

Corollary 4.16. *Let $M(E, X)$ be the minor of the determinant (4.6.1) obtained by omitting the first q rows and the columns $k_1, \ldots, k_q$. Then*

$$M(E, X) = \frac{Cx^\alpha}{\alpha!} > 0 \qquad for \quad 0 < x. \qquad (4.6.3)$$

Proof. Expanding $D(E, X)$ successively about its first q rows, we have $M(E, X) = (-1)^\rho D(E, X)$. The contribution of the row numbers to ρ is $1 + 2 + \cdots + q = q(q+1)/2$, while the contribution from the column numbers is $(k_1 + 1) + \cdots + (k_q + 1) = \sum_{j=1}^q k_j + q$. Thus, $\rho = q(q+3)/2 + k_1 + \cdots + k_q$ and (4.6.3) follows from (4.6.1).   □

For $0 \leqslant j$ we shall adopt the notation

$$(m)_j = j! \binom{m}{j} = m(m-1) \cdots (m - j + 1); \qquad (m)_0 = 1. \quad (4.6.4)$$

Then we can write

$$M(E, X) = \det\left[(l_1')_{l_j} \frac{x^{l_1' - l_j}}{l_1'!}, \ldots, (l_p')_{l_j} \frac{x^{l_p' - l_j}}{l_p'!}; j = 1, \ldots, p \right]$$

$$= \frac{x^\alpha}{l_1'! \cdots l_p'!} \det\left[(l_1')_{l_j}, \ldots, (l_p')_{l_j}; j = 1, \ldots, p \right],$$

where $\alpha = \sum_{j=1}^p (l_j' - l_j)$. This leads to two interesting consequences.

The first gives an explicit formula for the constant C (see [103]):

Corollary 4.17. *We have*

$$C(l_1, \ldots, l_p; l_1', \ldots, l_p') = \frac{\alpha!}{l_1'! \cdots l_p'!} \det\left[(l_1')_{l_j}, \ldots, (l_p')_{l_j}; j = 1, \ldots, p \right].$$

where $\alpha = \sum_{j=1}^p (l_j' - l_j)$.

The second corollary will be very useful in Chapter 11.

Corollary 4.18. [192]. *For any positive integer q and for integers l_j, l_j', $j = 1, \ldots, p$, satisfying $0 \leqslant l_1 < \cdots < l_p$, $0 \leqslant l_1' < \cdots < l_p'$, $l_j \leqslant l_j'$, $j = 1, \ldots, p$, the determinant*

$$\det\left[\binom{l_1'}{l_j}, \ldots, \binom{l_p'}{l_j}; j = 1, \ldots, p \right]$$

$$= \frac{l_1! \cdots l_p!}{l_1'! \cdots l_p'!} \det\left[(l_1')_{l_j}, \ldots, (l_p')_{l_j}; j = 1, \ldots, p \right]$$

$$(4.6.5)$$

is greater than 0.

Exercise. If $0 \leqslant m_0 < \cdots < m_p$ are integers and $0 < x_1 < \cdots < x_{p+1}$, then

$$\det\left[x_i^{m_0}, \ldots, x_i^{m_p}; \, i = 1, \ldots, p+1\right] > 0. \qquad (4.6.6)$$

This determinant is a well-known generalization of the Vandermonde determinant (see [H]).

Chapter 5

Rolle Extensions and Independent Sets of Knots

§5.1. ROLLE EXTENSIONS

We have used Rolle's theorem in the proof of the basic Theorem 1.5 of Atkinson and Sharma. The proof worked—Rolle's theorem produced a zero—because we assumed that the matrix E has no odd supported sequences. This argument may be applicable in other situations. The main purpose of this chapter is to show that for each system $\mathcal{G}$ there is a set of knots X (called *independent* knots) with the property that for polynomials in $\mathcal{G}$, Rolle's theorem produces zeros at each step, *whatever the structure of the* (*Pólya*) *matrix E may be*. The method of independent knots was formulated by Lorentz and Zeller [107] and developed further by Lorentz ([90,92]) in order to study singular interpolation matrices. This chapter is based on the paper [96].

Let E be an $m \times (n+1)$ interpolation matrix, and f be an n times differentiable function on $[a, b]$ that is annihilated by E and $X = \{x_1, \ldots, x_m\} \subset [a, b]$; that is, let f satisfy

$$f^{(l)}(x_i) = 0 \quad \text{whenever} \quad e_{i,l} = 1 \text{ in } E. \qquad (5.1.1)$$

The pair E, X defines the equations (5.1.1) and conversely. We shall often identify this pair with the equations. From the zeros of f and its derivatives specified by (5.1.1) we can derive further zeros by means of Rolle's theorem. A selection of a complete set of such zeros is called a *Rolle extension of the equations* (5.1.1). This is a pair $\tilde{E}$, $\tilde{X}$ with the corresponding equations

$$f^{(l)}(\tilde{x}_i) = 0, \qquad \tilde{e}_{i,l} = 1 \text{ in } \tilde{E}, \tag{5.1.2}$$

which contain all of the equations (5.1.1), but in general also some additional equations. The extension is not unique. The formal definition is as follows:

A *Rolle extension* $\mathcal{R} = \tilde{E}, \tilde{X}$ for a function f annihilated by the pair E, X [or for the equations (5.1.1)] is obtained by induction by selecting *Rolle extensions* $\mathcal{R}_k = E^k, X^k$ for each $k = 0, 1, \ldots, n$. Here, E^k is an $m_k \times (n - k + 1)$ matrix (with columns numbered $k, k + 1, \ldots, n$), and the equations of $\mathcal{R}_k$ contain all equations (5.1.1) with $l \geqslant k$.

The equations of $\mathcal{R}_0$ are simply the set (5.1.1). If $\mathcal{R}_0, \ldots, \mathcal{R}_k$ have been already selected, we choose a pair $E^{k+1}, X^{k+1} = \mathcal{R}_{k+1}$ according to the following prescriptions:

1. It contains all equations of E^k, X^k for derivatives $f^{(l)}, l \geqslant k + 1$.
2. Between any two *adjacent* zeros $\alpha < \beta$ of $f^{(k)}$ belonging to $\mathcal{R}_k$, we select, if possible, a zero of $f^{(k+1)}$ not listed in $\mathcal{R}_k$. This could be (a) a new zero ξ of $f^{(k+1)}$, or (b) a new zero ξ of $f^{(l)}, l > k + 1$, if equations $f^{(k+1)}(\xi) = 0, \ldots, f^{(l-1)}(\xi) = 0$ are contained in $\mathcal{R}_k$. In this case the multiplicity of ξ as a zero of $f^{(k+1)}$, which is acknowledged by $\mathcal{R}_k$, is increased by at least one.
3. If for a pair $\alpha < \beta$ this is impossible, we register a *loss* and do not add a new equation to $\mathcal{R}_k$ for the pair α, β.

The Rolle extension $\mathcal{R}$ consists of all equations contained in all $\mathcal{R}_k$, $k = 0, 1, \ldots, n$. In other words, the equations (5.1.2) of $\mathcal{R}$ for a given l, $l = 0, \ldots, n$, consist of the equations for $f^{(l)}$ that belong to the extension $\mathcal{R}_l$. A Rolle extension constructed without losses at any of its steps is called *maximal*. An extension $\mathcal{R}$ in which 2(b) has never been used is called an extension *without duplication*. A function f may have many Rolle extensions $\mathcal{R}$; some of them may be maximal, while others are not maximal.

Some properties of Rolle extensions are immediate consequences of the selection procedure. A zero η of $f^{(k)}$ in $\mathcal{R}_k$ of multiplicity σ is also a zero of $f^{(k+1)}$ in $\mathcal{R}_{k+1}$, of multiplicity exactly $\sigma - 1$. A new zero ξ of $f^{(k+1)}$ selected by 2(a) or 2(b) has in $\mathcal{R}_{k+1}$ a multiplicity not less than $\tau + 1$ if τ is the multiplicity of $f^{(k+1)}(\xi) = 0$, acknowledged by $\mathcal{R}_k$ [$\tau = 0$ in case 2(a)]. This multiplicity will be greater than $\tau + 1$ exactly when $\mathcal{R}_k$ contains the equation $f^{(l+1)}(\xi) = 0$ [see 2(b)] as well.

It follows also that the matrix E^{k+1} contains as a submatrix the last $n - k$ columns of E^k (hence also the last $n - k$ columns of E). In part 2 of the construction, for given $\alpha < \beta$, ξ can be found (there is no loss) if we assume that rows of E for which $\alpha < x_i < \beta$ have no odd supported sequences. This follows at once from the extended form of Rolle's theorem, Lemma 1.6. In particular:

Lemma 5.1. *If the matrix E has no odd supported sequences, then all Rolle extensions of a function f annihilated by E, X are maximal.*

We also have

Lemma 5.2. *A maximal Rolle extension $\mathcal{R}$ of a pair E, X has the properties:* (i) *If E satisfies the Pólya condition $M_l \geq l+1$ for $0 \leq l \leq k_0$, then all matrices E^k, $k \leq k_0$, also satisfy this condition for $l \leq k_0$;* (ii) *if E is a Pólya matrix, then all E^k are Pólya matrices.*

Under certain conditions we can find a simple formula for the number of equations for $f^{(k)}$ in $\mathcal{R}$. Let m_k, M_k, $k = 0, 1, \ldots, n$, be the Pólya functions of E; let $\mu_{-1} = 0$ and

$$\mu_k = \left(\cdots \left((m_0 - 1)_+ + m_1 - 1 \right)_+ \cdots + m_{k-1} - 1 \right)_+ + m_k, \qquad k = 0, \ldots, n.$$

$$(5.1.3)$$

We have then

$$\mu_0 = m_0, \quad \mu_k = \left(\mu_{k-1} - 1 \right)_+ + m_k, \qquad k = 1, \ldots, n. \qquad (5.1.4)$$

In particular, if E is a Pólya matrix, we can drop all the subscripts $+$ in these formulas; then

$$\mu_k = M_k - k, \qquad k = 0, \ldots, n. \qquad (5.1.5)$$

Lemma 5.3. *Let $\mathcal{R}$ be a maximal Rolle extension of the equations* (5.1.1) *(or, of E) obtained without duplication. Then the number of equations for $f^{(k)}$ in $\mathcal{R}_k$ is exactly μ_k. Moreover, $\mathcal{R}$ contains a Rolle zero ξ of $f^{(k)}$, $f^{(k)}(\xi) = 0$, if and only if $e^0_{i,k} = 1$, where E^0 is the coalescence of E to a one-row matrix.*

Proof. Let this be true for some k. Then the number of adjacent pairs of zeros $\alpha < \beta$ of $f^{(k)}$ in $\mathcal{R}_k$ is $(\mu_k - 1)_+$, hence the number of different zeros of $f^{(k+1)}$ in the construction of $\mathcal{R}_{k+1}$ is $(\mu_k - 1)_+ + m_{k+1} = \mu_{k+1}$. The last statement follows from the definition of the level function for E^0 (see §3.1).   □

Corollary 5.4. *Let f, annihilated by E, X, be such that at each step $k = 0, \ldots, n$, the Rolle zeros of f can be chosen so as to avoid all x_i, $i \neq i_0$. If row i_0, $1 < i_0 < m$, of E has no odd supported sequences, and E satisfies the Birkhoff conditions for all k with $0 \leq k < k_0$, then there is a Rolle extension $\mathcal{R}$ of E, X having for each $k \leq k_0$ either μ_k or $\mu_k - 1$ Rolle zeros. The last case can happen only if $e_{i_0,k} = 1$ belongs to an (even) supported sequence.*

Proof. Indeed, in constructing $\mathcal{R}_k$, we always have $\mu_k = (\mu_{k-1} - 1)_+ + m_k$ zeros until there is duplication at some level k. This means that in prescription 2 we have $f^{(k-1)}(\alpha) = f^{(k-1)}(\beta) = 0$, $\alpha < x_{i_0} < \beta$, and that

Rolle's theorem produces a zero $f^{(l)}(x_{i_0}) = 0$, $l > k + 1$, according to 2(b). Then for $k \leqslant j < l$ we have $\mu_j - 1 = (\mu_{j-1} - 2)_+ + m_j$ Rolle zeros; for $j = l$ again μ_l zeros, and so on.   □

§5.2. AN AUXILIARY THEOREM

In Chapter 6 we shall need relations between the number of Rolle zeros of E and of different matrices derived from E. Let $1 < i_0 < m$ be fixed. Let E' be the $i_0 \times (n + 1)$ matrix consisting of the rows $i = 1, 2, \ldots, i_0$ of E, and E'' be the $(m - i_0 + 1) \times (n + 1)$ matrix consisting of the rows $i = i_0, \ldots, m$ of E. If m_k, m'_k, and m''_k are the respective Pólya functions for E, E', and E'', then

$$m_k = m'_k + m''_k - e_{i_0, k}, \qquad k = 0, 1, \ldots, n. \tag{5.2.1}$$

Let μ_k, μ'_k, and μ''_k be the numbers defined by (5.1.3) for the matrices E, E', and E'', respectively.

Theorem 5.5. (i) *Assume that the matrix E satisfies the Birkhoff condition for its columns $k = 0, 1, \ldots, k_0 - 1$. Then*

$$\mu_k \geqslant \mu'_k + \mu''_k - e_{i_0, k}, \qquad k = 0, \ldots, k_0. \tag{5.2.2}$$

(ii) *Moreover, if $e_{i_0, k_0 - 1} = 0$, then equality holds in (5.2.2) if and only if $e_{i, k} = 0$, $0 \leqslant k < k_0$, for either* (a) *all $i < i_0$, or else* (b) *all $i > i_0$.*

Proof. By the Birkhoff condition and (5.1.5), $\mu_k \geqslant 2$ for $0 \leqslant k < k_0$. Thus, $(\mu_k - 1)_+ = \mu_k - 1$, $k = 0, \ldots, k_0$. The proof is carried out by induction. It is clear that (5.2.2) holds for $k = 0$. Let

$$\sigma_k = \mu_k - \mu'_k - \mu''_k + e_{i_0, k}. \tag{5.2.3}$$

By (5.1.4) and (5.2.1) we have

$$\sigma_k = \mu_{k-1} - 1 - (\mu'_{k-1} - 1)_+ - (\mu''_{k-1} - 1)_+, \qquad k = 0, \ldots, k_0 - 1. \tag{5.2.4}$$

Lemma 5.6. (i) *We have $\sigma_k \geqslant 0$, $k = 0, \ldots, k_0$.* (ii) *If for some $k \leqslant k_0$, $\sigma_k = 0$, then $\sigma_l = 0$ for all $l \leqslant k$.*

Proof. Clearly we have one of the three (not mutually exclusive) cases:

Case 1. $\mu'_{k-1}, \mu''_{k-1} \leqslant 1$. Then from (5.2.4), $\sigma_k = \mu_{k-1} - 1 \geqslant 1$, and we have (i).

Case 2. One of the μ'_{k-1}, μ''_{k-1} is $= 0$, and the other is $\geqslant 1$; for example, let $\mu'_{k-1} \geqslant 1$, $\mu''_{k-1} = 0$. Then, again by (5.2.4), $\sigma_k = \mu_{k-1} - \mu'_{k-1}$. Since $e_{i_0, k-1} \leqslant \mu''_{k-1}$, we have $e_{i_0, k-1} = 0$, and by (5.2.3), $\sigma_{k-1} = \mu_{k-1} - \mu'_{k-1}$; hence we have $\sigma_{k-1} = \sigma_k$.

Case 3. Let $\mu'_{k-1}, \mu''_{k-1} \geqslant 1$. Then from (5.2.3), (5.2.4),

$$\sigma_{k-1} = \mu_{k-1} - \mu'_{k-1} - \mu''_{k-1} + e_{i_0, k-1},$$

$$\sigma_k = \mu_{k-1} - \mu'_{k-1} - \mu''_{k-1} + 1. \tag{5.2.5}$$

In this case, $\sigma_{k-1} \leqslant \sigma_k$.

Now (i) follows by induction, from $\sigma_0 = 0$. After (i) has been established, if $\sigma_k = 0$, case 1 cannot happen, and in the other two cases we have $\sigma_{k-1} = 0$. Thus $\sigma_l = 0$, $l \leqslant k$. □

Lemma 5.7. *Let $\sigma_k = 0$, $k > 0$, and let the Birkhoff condition be satisfied for columns $0, \ldots, k$ of E. (i) If $e_{i_0, k-1} = 0$, then either $\mu'_{k-1} = \mu_{k-1} \geqslant 2$, $\mu''_{k-1} = 0$, or $\mu'_{k-1} = 0$, $\mu''_{k-1} = \mu_{k-1} \geqslant 2$. (ii) If $\mu''_k = e_{i_0, k} = m''_k$ ($= 0$ or 1), then $\mu'_{k-1} = \mu_{k-1} \geqslant 2$, $\mu''_{k-1} = e_{i_0, k-1} = m''_{k-1}$ ($= 0$ or 1).*

Proof. (i) Case 1 of Lemma 5.6 is impossible, as is case 3, since equations (5.2.5) together with $\sigma_{k-1} = \sigma_k = 0$ would imply $e_{i_0, k-1} = 1$. Hence one of the numbers μ'_{k-1}, μ''_{k-1} is $\geqslant 2$, and the other is 0. Let, for example, $\mu'_{k-1} \geqslant 2$. Then (5.2.4) gives $0 = \sigma_k = \mu_{k-1} - \mu'_{k-1}$, and (i) follows.

(ii) We need to consider only the case in which $e_{i_0, k-1} = 1$. From $e_{i_0, k} = (\mu''_{k-1} - 1)_+ + m''_k$ we derive $(\mu''_{k-1} - 1)_+ = 0$; that is, $\mu''_{k-1} = 0$ or 1. Then

$$\mu''_{k-1} = (\mu''_{k-2} - 1)_+ + m''_{k-1}$$

and $m''_{k-1} \geqslant 1$ imply $\mu''_{k-1} = 1 = m''_{k-1}$, which is the second part of assertion (ii). The first part follows from $0 = \sigma_k = \mu_k - \mu'_k$. □

It is clear that from Lemma 5.7 (i), (ii), and a statement symmetric to (ii), the proof of Theorem 5.5 can be completed by induction. □

§5.3. MARKOV'S INEQUALITY AND APPLICATIONS

The construction of independent sets of knots in §5.4 will be based on a weak form of Markov's inequality. This inequality makes it possible to guarantee (Theorem 5.11) the existence of a Rolle zero of a derivative that is not too close to the given zeros of the function.

Let $\mathcal{G} = \{g_0, \ldots, g_n\}$ be a system of n times continuously differentiable functions on $[a, b]$. In the remainder of this chapter we shall always assume that the functions g_k are *linearly independent on each subinterval* $[a_1, b_1]$ of $[a, b]$. For example, Birkhoff systems (see §5.5) have this property.

But we need more. For each $k = 1, \ldots, n$, let the *reduced set of derivatives* $\mathcal{G}^{(k)}$ *for* $[a_1, b_1]$ consist of those $g_j^{(k)}$ that are not identically 0 on $[a_1, b_1]$. We shall assume that *the reduced sets* $\mathcal{G}^{(k)} k = 1, \ldots, n$, *for* $[a, b]$

consist of linearly independent functions on each subinterval of $[a, b]$. This assumption is less restrictive than it might appear; each set $\mathcal{G}$ has this property locally. More exactly:

Proposition 5.8. *Let the functions of $\mathcal{G}$ be linearly independent on each subinterval of $[a, b]$. Then there exists a new basis (also denoted by $g_0, \ldots, g_n$) in the linear hull of $\mathcal{G}$ and an interval $[a_0, b_0] \subset [a, b]$ for which all reduced sets $\mathcal{G}^{(k)}$ are linearly independent on each subinterval of $[a_0, b_0]$.*

Proof. We construct the basis $g_0, \ldots, g_n$ and the interval $[a_0, b_0]$ by induction. Let the required conditions be satisfied for $\mathcal{G}^{(1)}, \ldots, \mathcal{G}^{(k-1)}$ on $I_{k-1} = [a_{k-1}, b_{k-1}]$. This means that there is a basis $g_0, \ldots, g_n$ for which $g_j^{(k-1)} \equiv 0, j = 0, \ldots, p-1$, on I_{k-1}, while $g_p^{(k-1)}, \ldots, g_n^{(k-1)}$ are linearly independent on each subinterval $[\alpha, \beta]$ of I_{k-1}. We consider two cases: (1) On no subinterval of I_{k-1} does the linear hull $\text{lin} \, \mathcal{G}^{(k-1)}$ contain constants. Then we take $I_k = [a_k, b_k] = I_{k-1}$. None of the functions $g_j^{(k)}, j \geq p$, can vanish identically on $[\alpha, \beta] \subset I_k$; hence $\mathcal{G}^{(k)} = \{g_p^{(k)}, \ldots, g_n^{(k)}\}$ is the reduced set of derivatives for $[\alpha, \beta]$. If there is a relation $a_p g_p^{(k)} + \cdots + a_n g_n^{(k)} \equiv 0$ on $[\alpha, \beta]$, then, integrating, we obtain $a + a_p g_p^{(k-1)} + \cdots + a_n g_n^{(k-1)} \equiv 0$. Here $a = 0$ since $\text{lin} \, \mathcal{G}^{(k-1)}$ does not contain constants, and by the inductive assumption $a_j = 0, j \geq p$. Thus, $g_j^{(k)}, j \geq p$, are linearly independent on $[\alpha, \beta]$. (2) If $\text{lin} \, \mathcal{G}^{(k-1)}$ contains constants on some subinterval of I_{k-1}, let I_k be the smallest subinterval. Changing $g_p, \ldots, g_n$ to some other basis, we can assume that $g_p^{(k-1)} = 1$ on I_k; then the linear hull of $g_{p+1}^{(k-1)}, \ldots, g_n^{(k-1)}$ does not contain constants on any subinterval $[\alpha, \beta]$ of I_k, and as before, $g_{p+1}^{(k)}, \ldots, g_n^{(k)}$ are linearly independent on $[\alpha, \beta]$. At the end, $[a_0, b_0] = I_n$, and the g_k are the elements of the last basis. $\square$

For sets $\mathcal{G}$ with the foregoing properties, we have

Theorem 5.9 ("Markov's inequality"). *For each $l > 0$ there is a constant C_l that depends on l and $\mathcal{G}$ and decreases as a function of l, with the property that for each linear combination P of the functions g_k, and each subinterval $[\alpha, \beta]$ of length $\geq l$ in $[a, b]$, there holds*

$$\|P'\|_{[a, b]} \leq C_l \|P\|_{[\alpha, \beta]}, \qquad \|P\|_{[\alpha, \beta]} = \max_{\alpha \leq x \leq \beta} |P(x)|. \qquad (5.3.1)$$

Proof. We can subdivide $[a, b]$ into intervals $I_j = [a + j\delta, \, a + (j+1)\delta]$, $j = 0, \ldots, p-1$, in such a way that each interval $[\alpha, \beta]$ of length $\geq l$ contains one of the I_j [it is sufficient to take $\delta = (b-a)/p \leq \frac{1}{2}l$]. The norm $\|P\|_{[\alpha, \beta]}$ of the restriction of P to $[\alpha, \beta]$ is not less than the norm $\|P\|_j$ of P in $C[I_j]$. Since the correspondence $\sum_{k=0}^{n} c_k g_k \to \sum_{k=0}^{n} c_k g_k'$ maps the $(n+1)$-dimensional linear space spanned by the g_k in $C[I_j]$ linearly into the space spanned by the g_k' in $C[a, b]$, it has a finite norm N_j. Therefore,

$$\|P'\|_{[a, b]} \leq N_j \|P\|_j \leq \max_j N_j \|P\|_{[\alpha, \beta]} = C_l \|P\|_{[\alpha, \beta]}.$$

The constants C_l will decrease as functions of l if we choose each of them to be best possible in (5.3.1).   □

Lemma 5.10. *For each $l > 0$ there is a number $d = d(l)$, $0 < d \leqslant \frac{1}{2}l$, with the property that if $P(\alpha) = P(\beta) = 0$, $\alpha, \beta \in [a, b]$, $\beta - \alpha \geqslant l$, then at least one point $\alpha + d < \xi < \beta - d$ satisfies $P'(\xi) = 0$. The function $d(l)$ is monotone increasing in l.*

Proof. We can assume that P is not identically 0 on $[\alpha, \beta]$. Let ξ be the point on (α, β), where $|P(x)|$ attains its maximum $M = \|P\|_{[\alpha, \beta]}$. Then $P'(\xi) = 0$. On the other hand, for some $\alpha < \eta < \xi$,

$$M = |P(\xi) - P(\alpha)| = (\xi - \alpha)|P'(\eta)| \leqslant C_l \|P\|_{[\alpha, \beta]}(\xi - \alpha) = C_l M(\xi - \alpha).$$

Therefore $\xi - \alpha \geqslant C_l^{-1}$, and likewise $\beta - \xi \geqslant C_l^{-1}$. We select $d(l) = \min(C_l^{-1}, \frac{1}{2}l)$.   □

Remarks. For algebraic polynomials of degree $\leqslant n$, the best value $d(l) = d_n(l)$ has been found by Turán [177]. If n is even, $d_n(l) = \frac{1}{2}l[1 - \cos(\pi/n)]$, and for any n, $d_n(l) \approx (l\pi^2/4)n^{-2}$.

For the system $\mathcal{G}^{(k)}$, Theorem 5.9 and Lemma 5.10 produce a number $d_k(l)$. Taking $\delta(l) = \min_{0 \leqslant k \leqslant n} d_k(l)$, we obtain, for given $\mathcal{G}$ and n:

Theorem 5.11. *There is a monotone increasing function $\delta(l)$, $0 \leqslant \delta(l) \leqslant \frac{1}{2}l$, such that if $\beta - \alpha \geqslant l$, $a \leqslant \alpha < \beta \leqslant b$ and $P^{(k)}(\alpha) = P^{(k)}(\beta) = 0$ for some P and k, $0 \leqslant k \leqslant n - 1$, then there exists a ξ, $\alpha + \delta(l) \leqslant \xi \leqslant \beta - \delta(l)$, for which $P^{(k+1)}(\xi) = 0$.*

§5.4. INDEPENDENT SETS OF KNOTS

Let $\mathcal{G}$ be a system of functions on $[a, b]$, satisfying the assumptions of §5.3. A set of knots $X = \{x_1, \ldots, x_m\} \subset [a, b]$ is called *independent with respect to* $\mathcal{G}$ if for each interpolation matrix E each polynomial P annihilated by E, X has a Rolle extension $\mathcal{R}$ with all new Rolle zeros ξ different from the x_k. As we know from §5.1, this $\mathcal{R}$ will be maximal and have no duplications. Lemma 5.3 yields then that the total number of (new and old) Rolle zeros of $P^{(k)}$ in $\mathcal{R}$ is exactly μ_k.

The construction of independent sets of knots is based on the technical lemma (Lemma 5.12) that follows. Without loss of generality, let $a = -1$, $b = 1$. We take $0 < y_1 < 1$ arbitrarily and choose y_j, $j = 2, 3, \ldots$, to increase rapidly to 1, with the following restrictions. Let $\Delta(u) = \frac{1}{2}\delta(u)$, where $\delta(u)$ is the function of Theorem 5.11, and let $\Delta^n(u)$ be defined by induction, $\Delta^n(u) = \Delta^{n-1}(\Delta u)$. We require that $0 < y_{j-1} < y_j$ and

$$1 - y_j \leqslant \Delta^n(y_j - y_{j-1}), \qquad j = 2, 3, \ldots. \tag{5.4.1}$$

Since $\Delta(u) = \frac{1}{2}\delta(u) \leqslant \frac{1}{4}u$ and $\Delta(u)$ is increasing, it follows that

$$1 - y_j \leqslant \Delta(y_j - y_{j-1}) < \delta(y_j - y_{j-1}). \tag{5.4.2}$$

We also select numbers l_j satisfying

$$\Delta^{n-1}(y_j - y_{j-1}) \leqslant l_j \leqslant y_j - y_{j-1}, \qquad j = 2, 3, \ldots, \qquad (5.4.3)$$

and put $l'_j = \Delta(l_j)$, $j = 2, 3, \ldots$; in general, l'_j are much smaller than l_j, for $\Delta(u) \leqslant \frac{1}{4} u$. We shall take X to contain some of the points $\pm y_j$, and will take care to select the new Rolle zeros ξ of X^k to be different from these points. This will follow because the ξ will be even outside of small intervals $(y_j - \varepsilon, y_j]$ and $[-y_j, -y_j + \varepsilon)$.

Lemma 5.12. *Let $0 < \rho < 1$, and let s be so large that $\rho \leqslant y_s$; let $s + 2 \leqslant t$. Assume that P is a polynomial in $\mathcal{G}$ annihilated by E, X, and that the knots X and X^k for a Rolle extension $\mathfrak{R}_k = E^k$, X^k are contained in*

$$[-\rho, \rho] \cup [y_t, 1] \cup [-1, -y_t] \qquad (5.4.4)$$

but miss all intervals

$$(y_j - l_j, y_j), \quad (-y_j, -y_j + l_j), \qquad j = t, t+1, \ldots.$$

Let $\rho' = y_{s+1}$. Then there is a Rolle extension E^{k+1}, X^{k+1} such that X^{k+1} is contained in the set

$$[-\rho', \rho'] \cup [y_t, 1] \cup [-1, -y_t] \qquad (5.4.5)$$

and that the Rolle zeros selected between adjacent zeros of X^k miss the intervals

$$(y_j - l'_j, y_j], \quad [-y_j, -y_j + l'_j), \qquad j = t, t+1, \ldots. \qquad (5.4.6)$$

Proof. Since $\rho' - \rho \geqslant y_{s+1} - y_s$, we have by (5.4.2)

$$1 - \delta(\rho' - \rho) < \rho'. \qquad (5.4.7)$$

Let $\alpha < \beta$ be two adjacent zeros of X^k. By means of Rolle's theorem, we shall find a zero ξ of $P^{(k+1)}$ of the required kind.

(a) If $\alpha \leqslant \rho$, $\beta > \rho'$, or $\alpha < -\rho'$, $\beta \geqslant -\rho$, then ξ can be found in $(-\rho', \rho')$. Indeed, the length of (α, β) is at least $\rho' - \rho$. By Theorem 5.11 we find a Rolle zero ξ for which

$$\xi < \beta - \delta(\beta - \alpha) < 1 - \delta(\rho' - \rho) < \rho';$$

similarly, ξ satisfies $\xi > -\rho'$.

(b) The zeros of X^k fall into three groups: zeros contained in $[-\rho, \rho]$, those in $[-1, -y_t]$, and those in $[y_t, 1]$. If one or both of α, β belong to the first group, then (a) shows that we can take $\xi \in [-\rho', \rho']$. This is still true, again by (a), if α, β belong to different groups. If α, β both belong to the second or the third interval, then ξ also belongs to this interval.

(c) In the last case, we still have to show that ξ can be selected so as to miss intervals (5.4.6). We can assume that $y_t \leqslant \alpha < \beta$. None of the intervals $(y_j - l_j, y_j)$ contains α or β; hence each of them is either contained in the interval (α, β) or disjoint with it. If the first possibility does not occur, we are finished. In the opposite case, let j be the smallest integer $j > t$

for which $(y_j - l'_j, y_j) \subset (\alpha, \beta)$. We shall find a $\xi < y_j - l'_j$, thus completing the proof. By Theorem 5.11 there is a ξ satisfying

$$\alpha < \xi \leqslant \beta - \delta(l_j) < y_j + (1 - y_j) - \delta(l_j).$$

From (5.4.1) and (5.4.3),

$$1 - y_j \leqslant \Delta(l_j) = \tfrac{1}{2}\delta(l_j).$$

Hence $\xi < y_j - \tfrac{1}{2}\delta(l_j) = y_j - l'_j$.   □

Corollary 5.13. *Let the knots X be only among the points $\pm y_j$, $j \geqslant s + 2$, or in $[-\rho, \rho]$, and suppose that the rows of E that correspond to knots x_i, $-\rho \leqslant x_i \leqslant \rho$, have no odd supported sequences. Then the construction of X^{k+1} of Lemma 5.12 will be without losses also in $[-\rho, \rho]$, with duplications possible only in this interval. Moreover, Rolle zeros in $I_1 = [-1, -y_{s+2}]$ or in $I_2 = [y_{s+2}, 1]$ will be produced by zeros only from the same interval; all other Rolle zeros will belong to $I' = [-y_{s+1}, y_{s+1}]$.*

We can apply Lemma 5.12, Corollary 5.13, and Lemma 5.2 to all derivatives $P^{(k)}$, $k = 0, \ldots, n - 1$. In this way we obtain the following formulation of the method of independent knots.

Theorem 5.14. *There exist numbers $\rho = y_s$, ρ', $\rho < \rho' < 1$, and an integer $t > s$ with the following properties. Let $I = [-\rho, \rho]$, $I' = [-\rho', \rho']$, $I_1 = [-1, -y_t]$, $I_2 = [y_t, 1]$. Let X be a subset of $I \cup \{\pm y_t, \pm y_{t+1}, \ldots\}$, and let E be an $m \times (n + 1)$ interpolation matrix with no odd supported sequences in the rows corresponding to knots $x_i \in I$. Then each polynomial P in $\mathcal{G}$ annihilated by E, X has a maximal Rolle set $\mathcal{R}$ with duplication possible only in I. Moreover, all Rolle zeros are contained in $I' \cup I_1 \cup I_2$; those in I_1 (or in I_2) are produced only by zeros of the same interval; Rolle zeros produced with participation of one of the zeros in I_1 (or I_2) lie in $[-1, \rho]$ (or in $[-\rho, 1]$). The total number of zeros of $P^{(l)}$ in $\mathcal{R}$ is equal to μ_l if X has no points in I.*

(If there is just one such point, Corollary 5.4 may apply.)

We can also assume that each P annihilated by E, Y, $Y \subset I \cup \{\pm y_t, \ldots\}$, has a maximal Rolle set if E has no odd supported sequences for knots in I.

This is proved by applying Lemma 5.12 and Corollary 5.13 in turn to $P, P', \ldots, P^{(n-1)}$. At the kth step, we select $l_j = l_{jk} = \Delta^k(y_j - y_{j-1})$ and have $l'_j = \Delta(l_{jk}) = l_{j, k+1}$. The integer t can be selected as $t = s + n + 1$.   □

Theorem 5.15. *For a given system $\mathcal{G}$ there exists an infinite sequence $Y \subset [a, b]$ with the property that each finite set $X \subset Y$ of m points is an independent set of knots for each $m \times (n + 1)$ interpolation matrix.*

Proof. For Y we can take each of the two sets

$$\{\pm y_i\}_{i \geqslant t}, \qquad \{y_i\}_{i \geqslant t}.$$ □

The three points $-1 < x < 1$ are independent if x is sufficiently close to -1 or 1.

§5.5. APPLICATIONS; BIRKHOFF SYSTEMS

Our first application of the Rolle extension method is to systems

$$\mathcal{G} = \{1, \dots, x^{k-1}, g_k, \dots, g_n\}, \qquad g_j \in C^k[a, b], \qquad j = k, \dots, n. \quad (5.5.1)$$

Let $\{g_0, \dots, g_p\}$ be a Chebyshev system. We call a point c, $a < c < b$, a *double zero* of the polynomial $P = \sum_0^p b_j g_j$ if P vanishes at c without change of sign.

Lemma 5.16 (see [H, p. 23]).   *A nontrivial polynomial P in $\{g_0, \dots, g_p\}$ cannot have more than p zeros, even if double zeros are counted twice.*

Proposition 5.17 (Ikebe [52]).   *Let E be a Pólya matrix with no 1's in columns $l > k$, where $k \leqslant n$, and without odd supported sequences. Then E is regular with respect to the system (5.5.1) if the derivatives $g_k^{(k)}, \dots, g_n^{(k)}$ form a Chebyshev system on $[a, b]$.*

Proof.   Let $P = \sum_{j=0}^{k-1} a_j x^j + \sum_k^n a_j g_j$ be annihilated by E, X. We imitate the proof of Lemma 5.3, showing that the Rolle extension $\mathcal{R}_k$ produces at least $\mu_k = n - k + 1$ zeros of $P^{(k)} = \sum_{j=k}^n a_j g_j^{(k)}$. At step l, between two zeros $\alpha < \beta$ of $P^{(l)}$, we can select an *odd* zero ξ of $P^{(l+1)}$. There are no losses. If there is duplication [case 2(b) of §5.1], then $\xi = x_i$ and the ith row of E has a sequence of 1's in columns $l+1, \dots, l_1$. If $l_1 < k$, this construction produces a zero of $P^{(l_1+1)}$ at x_i. If $l_1 = k$, then, because the sequence is even, the zero x_i of $P^{(k)}$ must be double. Including these double zeros, $P^{(k)}$ has μ_k zeros. Lemma 5.16 implies $a_j = 0$, $j = k, \dots, n$. Then also $a_j = 0$, $j = 0, \dots, k - 1$, for E is regular for polynomials of degree $k - 1$.   □

It is interesting to investigate systems $\mathcal{G}$ for which Theorem 1.5, the Atkinson–Sharma theorem, remains valid (see [96]). We call $\mathcal{G} = \{g_0, \dots, g_n\}$ a *Birkhoff system* if $g_j \in C^n[a, b]$, $j = 0, \dots, n$, and if each Pólya matrix E that has no essential odd supported sequences is regular with respect to $\mathcal{G}$.

Theorem 5.18.   *A system $\mathcal{G} = \{g_0, \dots, g_n\}$ is a Birkhoff system if and only if for arbitrary x_k, $a \leqslant x_k \leqslant b$, $k = 0, \dots, n$, equations*

$$P^{(k)}(x_k) = 0, \qquad k = 0, \dots, n, \qquad (5.5.2)$$

for a polynomial P in $\mathcal{G}$ imply $P \equiv 0$.

Proof.   This is sufficient. Let P be annihilated by E, X, where E is a Pólya matrix, and let $E = E_1 \oplus \cdots \oplus E_\mu$ be its canonical decomposition into matrices without odd supported sequences. If the last column of E_λ is n_λ, we obtain from Lemmas 5.1 and 5.2 that (5.5.2) is satisfied for $0 \leqslant k \leqslant n_1$.

Next, $P^{(n_1+1)}$ is annihilated by E_2, and in the same way we get (5.5.2) for $n_1 \leqslant k \leqslant n_2$, and so on. Thus $P \equiv 0$.

The condition is necessary, for equations (5.5.2) mean that P is annihilated by a matrix (an Abel matrix) whose canonical decomposition consists of one-column matrices. $\qquad\Box$

Another form of the condition (5.5.2) is that none of the determinants

$$V(x_0,\ldots,x_n) = \det\big[g_0^{(k)}(x_k),\ldots,g_n^{(k)}(x_k); k = 0,\ldots,n\big], \qquad a \leqslant x_k \leqslant b,$$
$$(5.5.3)$$

should vanish. As a simple application of this, the system (5.5.1) is a Birkhoff system exactly when $\{g_k^{(k)},\ldots,g_n^{(k)}\}$ is a Birkhoff system.

There are relations between Chebyshev and Birkhoff systems.

Proposition 5.19. (i) *A Birkhoff system is an extended Chebyshev system.*

(ii) *If for a system of functions $\mathcal{G}$ the Wronskian*

$$W(x) = \det\big[g_0^{(k)}(x),\ldots,g_n^{(k)}(x); k = 0,\ldots,n\big] \qquad (5.5.4)$$

does not vanish identically, in particular if $\mathcal{G}$ is an extended Chebyshev system, then $\mathcal{G}$ is a Birkhoff system locally, that is, a Birkhoff system on some closed subinterval $[\alpha, \beta]$ of $[a, b]$.

Proof. (i) If a polynomial P in $\mathcal{G}$ has $n+1$ zeros, counting their multiplicities, then by Rolle's theorem we obtain (5.5.2); hence $P \equiv 0$, since $\mathcal{G}$ is a Birkhoff system. (ii) If $W(\bar{x}) \neq 0$ for some $\bar{x} \in [a, b]$, then the determinants (5.5.3) are different from 0 if all x_k are close to $\bar{x}$. $\qquad\Box$

Example. $\mathcal{G} = \{x^\alpha, x^\beta\}$, where $0 < \alpha \leqslant \beta$, is a Chebyshev system on $[a, b]$, $0 < a < b$, if and only if $\alpha < \beta$, and this is equivalent to regularity of 2×2 Birkhoff matrices with respect to $\mathcal{G}$, but $\mathcal{G}$ is a Birkhoff system exactly when $(b/a)^{\beta-\alpha} < \beta/\alpha$.

A matrix E is *conditionally regular* for a system $\mathcal{G}$ on $[a, b]$ if we can find a set of knots $X \subset [a, b]$ for which the pair E, X is regular.

Remark 5.20. For Birkhoff systems, we can complete the statements of Lemma 5.2 as follows:

(a) If P is annihilated by a pair E, X that has a maximal Rolle extension, and if E satisfies the Pólya condition $M_l \geqslant l+1$, $0 \leqslant l \leqslant k_0$, then $P^{(l)}(z_l) = 0$ for some z_l, $0 \leqslant l \leqslant k_0$. If $k_0 = n$, then $P \equiv 0$.

(b) The same happens if in addition to the assumption, also $P^{(k_0)}$ is annihilated by a pair F, Y with similar properties for $k_0 < l \leqslant n$.

(c) For a Pólya matrix E, we have the regularity of the pair E, X in Theorem 5.14.

We can apply independent knots to the study of conditional regularity (see Windauer [189]).

Theorem 5.21. *Each Pólya matrix E is conditionally regular with respect to a Birkhoff system $\mathcal{G}$, as well as with respect to a system $\mathcal{G}$ for which the Wronskian $W(x)$ is not identically 0.*

Proof. Let $\mathcal{G}$ be a Birkhoff system on a subinterval $[\alpha, \beta]$ of $[a, b]$. By Theorem 5.15 we can find independent knots $U: u_1 < \cdots < u_m$ in $[\alpha, \beta]$. If P is annihilated by E, U, then this pair has a maximal Rolle extension, and $P^{(k)}(x_k) = 0$, $k = 0, \ldots, n$, for some $x_k \in [\alpha, \beta]$. Then $P \equiv 0$. Consequently, U is regular with respect to E. The second statement follows from Proposition 5.19 (ii).   □

This gives another proof of Corollary 4.2 of Chapter 4. It is interesting to note that now we can indicate explicitly some regular sets U.

§5.6. CONTINUITY OF INTERPOLATION

Let E be a matrix that is regular for the powers $x^j/j!$, $j = 0, \ldots, n$, on $[a, b]$. All different sets of knots $X: a \leqslant x_1 < \cdots < x_m \leqslant b$ form the "open" m-dimensional simplex S; its closure $\bar{S}$ consists of all points $X: a \leqslant x_1 \leqslant \cdots \leqslant x_m \leqslant b$. For a point X on the boundary of $\bar{S}$ the x_i take $t < m$ different values. We partition $1, \ldots, m$ into t disjoint groups of integers $[i_s, j_s]$, $s = 1, \ldots, t$, so that $x_i = \bar{x}_s$ is constant for $i_s \leqslant i \leqslant j_s$. We call $\bar{X}: a \leqslant \bar{x}_1 < \cdots < \bar{x}_t \leqslant b$ a *coalescence* of X; the corresponding $t \times (n + 1)$ matrix $\bar{E}$ has groups of rows i, $i_s \leqslant i \leqslant j_s$, of E coalesced each to one row.

The interpolating polynomial of a function $f \in C^n[a, b]$ is the unique polynomial $P(t) = P(f, E, X; t)$ that satisfies $P^{(k)}(x_i) = f^{(k)}(x_i)$, $e_{i,k} = 1$. We want to study the dependence of this polynomial on X. From the formula (1.6.1) it is clear that P is a continuous function of X for $X \in S$ (continuity for each $t \in [a, b]$ or continuity in the uniform norm is the same for polynomials). If all coalescences $\bar{E}$ of E are regular, there is a natural extension of P onto $\bar{S}$, given by $P(f, E, X; t) = P(f, \bar{E}, \bar{X}; t)$. This extension is continuous on $\bar{S}$ if it has the property

$$\lim_{X' \to X} P(f, E, X'; t) = P(f, \bar{E}, \bar{X}; t) \tag{5.6.1}$$

for each $X \in \bar{S}$ and $X' \in S$, $X' \to X$. We have the interesting and useful (see §10.5) fact:

Theorem 5.22 [33]. *For a conservative Pólya matrix E and $f \in C^n[a, b]$, the natural extension of $P(f, E, X; t)$ onto $\bar{S}$ is continuous.*

Proof. Because E is conservative, all coalescences $\bar{E}$ are regular. The proof will be by "de-coalescence." Let Q be the closed $(n + 1)$-dimensional cube with points $Y = (y_1, \ldots, y_{n+1})$, $a \leqslant y_i \leqslant b$, $i = 1, \ldots, n + 1$. A point $X =$

$(x_1,\ldots,x_m) \in \bar{S}$ we shall identify with $X^* = (x_1,\ldots,x_1, x_2,\ldots,x_2,\ldots)$, where each x_i is repeated as often as there are 1's in row i of E.

Let X be a fixed point on the boundary of $\bar{S}$. This point defines groups of integers $[i_s, j_s]$, $s = 1,\ldots,t$, and coalescences $\bar{E}, \bar{X}$. We "decoalesce" $\bar{E}$ into an $(n+1) \times (n+1)$ matrix E^*. For example:

$$\bar{E} = \begin{bmatrix} 1 & 0 & 1 & 1 & 0 \\ 0 & 1 & 0 & 0 & 0 \\ 1 & 0 & 0 & 0 & 0 \end{bmatrix}, \qquad E^* = \begin{bmatrix} 1 & 0 & 0 & 0 & 0 \\ 0 & 0 & 1 & 0 & 0 \\ 0 & 0 & 0 & 1 & 0 \\ 0 & 1 & 0 & 0 & 0 \\ 1 & 0 & 0 & 0 & 0 \end{bmatrix}.$$

Formally, $e^*_{j,k} = 1$ if and only if there is a pair (i, k) with $\bar{e}_{i,k} = 1$ whose lexicographic number is j. The determinant $D(E^*, Y)$ is a continuous function of Y and

$$D(E^*, Y) \to D(\bar{E}, \bar{X}) \qquad \text{for} \quad Y \to X^*. \tag{5.6.2}$$

Since all coalescences $\bar{E}$ of E are regular, $D(\bar{E}, \bar{X}) \neq 0$, and there is a neighborhood U of X^* in Q, given by inequalities $|y_i - \bar{x}'_s| < \varepsilon$, $i_s \leqslant i \leqslant j_s$, $s = 1,\ldots,t$, with some $\varepsilon > 0$ where $D(E^*, Y) \neq 0$. For $Y \in U$, the pair E^*, Y is regular (although E^* is likely to be singular), hence there exists the interpolating polynomial $P(f, E^*, Y; t)$. Writing it in the form of (1.6.1) we see that

$$\lim_{Y \to X^*} P(f, E^*, Y; t) = P(f, \bar{E}, \bar{X}; t). \tag{5.6.3}$$

The relation (5.6.1) will follow if we can show that for each $X' \in U \cap S$, there is a $Y \in U$ with the property

$$P(f, E, X'; t) = P(F, E^*, Y; t) \qquad \text{for all } t. \tag{5.6.4}$$

Let $P(t) = P(f, E, X'; t)$, $F = P - f$. The function F is annihilated by E, X': it satisfies

$$F^{(k)}(x'_i) = 0. \tag{5.6.5}$$

Let $[i_s, j_s]$ be one of the groups of integers defining the coalescence. We apply Rolle's theorem to equations (5.6.5) with $i_s \leqslant i \leqslant j_s$; by Lemma 5.3 we obtain, for each $\bar{e}_{s,k} = 1$ in $\bar{E}$, a point $x'_{i_s} \leqslant \xi_{s,k} \leqslant x'_{j_s}$ with $F^{(k)}(\xi_{s,k}) = 0$. Listed lexicographically, the $\xi_{s,k}$ give a point $Y \in U$.

Since F is annihilated by the regular pair E^*, Y, the polynomial $P(t)$ is identical with $P(f, E^*, Y; t)$. $\square$

Remarks.   1. The theorem remains true if the powers $x^j/j!$ are replaced by functions g_j of an arbitrary Birkhoff system on $[a, b]$ (see §5.5).

2. There is a companion result to Theorem 5.22. In various ways we can prove that the interpolating polynomial $P(f, E, X; t)$ is an entire function of $x_1,\ldots,x_m$ if f is entire and E is a Hermitian matrix. When is this true for an arbitrary E? Lorentz [98] proves that this is the case for any entire function f if and only if E decomposes into Hermitian and two-row matrices.

Chapter 6

Singular Matrices

§6.1. INTRODUCTION; SUPPORTED SINGLETONS

In the first three sections of this chapter we discuss two large classes of singular matrices. The Atkinson–Sharma theorem provides only a sufficient condition for the regularity of matrices; the condition is not necessary. However, a good guiding principle is that the condition is necessary in most cases, or necessary under some mild additional assumptions. Sections 6.1–6.3 confirm this idea. The condition is also necessary for "most" large matrices. We use probabilistic tools to prove this in §6.4. Main theorems of this chapter deal with the order singularity of Pólya or Birkhoff matrices; Theorems 6.1, 6.2 and 6.9 for Birkhoff systems $\mathscr{G}$ and the rest for algebraic interpolation.

A *singleton* is a 1 in a row of E that does not contain other 1's. A singleton is an odd sequence, which may be supported in E or not. We begin with the simplest theorem:

Theorem 6.1 (Lorentz and Zeller [107]). *A Birkhoff matrix is strongly singular if it contains a supported singleton.*

There is an immediate generalization:

Theorem 6.2 (Lorentz [90]). *A Birkhoff matrix is strongly singular if it contains a row with precisely one odd supported sequence (all other sequences of this row being even or not supported).*

By the localization theorem (Proposition 5.19(ii)), both theorems also hold for extended Chebyshev systems. Thus, two localization theorems, Propositions 5.8 and 5.19, were used to obtain this conclusion.

Theorem 6.2 also appears in Karlin and Karon [70] as an application of the method of coalescence, but the proof there is not correct (see also §6.5). Nevertheless, we use some ideas of this paper to prove strong singularity in §6.2. The proof of Theorem 6.2 can be based on coalescence (one coalesces the matrix to three rows by means of Theorem 4.5) or on the method of independent knots (this method was used for the proofs in [107] and [90]). The first is perhaps simpler (the main simplification being that the intervals I_1 and I_2 of §6.2 become simply points -1 and $+1$). Its disadvantage is that it requires the existence of derivatives of g_k of orders higher than n (see the end of §3.5).

We prefer to prove Theorem 6.1 first, because this proof is simpler, illustrates the method, and is (at least formally) not contained in the proof of the general Theorem 6.2. Throughout the proof, the system $\mathcal{G} = \{g_0, g_1, \ldots, g_n\}$ will be a Birkhoff system on $[a, b]$.

Proof of Theorem 6.1.   Let E be an $m \times (n + 1)$ Birkhoff matrix that has a supported singleton $e_{i_0, q} = 1$ in the interior row i_0. We denote by E_0 or E_1 matrices derived from E by omitting the row i_0 or by replacing it by $(1, 0, \ldots, 0)$, respectively. We use Theorem 5.14 and place knots x_i, $i < i_0$, into fixed independent positions in the interval $I_1 = [-1, -y_t]$, knots x_i, $i > i_0$, into similar positions in $I_2 = [y_t, 1]$. To this set of knots X_0 we add a variable knot $x \in I = [-\rho, \rho]$ in order to obtain the set X.

According to Theorem 5.14 and Remark 5.20 the pair E_1, X is regular; hence $P(x) = D(E_1, X) \neq 0$, $x \in I$. The function $P(x)$ is a polynomial in x and it is clearly annihilated by E_0, X_0. This pair has a maximal Rolle extension $\mathcal{R}$. We want to find a point $x = \xi \in I$ of $\mathcal{R}_q$ for which

$$P^{(q)}(\xi) = 0. \tag{6.1.1}$$

It is sufficient to find an $l < q$ for which $P^{(l)}$ has zeros of $\mathcal{R}$ in both I_1 and I_2; then by Theorem 5.14 the Rolle extension would produce a required ξ. Should an l of this type not exist, then for each $l < q$, Rolle zeros would either all be in I_1 or all be in I_2. The Birkhoff condition, which is satisfied for $l < q$ in E_0, yields at least two zeros of $P^{(l)}$. Suppose that for $l = 0$ all of these zeros are in I_1. Then Rolle's theorem produces a zero of P' in I_1; hence all Rolle zeros of P', and similarly for $P^{(l)}$, $l < q$, lie in this interval. This is impossible, since the one that is supporting the sequence from below gives a zero in I_2.

Since $D(x) = D(E, X) = P^{(q)}(x)$, we see from (6.1.1) that E is singular. To establish the strong singularity, we have to show that $D(x)$ changes sign at ξ. This is so because ξ is a simple zero of $P^{(q)}(x) = D(x)$. Indeed, otherwise we could add the one $e_{i_0, q+1} = 1$ to E, omitting a 1 in another row. By Remark 5.20, the new matrix and X would be a regular pair, and we would obtain $P \equiv 0$, for a contradiction. $\square$

§6.2. PROOF OF THEOREM 6.2

After proving Theorem 6.1, we can assume that the row i_0 containing the single supported odd sequence in the Birkhoff matrix E has at least two 1's. Let $e_{i_0, q} = 1$ be the first 1 of the odd supported sequence. Let E_0 denote the matrix obtained from E by replacing this 1 by 0; let $\bar{E}', E', \bar{E}'', E''$ be the matrices consisting of rows $i < i_0$, $i \leqslant i_0$, $i > i_0$, $i \geqslant i_0$ of E_0, respectively; and let $\mu_l, \bar{\mu}'_l, \mu'_l, \bar{\mu}''_l, \mu''_l$ be the functions μ of §5.2 for the last five matrices. Obviously, $\bar{\mu}'_l \leqslant \mu'_l$, $\bar{\mu}''_l \leqslant \mu''_l$, and $\mu_l = \bar{\mu}'_l + \mu''_l$.

To be able to use Theorem 5.14, we assign to x_i, $i < i_0$ and $i > i_0$, independent positions $\pm y_j$ in the intervals I_1, I_2. To this set X_0 we add the knot $x_{i_0} = x$ in $I = [-\rho, \rho]$; let $X = X_0 \cup \{x\}$. In addition, let $-\rho' \leqslant y \leqslant \rho'$ and $Y = X \cup \{y\}$. Further, let E_1 be the matrix obtained from E_0 by adding the row $(1, 0, \ldots, 0)$ between rows i_0 and $i_0 + 1$. Then $P(x, y) = D(E_1, Y)$ is a polynomial in x and y. For a fixed x, it is a polynomial $P(y)$ of degree $\leqslant n$ in y, and the structure of the determinant shows that $P(y)$ is annihilated by E_0, X. The coefficient of $y^n / n!$ in $D(E_1, Y)$ is $\pm D(E'_0, X)$ where E'_0 is E_0 without its last column of 0's. Since E'_0, X is regular for each choice of x in $[-\rho, \rho]$, the polynomial $P(y)$ is nontrivial.

We consider the derivative $\partial^q P / \partial y^q = P^{(q)}(y) \stackrel{\perp}{=} P^{(q)}(x, y)$. Since

$$P^{(q)}(x, x) = \pm D(E, X), \tag{6.2.1}$$

the singularity of E will be established if we show that for some x, the equation

$$P^{(q)}(x, y) = 0 \tag{6.2.2}$$

is satisfied for $y = x$. We are thus led to consider solutions y of (6.2.2) for fixed x. We can say at once that for $x = \rho$ (or $x = -\rho$) this equation has no solution $y = \rho$ (or, correspondingly, $y = -\rho$). For if $x = \rho$, X is independent, and $D(E, X) = \pm P^{(q)}(\rho, \rho) \neq 0$ by Remark 5.20(c).

By Theorem 5.14, there is a maximal Rolle extension $\mathcal{R}$ of the pair E_0, X that annihilates P. For this extension, Rolle zeros y produced by pairs of knots other than those confined to I_1 or to I_2 lie in $[-\rho', \rho']$. This explains our choice of the domains of the variables, $-\rho \leqslant x \leqslant \rho$, $-\rho' \leqslant y \leqslant \rho'$.

Let t be the number of solutions of (6.2.2) for a given x, in other words, the number of Rolle zeros of $P^{(q)}$ in $[-\rho', \rho']$. Rolle zeros of $P^{(l)}$ of $\mathcal{R}$ in the intervals I_1, I_2 are produced (by Theorem 5.14) by knots in these intervals and by matrices $\bar{E}', \bar{E}''$; their numbers are $\bar{\mu}'_l, \bar{\mu}''_l$, $l = 0, \ldots, n$, respectively. We cannot claim that μ_l is the total number of Rolle zeros of

$P^{(l)}$. For $l = q$, however, this is true by Corollary 5.4 Hence

$$t = \mu_q - \bar{\mu}'_q - \bar{\mu}''_q. \qquad (6.2.3)$$

We see that t is independent of x. We also have $t \geqslant 1$. This can be deduced by the argument in the proof of Theorem 6.1. Alternatively, we have

$$t \geqslant \sigma = \mu_q - \mu'_q - \mu''_q, \qquad (6.2.4)$$

and by Theorem 5.5, $\sigma \geqslant 1$.

Let

$$-\rho' \leqslant y_1(x) < \cdots < y_t(x) \leqslant \rho', \qquad -\rho \leqslant x \leqslant \rho, \qquad (6.2.5)$$

be all Rolle zeros of $P^{(q)}(y)$ contained in $[-\rho', \rho']$. We claim that (a) $P^{(q)}(y)$ has no other zeros in $[-\rho, \rho]$; (b) each of the zeros (6.2.5) is simple except perhaps the zero $y_s(x) = x$; (c) if there is a zero $y_s(x) = x$, it has an odd multiplicity (equal to the length of the sequence containing $e_{i_0, q} = 1$). Indeed, E_0 is an $m \times n$ Pólya matrix, and in the Rolle extension E_0^q, X^q, E_0^q is a Pólya matrix with $n - q$ columns (and zero column numbered $n + 1 - q$). If one of the foregoing statements were not true, we would be able to add to E_0^q an additional 1, obtaining a new Pólya matrix with $n + 1 - q$ ones and columns, which annihilates $P^{(q)}$ and has no odd supported sequences for knots in $[-\rho', \rho']$. By Remark 5.20(b) we would obtain $P \equiv 0$, a contradiction.

Since the function $P(x, y)$ is continuous, it is now easy to prove the continuity of the zeros (6.2.5).

To prove the singularity of E, we have to show that for some s we have $y_s(x) = x$ for a certain $x \in I$; in other words, that one of the curves (6.2.5) in the rectangle $-\rho \leqslant x \leqslant \rho$, $-\rho' \leqslant y \leqslant \rho'$ intersects the line $y = x$ (see Figure 6.1). It is not obvious that this intersection exists, for there are intervals on the lines $x = \pm \rho$ through which the curves could escape. (This remark applies also to the proof that uses coalescence to three rows. If x, y change in the open interval $(-1, 1)$, the intersection must lie in the open square. The curves could still escape through the corners of the square—a point missed in [70].)

Let $N(x)$, $-\rho \leqslant x \leqslant \rho$, be the number of $y_s(x)$ that satisfy $y_s(x) > x$. We can find $N(\rho)$: This is the number of Rolle zeros ξ of $P^{(q)}(\rho, y)$ that satisfy $\rho < \xi \leqslant \rho'$, or equivalently $\rho \leqslant \xi \leqslant \rho'$. Now Rolle zeros ξ in $[\rho, 1]$ are produced by the matrix E'' and the independent knots $\rho = x_{i_0}, \ldots, x_m$. Their number (by Theorem 5.14) is μ''_q. The ξ with $\xi > \rho'$ are produced by $\bar{E}''$ and the knots $x_{i_0 + 1}, \ldots, x_m$; there are $\bar{\mu}''_q$ of them. Hence

$$N(\rho) = \mu''_q - \bar{\mu}''_q.$$

Next let $x = x_{i_0} = -\rho$. The number of $y_s > -\rho$ is equal to the number of Rolle zeros ξ of $P^{(q)}(-\rho, y)$ with $-\rho \leqslant \xi \leqslant \rho'$. Here again, the

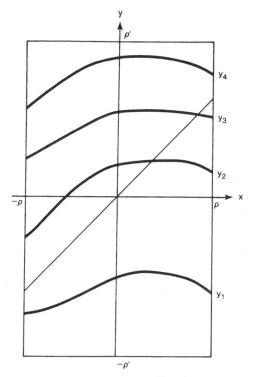

FIGURE 6.1 Zeros of $P^{(q)}(x, y)$

knots $x_1, \ldots, x_m$ are in independent positions, and we can find the numbers of different types of ξ by means of Theorem 5.14. Their total number is μ_q. The zeros $\xi < -\rho$ are produced exclusively by the knots $x_1, \ldots, x_{i_0} = -\rho$ and the matrix E', and their number is μ'_q. The number of ξ with $\xi > \rho'$ is $\bar{\mu}''_q$. It follows that

$$N(-\rho) = \mu_q - \mu'_q - \bar{\mu}''_q.$$

This yields

$$N(-\rho) - N(\rho) = \mu_q - \mu'_q - \mu''_q = \sigma. \qquad (6.2.6)$$

By Theorem 5.5, $\sigma \geqslant 1$. Thus, at least σ curves $y_s(x)$ cross the line $y = x$ inside the interval $[-\rho, \rho]$. The curves (6.2.5) divide the rectangle $-\rho \leqslant x \leqslant \rho$, $-\rho' \leqslant y \leqslant \rho'$ into $t + 1$ regions with $P^{(q)}(x, y)$ alternating in sign from region to region. The point (x, x) moving on the line $y = x$ crosses $\sigma + 1$ of the regions; the points $(-\rho, -\rho), (\rho, \rho)$ are not on the curves. This means that $D(E, X)$ changes sign σ times as x moves on $[-\rho, \rho]$. $\qquad \square$

§6.3. ALMOST-SIMPLE MATRICES

A row, i, of the interpolation matrix E is called *simple* if it contains exactly one entry $e_{i,k} = 1$. A matrix E is *almost simple* if all of its rows except at most one are simple. For this class of matrices, the order regularity problem can again be decided by means of the conditions of the Atkinson–Sharma theorem.

In the canonical decomposition

$$E = E_1 \oplus \cdots \oplus E_\mu \qquad (6.3.1)$$

of an almost-simple matrix each component matrix is also almost simple. More interestingly, the following lemma holds:

Lemma 6.3. *A normal almost-simple interpolation matrix E is strongly order singular if and only if one of the components in its canonical decomposition (6.3.1) is strongly order singular.*

Proof. (Note that Theorem 1.4 is not sufficient here.) Suppose that E_τ is strongly order singular. We have

$$D(E, X) = \pm \prod_{\lambda=1}^{\mu} D(E_\lambda, X_\lambda) \qquad (6.3.2)$$

where X_λ is the subset of X corresponding to the nonzero rows of E_λ. Let x_{i_0} correspond to the nonsimple row i_0 of E (if such a row exists). Because the determinants are translation invariant, we may fix x_{i_0} and still $D(E_\tau, X_\tau)$ will change sign. This determinant will become a function of the remaining variables $y_1, \ldots, y_s$, of X_τ. It is easy to find (e.g., by Lemma 1.1) a set of points $y_1^*, \ldots, y_s^*$ and disjoint neighborhoods V_j of the y_j^*, not containing x_{i_0}, so that $D(E_\tau, X_\tau)$ will change sign even for $y_j \in V_j$, $j = 1, \ldots, s$. Since X_τ and X_λ, $\lambda \neq \tau$, can intersect in at most x_{i_0}, the nontrivial polynomials $D(E_\lambda, X_\lambda)$, $\lambda \neq \tau$, do not depend on the variables of X_τ. Therefore, we can fix all variables in X_λ, $\lambda \neq \tau$, avoiding the neighborhoods V_j, $j = 1, \ldots, s$, in such a way that $D(E_\lambda, X_\lambda) \neq 0$ and the order relation on X is satisfied for all x_i whenever $y_j \in V_j$, $j = 1, \ldots, s$. By (6.3.2), $D(E, X)$ changes sign for this choice of X_λ, $\lambda = 1, \ldots, \mu$. Hence, E is strongly order singular. $\square$

Remark. For an arbitrary matrix E it can happen that one of the factors $D(E_\lambda, X_\lambda)$ in (6.3.2) changes sign while $D(E, X)$ remains of constant sign. For example, another factor of the product may equal $D(E_\lambda, X_\lambda)$, thus cancelling the change of sign.

Theorem 6.4 (Lorentz [91]). *An almost-simple Pólya matrix E is regular if it has no essential odd supported sequences, and is strongly order singular otherwise.*

Proof. By Lemma 6.3 it is enough to prove the theorem for Birkhoff matrices. Because of the Atkinson–Sharma theorem, we need only show that an almost-simple Birkhoff matrix E with an odd supported sequence is necessarily strongly order singular. Assume to the contrary that there exists an almost-simple Birkhoff matrix E, having an odd supported sequence in a certain row i_0, that is regular or weakly singular. We consider matrices of this type with the smallest number of columns n, and among such E, a matrix with the smallest number of rows m. We let E_0 be one of these minimal matrices.

By Theorem 6.2, since E_0 is not strongly singular, none of the entries in the simple rows in E_0 is supported, and there must be at least two odd supported sequences in the row i_0 of E_0 (which proves to be nonsimple). Let

$$I_1 = \left\{ e_{i_0, q_1} = \cdots = e_{i_0, r_1} = 1 \right\}, \qquad I_2 = \left\{ e_{i_0, q_2} = \cdots = e_{i_0, r_2} = 1 \right\}$$

be the first two odd supported sequences in row i_0.

We first prove the following property of E_0: There are at least three elements $e_{i,k} = 1$ of E_0 outside of row i_0, and all such elements support I_1 (in other words, satisfy $k < q_1$).

The number of 1's outside i_0 is equal to the number of zeros in row i_0. This row contains at least three zeros, in columns $q_1 - 1 < r_1 + 1 \leqslant q_2 - 1 < r_2 + 1$. Hence there are at least three simple rows $i \neq i_0$.

Let k_0 be the maximum value of k with $e_{i,k} = 1$, $i \neq i_0$. We must have either $e_{i, k_0} = 1$ or $e_{m, k_0} = 1$; for if not, then the elements in the simple rows 1 and m would support the entry $e_{i, k_0} = 1$ in the simple row i, $i \neq 1, i_0, m$.

We shall show that all 1's in simple rows support the sequence I_1, in other words that $k_0 \leqslant q_1 - 1$. We can assume that $e_{1, k_0} = 1$. If $k_0 \geqslant q_1$, then we consider the maximal coalescence of row 1 in E_0. This moves the 1 in position $(1, k_0)$ to the position $(1, n)$, generating a decomposable matrix E_0^*. By Theorem 4.5, E_0^* cannot be strongly order singular. Clearly, E_0^* is an almost-simple matrix, and by Lemma 6.3, each of the components in its canonical decomposition is order regular or weakly singular. Now the Pólya functions of E_0 and E_0^* agree for $k = 0, 1, \ldots, q_1 - 1$ and $e_{i_0, q_1} = 1$. Therefore, the first component, E_1, of the canonical decomposition of E_0^* is an almost-simple Birkhoff matrix and it contains the odd supported sequence I_1 (since $k_0 \geqslant q_1$, we have not destroyed the support of I_1 by removing $e_{1, k_0} = 1$). Since E_1 has fewer columns than E_0, this contradicts the assumption that E_0 is minimal. Hence, $k_0 \leqslant q_1 - 1$.

Without loss of generality, we may assume that E_0 contains two rows above the row i_0. Then we coalesce row $i_0 - 1$ to row i_0, obtaining a new matrix E'. The one $e_{i_0 - 1, k} = 1$ moves into position (i_0, k'), with $k' \leqslant q_1 - 1$ since $k \leqslant k_0 \leqslant q_1 - 1$. Thus, the sequence I_2 remains intact and is supported in E' (e.g., by the 1's in rows $i_0 - 2$ and $i_0 + 1$). We apply Proposition 3.7. In the general case of the proposition, E' is an almost-simple Birkhoff matrix with $m - 1$ rows. It contains an odd supported sequence and is order regular or weakly singular by Theorem 4.5. Again this contradicts the minimal nature of E_0.

In the exceptional case of Proposition 3.7, the matrix E' decomposes $E' = E_1' \oplus \cdots \oplus E_\lambda'$. The last component E_λ' of the decomposition contains the sequence I_2, and two 1's that support it. One of them is $e_{i_1, \lambda} = 1$ of the lemma, $i_1 \neq i_0 - 1, i_0$. The other is any 1 located on the other side of i_0. This exists because there are $e_{i, k} = 1$ both with $i < i_0 - 1$ and $i > i_0$.

Since E_λ' has fewer columns than E_0, it must be strongly singular in contradiction to Lemma 6.3.   □

The simplest special case of Theorem 6.4 is the following. Let E be a Pólya matrix with at least three rows that has nonzero entries only in an interior row i_0 or in the column 0. Let k_j, $0 \leqslant k_0 < \cdots < k_p \leqslant n$, be the positions of all 0's in row i_0. Each interval (k_{j-1}, k_j) contains a sequence of 1's in row i_0. As a corollary to Theorem 6.4, we have

Theorem 6.5. *A matrix E of type just described is order regular if and only if all the differences $k_j - k_{j-1}, j = 1, \ldots, p$, are odd.*

This theorem was obtained by K. Zeller in 1969 with a very different proof (the original ideas of Zeller are reproduced in [92, p. 75]).

§6.4. PROBABILITY OF REGULARITY

We wish to show that the singularity of Birkhoff or of Pólya matrices E is the rule rather than exception, at least when the number of 1's in E is large. In fact, we will show that almost all $m \times n$ Birkhoff matrices with large n have supported singletons. Then our statement about Birkhoff matrices will follow from Theorem 6.1. This section is based on the paper [100].

The approach outlined requires a large number of rows in the matrices; we shall need the assumption

$$(1 + \delta)\frac{n}{\log n} \leqslant m \tag{6.4.1}$$

where $\delta > 0$ is a constant.

Let $B(m, n)$ denote both the class and the number of $m \times n$ Birkhoff matrices. By $B^p(m, n)$ or $B_p(m, n)$ we mean the subclasses of $B(m, n)$ of those matrices having at least p or exactly p simple rows, respectively, as well as the numbers of such matrices.

Theorem 6.6. *Let m satisfy (6.4.1). Then for each p and all large n, almost all $m \times n$ Birkhoff matrices have at least p simple rows:*

$$\lim_{n \to \infty} \frac{B^p(m, n)}{B(m, n)} = 1. \tag{6.4.2}$$

Proof. Let $E \in B_{k-1}(m, n)$ be an $m \times n$ Birkhoff matrix with exactly $k - 1$ simple rows, and let F be an $(m + 1) \times n$ interpolation matrix containing only a single 1. There are $(m + 1)n$ such matrices F. By combining E and F, we construct an $(m + 1) \times (n + 1)$ matrix E' as follows: The m rows of E are placed in their natural order into the m empty rows of F, and then an $(n + 1)$st column of 0's is added. Clearly, $E' \in B_k(m + 1, n + 1)$. In this way, we obtain some of the matrices in $B_k(m + 1, n + 1)$ with at most k duplications, for we can use at most k simple rows of E' for F. Therefore,

$$kB_k(m + 1, n + 1) \geqslant (m + 1)nB_{k-1}(m, n), \qquad k = 1, 2, \ldots .$$

Summing these inequalities for $k = 1, \ldots, p$ yields

$$B_1(m + 1, n + 1) + \cdots + B_p(m + 1, n + 1)$$

$$\geqslant \frac{(m + 1)n}{p} [B(m, n) - B^p(m, n)].$$

Therefore,

$$1 - \frac{B^p(m, n)}{B(m, n)} \leqslant \frac{p}{(m + 1)n} \cdot \frac{B(m + 1, n + 1)}{B(m, n)} . \qquad (6.4.3)$$

Using equation (2.2.3) of Theorem 2.6, we have

$$\frac{B(m + 1, n + 1)}{B(m, n)} \leqslant \frac{\binom{mn + n}{n + 1}}{\binom{mn - m}{n}} = \frac{(m + 1)n}{n + 1} \prod_{k=1}^{n-1} \frac{mn + n - k - 1}{mn - m - k}$$

$$\leqslant (m + 1) \left(\frac{mn}{mn - m - n + 1} \right)^n$$

$$= (m + 1) \left[1 + \frac{m + n - 1}{(m - 1)(n - 1)} \right]^n$$

$$\leqslant (m + 1) \left(1 + \frac{2}{n - 1} + \frac{1}{m - 1} \right)^n .$$

Putting this in (6.4.3), we obtain

$$1 - \frac{B^p(m, n)}{B(m, n)} \leqslant \frac{p}{n} \left(1 + \frac{2}{n - 1} + \frac{1}{m - 1} \right)^n . \qquad (6.4.4)$$

Because of the assumption (6.4.1),

$$\log \left(1 + \frac{2}{n - 1} + \frac{1}{m - 1} \right)^n \leqslant (1 - \delta') \log n$$

for some $\delta' > 0$ and all large n. Thus, the right-hand side of (6.4.4) does not exceed $pn^{-\delta'}$. □

A similar argument shows that for each p, and m satisfying (6.4.1),

$$\lim_{n \to \infty} \frac{P^p(m,n)}{P(m,n)} = 0.$$

However, we will not use this fact.

We want to show that most matrices in $B^p(m,n)$ have simple rows with *supported* singletons. In particular, let $E \in B_p(m,n)$. Then the matrix E has exactly p simple rows. Let $A: i_1 < \cdots < i_p$ be their positions. The rows $i_j, j = 1,\ldots,p$, of the matrix E can be identified with a function $k = f(j)$, with values $k = 0,\ldots,n-1$, that gives the position of the singleton $e_{i_j,k}$ of the row i_j. If the function f is strongly nonmonotone, then the matrix E will have supported singletons. To carry out this proof, we also have to consider cyclic permutations of columns, because Birkhoff matrices are best understood probabilistically as equivalence classes produced by cyclic permutations (see §2.2).

Thus, we are led to consider functions $f(j), j = 1,2,\ldots,p$, with values in the set $\{0,1,\ldots,n-1\}$, and their monotonicity properties. There are altogether n^p such functions. We say f has at most four monotone branches if there exist four (perhaps degenerate) intervals $p_s \leqslant j < p_{s+1}, p_0 = 1, p_4 = p$, on each of which f is monotone.

Proposition 6.7. (i) *There are* $\dbinom{n+p-1}{p}$ *monotone increasing functions f.*

(ii) *The number of functions with at most four monotone branches does not exceed*

$$\frac{2^{3p+4}}{p!} n^p \leqslant \varepsilon n^p \tag{6.4.5}$$

for each $\varepsilon > 0$ and all sufficiently large p and $n \geqslant p$.

Proof. The monotone increasing functions f are in one-to-one correspondence with paths connecting the point $(1,0)$ with $(p, n-1)$ formed by moving one unit upward or one unit to the right at each stage, and also marking a special point (p, n'), $n' \leqslant n-1$ (for the value of $n' = f(p)$). On the other hand, these paths can be described as follows. On the horizontal axis we mark $(p-1)+(n-1)+1 = p+n-1$ points. We select p of the points. The first $p-1$ points serve as the left end points of the horizontal stages of the graph, and the last point serves as the point (p,n'). This selection determines the graph uniquely, and shows that there are $\dbinom{n+p-1}{p}$ graphs and increasing functions.

A function f with at most four monotone branches is determined by the selection of four intervals of lengths l_s, $\sum_1^4 l_s = p$, and of the sense of increase or decrease of f on each of them. The range of f on each interval is at most $k = 0,\dots,n-1$. Therefore, an upper bound for the number of functions is

$$2^4 \sum_{l_1 + \cdots + l_4 = p} \prod_{s=1}^{4} \binom{n + l_s - 1}{l_s}$$

$$\leqslant 2^4 (n+p)^p \sum_{l_1 + \cdots + l_4 = p} \frac{1}{l_1! \cdots l_4!} = 2^4 (n+p)^p \frac{4^p}{p!}$$

and (6.4.5) follows if $p \leqslant n$.   □

Corresponding to a function f, we form the $p \times n$ matrix with entries $e_{i,k} = 1$ if $f(i) = k$ and $e_{i,k} = 0$ otherwise. We generate new functions f' from f by cyclic permutations of the columns of the matrix. For a given f, we say that the n distinct functions f' obtained in this way are equivalent. The following fact is essential: If f has at most two monotone branches, then all of the equivalent functions f' have at most four monotone branches.

If a function f has the property that $[a, b]$ cannot be partitioned into two intervals on each of which f is monotone, then there are three adjacent intervals on which f is monotone in alternating sense and not constant. This implies the existence of $j_1 < j < j_2$ for which

$$f(j_1) < f(j), \qquad f(j_2) < f(j). \tag{6.4.6}$$

We now return to the matrices $B(m, n)$. Let $\overline{M}(m, n-1)$ denote the set of all $m \times (n-1)$ matrices of 0's and ones with exactly n 1's. According to Theorem 2.6, the set $B(m, n)$ can be identified with the set of equivalence classes of $\overline{M}(m, n-1)$, two matrices being equivalent if one can be obtained from the other by a cyclic permutation of the columns. In particular, let $\overline{M}_A = \overline{M}_A(m, n-1)$ for a given set A of integers i_j, $1 \leqslant i_1 < \cdots < i_p \leqslant m$, be the subset of $\overline{M}(m, n-1)$ consisting of matrices having their simple rows exactly in the positions $i_j \in A$. Let $M_A^* = M_A^*(m, n)$ be the set of equivalence classes of $\overline{M}_A$. (We use the same notation also for the number of their elements.)

Lemma 6.8. *For a given set A, all but at most εM_A^* classes of M_A^* have the property that they consist of matrices with a supported singleton.*

Proof. The matrices $E \in \overline{M}_A(m, n-1)$ are in one-to-one correspondence with pairs (f, E'), where $f(j)$, $j = 1,\dots,p$, is a function with values $0,\dots,n-2$, and E' is an arbitrary $(m-p) \times (n-1)$ matrix with exactly $n - p$ ones, formed by the rows $i \neq i_j, j = 1,\dots,p$, of E. Let β be the number of such matrices E'. Then there are $\beta(n-1)^p$ matrices in $\overline{M}_A$, and consequently, $\beta(n-1)^{p-1}$ equivalence classes M_A^*.

By Proposition 6.7 (ii), there are at most $\varepsilon(n-1)^p \beta$ matrices $E = (f, E')$ in $\overline{M}_A$ with f having at most four monotone branches. Since each equivalence class of M_A^* consists of $n-1$ matrices, there can be at most $\varepsilon(n-1)^{p-1}\beta = \varepsilon M_A^*$ equivalence classes consisting entirely of matrices $E = (f, E')$ with f having at most four monotone branches.

Suppose $E = (f, E')$ has no supported singletons. Then f fails to satisfy (6.4.6) and has at most two monotone branches. Let $E^* = (f', E'')$ be any matrix in the same equivalence class as E. Then f' is a cyclic permutation of the function f and has at most four monotone branches. Therefore, there are at most εM_A^* equivalence classes containing a matrix E having no supported singletons. $\square$

Theorem 6.9. *Let $\varepsilon > 0$ be given; let m satisfy (6.4.1); then for all large n, $n \geqslant n_0$, all but $\varepsilon B(m, n)$ of the $B(m, n)$ Birkhoff $m \times n$ matrices are order singular; even more, all but $\varepsilon B(m, n)$ of them have supported singletons.*

Proof. Because of (6.4.2), it is sufficient to prove that, for some p, all but $\varepsilon B^p(m, n)$ matrices of the class $B^p(m, n)$ have supported singletons. We take p to satisfy (6.4.5). The set $B^p(m, n)$ is the disjoint union of sets of matrices $E \in B_A(m, n)$, where $A = \{i_1 < \cdots < i_q\}$, $q \geqslant p$, and the matrices E of $B_A(m, n)$ have $i_j, j = 1, \ldots, q$, as their simple rows. It is sufficient to prove our result for each $B_A(m, n)$, for a fixed $q \geqslant p$ and fixed A.

In the identification of $B(m, n)$ with $\overline{M}(m, n-1)$ described above, the matrices $E \in B_A(m, n)$ will correspond to the equivalence classes of $M_A^*(m, n-1)$. If we exclude at most εM_A^* classes from M_A^* by Lemma 6.8, that is, if we exclude at most εB_A matrices from B_A, all the remaining matrices will have supported singletons. $\square$

Theorem 6.10. *Let $\varepsilon > 0$ be given; let m satisfy (6.4.1). Then for all large n, all but $\varepsilon P(m, n)$ of the $P(m, n)$ Pólya matrices are order singular.*

Proof. We consider the canonical decompositions

$$E = E_1 \oplus \cdots \oplus E_\mu, \qquad n = n_1 + \cdots + n_\mu \qquad (6.4.7)$$

of the $m \times n$ Pólya matrices E into $m \times n_\lambda$ matrices E_λ that are either Birkhoff matrices or have one column. We prove that for most such E, at least one of the E_λ is a Birkhoff matrix with supported singletons.

For given $\varepsilon > 0$, we select a sufficiently large integer p_0. The first requirement is that $p_0 \geqslant n_0$, where n_0 is given by Theorem 6.9. The second requirement for p_0 will be given later.

We first consider matrices (6.4.7) with fixed numbers $\mu, n_1, \ldots, n_\mu$, and with the property that $n_{\lambda_0} \geqslant p_0$ for some λ_0. There are $\prod_{\lambda=1}^{\mu} B(m, n_\lambda)$ such matrices, and all except $\varepsilon \prod_{\lambda=1}^{\mu} B(m, n_\lambda)$ of them have supported simple rows in the λ_0 component. The representations (6.4.7) for different $n_1, \ldots, n_\mu$ give rise to disjoint sets of matrices E. Hence the conclusion of the theorem is valid for the set of all E of form (6.4.7) with $n_{\lambda_0} \geqslant p_0$ for some λ_0.

Next we consider matrices E that satisfy $n_\lambda \leqslant p_0$ for all $\lambda = 1, \ldots, \mu$ in (6.4.7). It is easier to handle matrices of approximately equal but not too small length. By combining some of the matrices E_λ together, we can obtain another decomposition of E,

$$E = F_1 \oplus \cdots \oplus F_{\mu'}, \qquad p_1 + \cdots + p_{\mu'} = n, \qquad (6.4.8)$$

where each F_λ, $\lambda = 1, \ldots, \mu'$, is an $m \times p_\lambda$ Pólya matrix with

$$p_0 \leqslant p_\lambda \leqslant 3 p_0. \qquad (6.4.9)$$

The number of matrices E of this kind does not exceed the number γ of all possible representations (6.4.8).

In order to evaluate γ, we start with an estimate of $P(m, p)$ from above. Using (2.2.7) and (2.2.1) we have

$$P(m, p) = \frac{1}{p+1} \binom{m(p+1)}{p}$$

$$\leqslant \frac{1}{(m-1)p} \binom{m(p+1)}{p+1}$$

$$\leqslant C_1 \frac{1}{p^{3/2}} m^p \left(1 + \frac{1}{m-1}\right)^{(m-1)(p+1)}$$

$$\leqslant C_2 \frac{1}{p^{3/2}} m^p \left(1 + \frac{1}{m-1}\right)^{(m-1)p}$$

where C_1 and $C_2 = C_1 e$ are absolute constants. For $p = p_\lambda$ we have $p_\lambda^{-3/2} \leqslant p_0^{-3/2}$.

If μ', $p_1, \ldots, p_{\mu'}$, are fixed, the number of representations (6.4.8) does not exceed

$$\prod_{\lambda=1}^{\mu'} P(m, p_\lambda) \leqslant C_2^{\mu'} \frac{1}{p_0^{3\mu'/2}} \prod_{l=1}^{\mu'} m^{p_\lambda} \left(1 + \frac{1}{m-1}\right)^{(m-1)p_\lambda}$$

$$= C_2^{\mu'} \frac{1}{p_0^{3\mu'/2}} m^n \left(1 + \frac{1}{m-1}\right)^{(m-1)n}.$$

For given μ', there are at most $2p_0 + 1 \leqslant 3p_0$ choices of p_1 (in the interval $[p_0, 3p_0]$); after that at most $3p_0$ choices of p_2; and so on. Altogether, there are no more than $(3p_0)^{\mu'}$ choices of $p_1, \ldots, p_{\mu'}$, and there are at most n choices of μ'. We see that the number γ of representations (6.4.8) satisfies

$$\gamma \leqslant n(3p_0)^{\mu'} \prod P(m, p_\lambda) \leqslant \left(\frac{3C_2}{\sqrt{p_0}}\right)^{\mu'} n m^n \left(1 + \frac{1}{m-1}\right)^{(m-1)n}.$$

We now fix p_0 so large that $C_2 \leqslant \frac{1}{6}\sqrt{p_0}$. Then because $\mu' \geqslant n/(3p_0)$,

$$\gamma \leqslant 2^{-n/(3p_0)} n m^n \left(1 + \frac{1}{m-1}\right)^{(m-1)n}. \qquad (6.4.10)$$

It is easy to estimate $P(m, n)$ from below by means of (2.2.7) and (2.2.1):

$$P(m,n) \geqslant \frac{1}{n+1}\binom{mn}{n} \geqslant C_3 n^{-3/2} m^n \left(1 + \frac{1}{m-1}\right)^{(m-1)n}. \quad (6.4.11)$$

From (6.4.10) and (6.4.11), we see that $\gamma \leqslant \varepsilon P(m, n)$ for n sufficiently large.

□

§6.5. NOTES

6.5.1. Our proof of Theorem 6.2 is borrowed from [92]. The advantage of this method of proof is that it introduces two new techniques, useful elsewhere, the method of independent knots (see §5.6, Chapter 6, §8.2, §10.5) and the method of the null curves used also in Chapter 8. Another advantage of the method is that it allows proofs for very general sets of functions $\mathcal{G}$.

6.5.2. For the history of Theorems 6.1, 6.2 see [96]. Recently, Scherer and Zeller [148] gave another proof of the last theorem, using coalescence to a three-row matrix and some ideas from [107] and [90].

6.5.3. Unfortunately, we were not able to include the probabilistic results of G. G. Lorentz and R. A. Lorentz [99], which improve Theorems 6.9 and 6.10. Instead of Theorem 6.1, these authors use Theorem 4.12 as their basic singularity criterion. This requires different techniques and leads to the following.

Theorem 6.11. *If $P_r(m, n)$ is the number of regular $m \times n$ matrices among all $P(m, n)$ Pólya matrices, then*

$$\lim_{m,n \to \infty} \frac{P_r(m,n)}{P(m,n)} = 0 \qquad (6.5.1)$$

and for fixed $m \geqslant 3$,

$$\overline{\lim_{n \to \infty}} \frac{P_r(m,n)}{P(m,n)} \leqslant \frac{1}{2^{[m/3]}}.$$

where $[m/3]$ is the integer part of $m/3$.

However, the question whether it is true that $\lim_{n \to \infty}[P_r(3, n)/ P(3, n)] = 0$ remains unsolved. A further progress in the regularity problem seems to require a radical improvement of the singularity theorems (such as Theorem 4.12) or of regularity theorems (of the Atkinson–Sharma theorem).

6.5.4. In §6.4 and in [99], matrices of classes $M(m, n)$, $P(m, n)$, $B(m, n)$ were allowed to contain zero rows. Perhaps more natural are the corresponding classes $M^*(m, n)$, $P^*(m, n)$, $B^*(m, n)$, for which this possibility is eliminated. R. A. Lorentz [111] proves that $P^*(m, n)/M^*(m, n) \to 0$ if $n \to \infty$ and $B_r^*(m, n)/B^*(m, n) \to 0$ if $n \to \infty$ and $m \geqslant (\frac{1}{2} + \varepsilon)n$. An analogue of Theorem 6.11 in this situation is not known.

Chapter 7

Zeros of Birkhoff Splines

§7.1. BIRKHOFF'S KERNEL

Birkhoff's kernel is intimately connected with Birkhoff interpolation by algebraic polynomials. It was introduced by Birkhoff in his famous paper [7], which is more concerned with remainder formulas than regularity theorems. But we shall see that the Atkinson and Sharma theorem is a corollary—although not an obvious one—of Birkhoff's results. Let

$$E = [e_{i,k}]_{i=1\,k=0}^{m\quad n} \qquad (7.1.1)$$

be a normal interpolation matrix with 0's *in the last column*, and let $X = \{x_1, \ldots, x_m\}$ be an arbitrary set of knots. For the sake of simplicity let us assume that $x_1 < x_2 < \cdots < x_m$. In the determinant $D(E, X)$ defined in (1.3.2) we replace the entries in the last column by the elements $(x_i - t)_+^{n-1-k}/(n-1-k)!$; as usual, u_+^r defines the truncated power function:

$$u_+^r = \begin{cases} u^r & \text{if} \quad u \geq 0, \\ 0 & \text{if} \quad u < 0. \end{cases} \qquad (7.1.2)$$

Here, u_+^0 is not defined for $u = 0$. The resulting real function

$$K_E(X; t) = K(t)$$

$$= \det\left[\frac{x_i^{-k}}{(-k)!}, \ldots, \frac{x_i^{n-1-k}}{(n-1-k)!}, \frac{(x_i - t)_+^{n-1-k}}{(n-1-k)!}; e_{i,k} = 1 \right] \qquad (7.1.3)$$

is *Birkhoff's kernel* associated with the pair E, X.

We derive some simple properties of the kernel, given essentially in
[7]. If $t > x_m$ then all entries in the last column vanish; thus $K(t) = 0$. In
order to see that $K(t) = 0$ also for $t < x_1$, we use an argument of §1.3. For
arbitrary real y the determinant

$$K_y(t) = \det\left[\frac{(x_i + y)^{-k}}{(-k)!}, \ldots, \frac{(x_i + y)^{n-1-k}}{(n-1-k)!}, \frac{(x_i - t)_+^{n-1-k}}{(n-1-k)!} \; ; e_{i,k} = 1 \right]$$

is independent of y, $K_y(t) = K(t)$. We choose $y = -t$ and see that the
determinant has two identical columns if $t < x_1$. Consequently, the kernel
has compact support, $\text{supp}[K_E(X; t)] \subseteq [x_1, x_m]$.

Let us compute the integral of the kernel:

$$\int_{-\infty}^{+\infty} K_E(X; t)\, dt$$

$$= \int_{x_1}^{x_m} K_E(X; t)\, dt$$

$$= \det\left[\frac{x_i^{-k}}{(-k)!}, \ldots, \frac{x_i^{n-1-k}}{(n-1-k)!}, \int_{x_1}^{x_m} \frac{(x_i - t)_+^{n-1-k}}{(n-1-k)!}\, dt ; e_{i,k} = 1 \right]$$

$$= \det\left[\frac{(x_i - x_1)^{-k}}{(-k)!}, \ldots, \frac{(x_i - x_1)^{n-1-k}}{(n-1-k)!}, \frac{(x_i - x_1)^{n-k}}{(n-k)!} \; ; e_{i,k} = 1 \right]$$

$$= D(E, X - x_1) = D(E, X)$$

where we have used Proposition 1.2.

Further properties follow from the representation

$$K_E(X; t) = \sum_{e_{j,k} = 1} D_{i,k}(X) \frac{(x_i - t)_+^{n-1-k}}{(n-1-k)!} \tag{7.1.4}$$

where — as in §1.3— $D_{i,k}(X)$ denotes the algebraic complements of the
entries of the last column of the determinant defining the kernel. We
conclude that the kernel K is a piecewise polynomial function with joints
(knots) $x_1, \ldots, x_m$. This polynomial spline function has discontinuities of its
$(n-1-k)$th derivative only at the knots x_i for which $e_{i,k} = 1$ and $D_{i,k}(X)$
$\neq 0$. The degree of the spline (i.e., the maximal degree of the polynomial
pieces) is evidently at most $n-1$.

If E is not a Birkhoff matrix, we can see that the degree of K is
actually less than $n-1$. We can prove more. Let us first assume that E is
not a Pólya matrix and let r, $0 \leqslant r \leqslant n-2$, be the first integer for which
$M_r < r + 1$; then on rearranging the rows (compare §1.4) we get

$$K(t) = \pm \det \begin{bmatrix} * & | & * \\ \hline 0 & | & * \end{bmatrix} \begin{matrix} \} M_r \\ \} n + 1 - M_r \end{matrix}.$$
$$\underbrace{}_{r+1} \underbrace{}_{n-r}$$

This shows that K vanishes identically. Similarly, if E is decomposable, $E = E_1 \oplus E_2$, then $K_E(X; t) = \pm D(E_1, X) K_{E_2}(X; t)$ and in this case all the properties of $K_E(X; t)$ essentially follow from the properties of the kernel associated with the pair E_2, X. We summarize

Theorem 7.1 (Properties of Birkhoff's kernel). *Let E be an $m \times (n + 1)$ normal interpolation matrix with 0's in the last column, and let X: $x_1 < x_2 < \cdots < x_m$. Then Birkhoff's kernel*

$$K_E(X; t) = \det\left[\frac{x_i^{-k}}{(-k)!}, \ldots, \frac{x_i^{n-1-k}}{(n-1-k)!}, \frac{(x_i - t)_+^{n-1-k}}{(n-1-k)!}; e_{i, k} = 1 \right]$$

has the following properties:

    a. *$K_E(X; t)$ is a polynomial spline function of degree $\leqslant n - 1$ with knots $x_1, \ldots, x_m$, and $K_E^{(q)}(X; t)$ is continuous at x_i if and only if $e_{i, n-1-q} = 0$ or $D_{i, n-1-q}(X) = 0$.*
    b. *$\mathrm{supp}[K_E(X; t)] \subseteq [x_1, x_m]$.*
    c. *$\int_{-\infty}^{+\infty} K_E(X; t)\, dt = D(E, X)$.*
    d. *If E is not a Pólya matrix, then $K_E(X; t) \equiv 0$. If E is decomposable, $E = E_1 \oplus E_2$, then $K_E(X; t) = \pm D(E_1, X) K_{E_2}(X; t)$.*

§7.2. BIRKHOFF'S IDENTITY

In this section we shall see that Birkhoff's kernel is the generating kernel of a linear functional that is orthogonal to the space $\mathcal{P}_{n-1}$ of algebraic polynomials of degree at most $n - 1$. For an interval $[a, b]$, let $A_n[a, b]$ be the linear subspace of $C^{n-1}[a, b]$ consisting of all functions f for which $f^{(n-1)}$ is absolutely continuous.

For $f \in A_1[a, b]$ and $x \in [a, b]$ we have

$$f(x) = f(a) + \int_a^x f'(t)\, dt = f(a) + \int_a^b (x - t)_+^0 f'(t)\, dt.$$

Replacing f by $f^{(k)}$ here and integrating by parts several times, we conclude that for $f \in A_n[a, b]$,

$$f^{(k)}(x_i) = \sum_{j=0}^{n-1-k} f^{(k+j)}(a) \frac{(x_i - a)^j}{j!} + \int_a^b \frac{(x_i - t)_+^{n-1-k}}{(n-1-k)!} f^{(n)}(t)\, dt$$

$$(7.2.1)$$

if $x_i \in [a, b]$ and $0 \leqslant k \leqslant n - 1$. Of course, this is Taylor's formula with integral remainder.

Theorem 7.2 (Birkhoff's identity [7]). *Let E be an $m \times (n + 1)$ normal interpolation matrix with 0's in the last column, and let X:*

$x_1 < x_2 < \cdots < x_m$. Then, for all $f \in A_n[x_1, x_m]$,

$$\sum_{e_{i,k}=1} D_{i,k}(X) f^{(k)}(x_i) = \int_{x_1}^{x_m} f^{(n)}(t) K_E(X;t) \, dt, \qquad (7.2.2)$$

where $K_E(X;t)$ is Birkhoff's kernel associated with the pair E, X.

Proof. We use (7.2.1) with $a = x_1$ and $b = x_m$:

$$\sum_{e_{i,k}=1} D_{i,k}(X) f^{(k)}(x_i)$$

$$= \det\left[\frac{x_i^{-k}}{(-k)!}, \ldots, \frac{x_i^{n-1-k}}{(n-1-k)!}, f^{(k)}(x_i); e_{i,k}=1 \right]$$

$$= \det\left[\frac{x_i^{-k}}{(-k)!}, \ldots, \frac{x_i^{n-1-k}}{(n-1-k)!}, \sum_{j=0}^{n-1-k} f^{(k+j)}(x_1) \frac{(x_i - x_1)^j}{j!}; e_{i,k}=1 \right]$$

$$+ \det\left[\frac{x_i^{-k}}{(-k)!}, \ldots, \frac{x_i^{n-1-k}}{(n-1-k)!}, \int_{x_1}^{x_m} \frac{(x_i - t)_+^{n-1-k}}{(n-1-k)!} f^{(n)}(t) \, dt; e_{i,k}=1 \right].$$

The first term vanishes, since it is equal to

$$\det\left[\frac{(x_i - x_1)^{-k}}{(-k)!}, \ldots, \frac{(x_i - x_1)^{n-1-k}}{(n-1-k)!}, \right.$$

$$\left. \sum_{q=0}^{n-1} f^{(n-1-q)}(x_1) \frac{(x_i - x_1)^{n-1-k-q}}{(n-1-k-q)!}; e_{i,k}=1 \right]$$

where the last column is a linear combination of the preceding columns. The second term gives

$$\int_{x_1}^{x_m} f^{(n)}(t) \det\left[\frac{x_i^{-k}}{(-k)!}, \ldots, \frac{x_i^{n-1-k}}{(n-1-k)!}, \frac{(x_i - t)_+^{n-1-k}}{(n-1-k)!}; e_{i,k}=1 \right] dt$$

$$= \int_{x_1}^{x_m} f^{(n)}(t) K_E(X;t) \, dt. \quad \square$$

Theorem 7.2 shows that the functional $L(f)$, defined on $A_n[x_1, x_m]$ by the formula (7.2.2), annihilates the space $\mathcal{P}_{n-1}$.

If a linear functional $L(f), f \in A_n[a, b]$, has the representation

$$L(f) = \int_a^b f^{(n)}(t) K(t) \, dt,$$

then $K(t)$ is called the *Peano kernel* of L. It is interesting to note that Peano's paper [133] from which this term originated dates to the year 1913, whereas Birkhoff published his identity in 1906. (See §7.7.)

Of course, Theorem 7.2 is meaningful only if at least some cofactor $D_{i,k}(X)$ does not vanish. Now $D_{i,k}(X) = \pm D(E_{i,k}, X)$ with the normal interpolation matrix $E_{i,k}$ obtained from E by replacing the entry $e_{i,k}$ by 0 and deleting the last column. Thus the theorem gives information if and only if some reduced interpolation problem $E_{i,k}, X$ is regular.

Theorem 7.3 (Birkhoff's existence theorem, [7]).   *Let E be an $m \times (n+1)$ normal interpolation matrix with 0's in the last column, and let X: $x_1 < x_2 < \cdots < x_m$. If at least one of the algebraic complements $D_{i,k}$ is not 0, and if the constants $c_{i,k}$ and the integrable function g on $[x_1, x_m]$ satisfy*

$$\sum_{e_{i,k}=1} c_{i,k} D_{i,k}(X) = \int_{x_1}^{x_m} g(t) K_E(X; t)\, dt, \tag{7.2.3}$$

then there is a unique function $f \in A_n[x_1, x_m]$ with the properties

$$f^{(k)}(x_i) = c_{i,k} \quad \text{for} \quad e_{i,k} = 1 \tag{7.2.4a}$$

and

$$f^{(n)} = g \quad a.e. \tag{7.2.4b}$$

Proof.   Using Taylor's expansion for $f \in A_n[x_1, x_m]$ at the point $a = x_1$, we see that conditions (7.2.4) are equivalent to the system

$$\sum_{j=0}^{n-1} f^{(n-1-j)}(a) \frac{(x_i - a)^{n-1-j-k}}{(n-1-j-k)!}$$

$$= c_{i,k} - \int_{x_1}^{x_m} g(t) \frac{(x_i - t)_+^{n-1-k}}{(n-1-k)!}\, dt \quad \text{for} \quad e_{i,k} = 1. \tag{7.2.5}$$

We have to show that these $n+1$ equations are uniquely solvable for the n unknowns $f^{(n-1)}(a), f^{(n-2)}(a), \ldots, f(a)$. Condition (7.2.3) means the vanishing of the determinant

$$\det\left[\frac{x_i^{-k}}{(-k)!}, \ldots, \frac{x_i^{n-1-k}}{(n-1-k)!}, c_{i,k} - \int_{x_1}^{x_m} g(t) \frac{(x_i - t)_+^{n-1-k}}{(n-1-k)!}\, dt; e_{i,k} = 1 \right].$$

Since one of the $D_{i,k}(X) = D_{i,k}(X - a)$ does not vanish, the system (7.2.5) has a unique solution. It remains to note that every function $f \in A_n[x_1, x_m]$ is uniquely defined by $f(a), f'(a), \ldots, f^{(n-1)}(a)$ and $f^{(n)} = g$ a.e. $\square$

Theorem 7.2 can be used to find Peano's kernel of Birkhoff interpolation. Let E, X be a regular pair, $E = [e_{i,k}]_{i=1\,k=0}^{m\ \ n}$, X: $x_1 < x_2 < \cdots < x_m$, and let $x \in [x_1, x_m]$ and $q = 0, 1, \ldots, n$ be for the moment fixed. We define the extended set of knots

$$\tilde{X}: x_1 < x_2 < \cdots < x_j \leqslant x < x_{j+1} < \cdots < x_m$$

and the extended interpolation matrix $\tilde{E}$ as follows: If $x_j < x < x_{j+1}$, we put

$$\tilde{E} = [\tilde{e}_{i,k}]_{i=1}^{m+1}{}_{k=0}^{n+1}$$

$$= \begin{bmatrix}
e_{1,0} & e_{1,1} & \cdots & & \cdots & e_{1,n} & \vdots & 0 \\
\vdots & \vdots & & & & \vdots & \vdots & \\
e_{j,0} & e_{j,1} & \cdots & & \cdots & e_{j,n} & \vdots & 0 \\
\hline
0 & 0 & \cdots & 0 \quad 1 \quad 0 & \cdots & 0 & \vdots & 0 \\
\hline
e_{j+1,0} & e_{j+1,1} & \cdots & & \cdots & e_{j+1,n} & \vdots & 0 \\
\vdots & \vdots & & & & \vdots & \vdots & \vdots \\
e_{m,0} & e_{m,1} & \cdots & & \cdots & e_{m,n} & \vdots & 0
\end{bmatrix}$$

where the 1 in row $j+1$ is $\tilde{e}_{j+1,q} = 1$.

If $x_j = x$ and $e_{j,q} = 0$, then $\tilde{E} = [\tilde{e}_{i,k}]_{i=1}^{m}{}_{k=0}^{n+1}$ is obtained from E by replacing the entry $e_{j,q} = 0$ by $\tilde{e}_{j,q} = 1$ and adding a zero column. If $x = x_j$ and $e_{j,q} = 1$ we do not define $\tilde{E}$ but put $K_{\tilde{E}}(\tilde{X}; t) \equiv 0$. With these definitions in mind we apply Theorem 7.2 to the extended pair $\tilde{E}$, $\tilde{X}$ and obtain

Theorem 7.4 (Birkhoff's remainder theorem, [7]). *Suppose that $P \in \mathcal{P}_n$ is the polynomial that interpolates $f \in A_{n+1}[x_1, x_m]$ with respect to the regular pair E, X. If $\tilde{E}$, $\tilde{X}$ and q are as described, then*

$$f^{(q)}(x) - P^{(q)}(x) = \frac{(-1)^{n+1+\varepsilon}}{D(E, X)} \int_{x_1}^{x_m} f^{(n+1)}(t) K_{\tilde{E}}(\tilde{X}; t) \, dt \quad (7.2.6)$$

where ε is the number of 1's in $\tilde{E}$ that precede (in the lexicographic order) the new 1.

This shows that Peano's kernel of interpolation is—essentially—Birkhoff's kernel associated with the extended pair $\tilde{E}$, $\tilde{X}$.

If we specialize E to be a Hermite matrix, but not a Taylor matrix, then Theorem 7.2 gives a representation for divided differences. There are several equivalent definitions for divided differences. For our considerations the following definition is the most convenient one: If E, X is a normal Hermite problem satisfying the assumptions of Theorem 7.2, and if $f \in A_n[x_1, x_m]$, then the leading coefficient of the polynomial $P \in \mathcal{P}_n$ that interpolates f is the divided difference of f with respect to the knots X, the multiplicity of the knots being indicated by E. Using Cramer's rule, we see that the divided difference is given by

$$\Delta^n(f; X, E) = \frac{1}{D(E, X)} \sum_{e_{i,k}=1} D_{i,k}(X) f^{(k)}(x_i). \quad (7.2.7)$$

This formula and Theorem 7.2 show that apart from the scalar factor $D(E, X)^{-1}$, the Birkhoff kernel of a normal Hermite matrix is the Peano kernel of the corresponding divided difference. Kernels of divided differences are called *B-splines* (or basic splines; see de Boor [B, p. 108]). Thus, for regular Birkhoff problems E, X, we may consider Birkhoff's kernel to be a straightforward generalization of the concept of B-splines.

We conclude this section by dealing with the important special case when $K_E(X; t)$ is of one sign.

Theorem 7.5. *Under the assumptions of Theorem 7.2, for all $f \in C^n[x_1, x_m]$ we have*

$$\sum_{e_{i,k}=1} D_{i,k}(X)f^{(k)}(x_i) = f^{(n)}(\xi)D(E, X) \quad \text{for some} \quad \xi \in (x_1, x_m)$$

(7.2.8)

if and only if Birkhoff's kernel $K_E(X; t)$ does not change sign.

Proof. Sufficiency follows from (7.2.2), Theorem 7.1(c), and the mean value theorem. On the other hand, if a piecewise continuous function K changes sign, there is a continuous f for which the formula

$$\int_a^b Kf\,dt = f(\xi)\int_a^b K\,dt$$

does not hold for any ξ, $a \le \xi \le b$. Indeed, assume for example that $\int_a^b K\,dt \le 0$. Then we can find a continuous function f that at the points t of continuity of K is > 0 if $K(t) \ge 0$, and $= 0$ if $K(t) \le 0$. □

Finally, we have, as a corollary of Theorem 7.1(c) and Theorem 7.5, respectively:

Corollary 7.6. *If Birkhoff's kernel associated with the pair E, X is not identically 0 and does not change sign, then*

a. E, X *is regular and*
b. $f \in C^n[x_1, x_m]$, $f^{(k)}(x_i) = 0$ *for* $e_{i,k} = 1$ *implies* $f^{(n)}(\xi) = 0$ *for some* $\xi \in (x_1, x_m)$.

§7.3. SPLINES; DIAGRAMS, BOUNDARY

The deepest theorem of Birkhoff in [7] gives an estimate for the number of possible changes of sign for the kernel $K_E(X; t)$ in terms of numbers that depend only on the structure of the incidence matrix E:

Theorem 7.7 (Birkhoff [7]). *If E is a Birkhoff matrix and X: $x_1 < x_2 < \cdots < x_m$ is an arbitrary set of knots, then the number of changes of sign of the kernel $K_E(X; t)$ does not exceed $\gamma(E)$, the number of odd supported sequences in E.*

The main purpose of this and the next sections is to extend this result (see Theorem 7.13) in two directions: The number of changes of sign is replaced by the number of zeros; the kernel K is replaced by an arbitrary

spline of a certain class. This is best achieved by means of a still more
general result (Theorem 7.11), which takes into account "the shape of S" by
means of the diagram of S and its boundary.

Let $E = [e_{i,k}]_{i=1,k=0}^{m,\ n}$ be an interpolation matrix; let X: $-1 = x_1$
$< \cdots < x_m = 1$. Instead of E, the inverted matrix $\tilde{E} = [\tilde{e}_{i,k}]$ with elements
$\tilde{e}_{i,k} = e_{i,n-k}$ is sometimes more convenient. The *spline space* $\mathbb{S} = \mathbb{S}_n(E, X)$
consists of the functions

$$S(x) = P_n(x) + \sum_{\tilde{e}_{i,k}=1} \alpha_{i,k} \frac{(x_i - x)_+^k}{k!} \tag{7.3.1}$$

where P_n is a polynomial of degree $\leqslant n$, and the real numbers $\alpha_{i,k}$ are
defined whenever $\tilde{e}_{i,k} = 1$. On each interval $[x_i, x_{i+1}]$, S is a polynomial, and
the highest degree of these polynomials is the degree of S. We agree that the
expression $(1/l!)(x_i - x)_+^l$ is not defined if $x = x_i$, $l = 0$; if $l < 0$, it is
interpreted to be 0.

The splines $S \in \mathbb{S}$ also have another (intrinsic) definition.

Proposition 7.8. *Splines $S \in \mathbb{S}(E, X)$ are precisely piecewise polynomials on* **R**, *which may have a discontinuity of the derivative $S^{(k)}$ only at
knots x_i and only if $\tilde{e}_{i,k} = 1$.*

Proof. Assume that a function f is equal to a polynomial P of
degree $\leqslant n$ on (a, c) and to another such polynomial on (c, b). Let the
derivatives of f be continuous on (a, b) except for $f^{(k_j)}, j = 1, \ldots, p, 0 \leqslant k_1 <
\cdots < k_p \leqslant n$, which have jumps $(-1)^{k_j}\sigma_j$ at c. Then

$$f(x) = P(x) + \sum_{i=1}^{p} \sigma_j \frac{(c - x)_+^{k_j}}{k_j!}, \qquad a < x < b. \tag{7.3.2}$$

Indeed, the function $f(x) - \sum_1^p \sigma_j(c - x)_+^{k_j}/k_j!$ is continuously differentiable,
and at c has the same derivatives as $P(x)$. Hence (7.3.2) is valid on (a, b).
Applying (7.3.2) inductively with $c = x_1, \ldots, x_m$ for $f = S$, we obtain (7.3.1)
for all x. $\qquad\square$

Birkhoff splines S of the class $\mathbb{S}^0 = \mathbb{S}_n^0(E, X) \subset \mathbb{S}$ will be defined as
the splines (7.3.1) that vanish outside of $[-1, +1]$. This in particular implies
$P_n = 0$ in (7.3.1). There is an alternative definition: $S \in \mathbb{S}^0$ are functions that
are polynomials on $(-\infty, x_1), (x_1, x_2), \ldots, (x_m, +\infty)$, 0 outside of (x_1, x_m),
and can have a discontinuity of $S^{(k)}$ at x_i only if $e_{i,n-k} = 1$.

Of considerable interest is the question of the dimension of the space
$\mathbb{S}_n^0(E, X)$.

Exercises

1. $\dim \mathbb{S}^0 > 0$ if and only if the polynomials $(1/k!)(x_i - x)^k$, $\tilde{e}_{i,k} = 1$,
are linearly dependent.

2. In particular, $\dim \mathbb{S}^0 > 0$ if $|E| > n + 1$.

Splines may have *interval zeros* (see §7.4). In the past, interval zeros of splines have been a source of trouble (see §7.5). It is apparently better to separate interval and point zeros, both in formulations and in proofs, in this chapter and in Chapter 14. It is not sufficient to restrict oneself to splines that have interval support—they will have no interval zeros, but their derivatives may have such zeros. We shall define point zeros of splines (and their multiplicities) here, interval zeros in §7.5.

A point c, $-1 < c < +1$, can be a *point zero* of the spline S only if S satisfies $S(c + h)S(c - h) \neq 0$ for all small h. A point of discontinuity c of S is a zero (of multiplicity 1) if and only if S changes sign on c, that is, if $S(c + h)S(c - h) < 0$ for small h. A continuous zero c of S is defined in the usual way. The *multiplicity* of a point zero c is defined as follows. Let $S, \ldots, S^{(l)}$ be the *maximal* sequence with the property that c is a continuous zero of $S^{(k)}$, $k = 0, \ldots, l - 1$, and that c is a continuous or discontinuous zero of $S^{(l)}$. Then the multiplicity of c is $l + 1$.

Useful will be the notion of the diagram of a spline $S \in \mathbb{S}_n(E, X)$. Rolle theorem considerations for S are to be conducted within the diagram. The diagram of S is the set of intervals on which S and its derivatives "live."

Let Γ be the lattice of points (i, k), $i = 1, \ldots, m$; $k = 0, \ldots, n$. For each k, let a set of disjoint subintervals $[j, j']$ of $[1, m]$ be defined; we call them intervals of the kth level. This set of intervals is a *lower set* Φ or an *upper set* Ψ in Γ if each interval $[j, j']$ of level k, $k \geqslant 1$, is contained in an interval of level $k - 1$, or correspondingly, if each interval of level k, $k < n$, is contained in an interval of level $k + 1$. Examples of lower or upper sets are provided by rectangles $i_1 \leqslant i \leqslant i_2$, $0 \leqslant k \leqslant r_1$ or $i_1 \leqslant i \leqslant i_2$, $r_2 \leqslant k \leqslant n$.

A lower set Φ defines a set of points (i, k) that belong to all intervals of Φ; we also denote it by Φ. However, to a set of points there may correspond several lower sets. The end points of the maximal intervals $[j, j']$ of Φ are the *horizontal boundary* of Φ. The number of boundary points is an even number $2L(\Phi)$. We also need the *vertical boundary* of Φ. An interval $[j, j']$ is part of the vertical boundary of Φ at level $k = 0, 1, \ldots, n$ if it is a maximal connected union of intervals $[i, i + 1]$ with this property: they are contained in intervals of level k of Φ, but not contained in those of higher level.

Outside and *inside corners* of Φ are easily defined. An outside corner is a point (i, k) on the horizontal boundary of Φ for which $(i, k + 1)$ does not belong to it. For an inside corner (i, k), the point $(i, k + 1)$ belongs to the horizontal boundary, but (i, k) does not. Similar definitions are possible for an upper set Ψ.

A spline $S \in \mathbb{S}_n(E, X)$ defines in a natural way a lower set. For each k, we take the maximal intervals $[x_j, x_{j'}]$ on which $S^{(k)}$ does not vanish, except at isolated points. We denote these intervals by X_S. The corresponding $[j, j']$ form a lower set, the *diagram* $\Phi(S)$ of S. For an interval $[j, j']$ of the vertical boundary of $\Phi(S)$, $S^{(k)}$ is piecewise constant and not 0 on $[x_j, x_{j'}]$; it is discontinuous at x_j if (j, k) is an outside corner.

For an arbitrary lower set Φ, the matrix $\tilde{E}_\Phi$ is the submatrix of $\tilde{E} = [\tilde{e}_{i,k}]$ restricted to the entries $\tilde{e}_{i,k}$, $(i, k) \in \Phi$.

A position $(i, k) \in \Phi$ is *supported from the right* (*or the left*) in $\tilde{E}_\Phi$ if there are $\tilde{e}_{i_1, k_1} = \tilde{e}_{i_2, k_2} = 1$ in $\tilde{E}_\Phi$ with $i_1 < i < i_2$, $k_1, k_2 > k$ (or, respectively, $k_1, k_2 < k$).

Lemma 7.9. *Let* $S \in \mathbb{S}_n(E, X)$, *then each interior position* $(i, k) \in \Phi(S)$ (*i.e., a position not on the horizontal or ‹the vertical boundary*) *is supported on the right in* $\tilde{E}_\Phi$.

Proof. Consider points (j, l) with $j \geq i$, $l > k$ in Φ. Among them there is a point with the smallest value i_1 of j, and among those with $j = i_1$, a point with the largest value k_1 of l. We see that $i < i_1$, $k_1 > k$, and that (i_1, k_1) is an outside corner of Φ. Since $\tilde{e}_{i_1, k_1} = 1$, (i, k) is supported from the right above. Similarly, it is supported from the right below. □

§7.4. COUNT OF ZEROS OF SPLINES

Let $E = [e_{i,k}]$ be an $m \times (n + 1)$ interpolation matrix, let X: $-1 = x_1 < \cdots < x_m = +1$ be a corresponding set of knots. For a Birkhoff spline $S \in \mathbb{S}_n^0(E, X)$, we would like to estimate the number of its zeros, and to obtain the basic Theorem 7.13, a generalization of Birkhoff's Theorem 7.7. We conduct our investigations within the diagram $\Phi(S)$ of S. In this way we obtain Theorem 7.11 (due to the present authors [61]), which is stronger than Theorem 7.13 because it takes into account the "shape" of S and not just its degree.

If S, defined on $[a, b]$, does not vanish on intervals, then for each $\xi \in (a, b)$, the product $\Pi = S(\xi + h_1)S(\xi - h_2)$ is either always > 0 or always < 0 for small $h_1, h_2 > 0$. In the first case, S *does not change sign* at ξ, in the second case it *does*. This definition applies also to arbitrary splines S, but there is a third possibility: $\Pi = 0$ for all small $h_1 > 0$ or $h_2 > 0$.

An essential tool is the following lemma (*Rolle's theorem for splines*), which appears (in a slightly incorrect form) in Lorentz [93].

Lemma 7.10. *Let S be a spline on (ξ_1, ξ_2) with $S(\xi_1+) = S(\xi_2-) = 0$. Suppose that both S and S' do not vanish on intervals. Then there is a point ξ, $\xi_1 < \xi < \xi_2$, such that either* (A): *ξ is a zero of S' with change of sign, but not a zero of S, or* (B): *ξ is a discontinuous zero of S (with change of sign) on which S' preserves sign.*

Proof. We can assume that (ξ_1, ξ_2) contains no other continuous zeros of S, for otherwise we could replace the interval by a smaller one. For all small $h > 0$, $S(\xi_1 + h)S'(\xi_1 + h) > 0$ and $S(\xi_2 - h)S'(\xi_2 - h) < 0$. Therefore S and S' have in (ξ_1, ξ_2) numbers of changes of sign of different parity.

There exists a point ξ at which S' but not S, or conversely, S but not S', changes sign. In the first case ξ cannot be a zero of S (continuous or discontinuous) and we have (A); in the second case ξ is a discontinuous zero of S as in (B).   □

For a spline S that does not vanish on intervals, we call $\xi \in (-1, 1)$ an *even point* of S if S does not change sign at ξ; an *odd point* if S changes sign at ξ. We say that S, S' *alternate* at ξ if ξ is a point of different parity for S, S'. Thus, Lemma 7.10 asserts the existence of a zero ξ of S, or of S', at which S, S' alternate.

For a Birkhoff spline $S \in \mathbb{S}_n^0(E, X)$ we define its diagram $\Phi(S)$ as in §7.3. A *sequence* in $\Phi(S)$

$$B: \tilde{e}_{i,s} = \cdots = \tilde{e}_{i,s+t} = 1 \qquad (7.4.1)$$

is a maximal sequence of 1's in row i of the matrix $\tilde{E}_\Phi$; it is *interior* if each (i, k), $s \leq k \leq s + t$, is interior to Φ. By $Z(S, (a, b))$ we denote the number of *point zeros* (counting multiplicities) of a spline S defined on $[a, b]$.

Theorem 7.11. *Let $S \in \mathbb{S}_n^0(E, X)$ have support $[a, b] = [x_{i_1}, x_{i_2}]$. If $\Phi = \Phi(S)$ is the diagram of S, then*

$$Z(S; (a, b)) \leq |\tilde{E}_\Phi| - L(\Phi) - 1 + \gamma(\tilde{E}_\Phi) \qquad (7.4.2)$$

where $\gamma(\tilde{E}_\Phi)$ is the number of odd interior sequences in $\tilde{E}_\Phi$.

Proof. Let the matrix $\tilde{E}$ and the spline S with all of its zeros in (a, b) be given. We introduce the following notations: $l_k := $ the number of intervals of X_S (or of Φ) of level k; $u_k := $ the number of continuous zeros of $S^{(k)}$ inside X_S, which are also zeros of S of multiplicity $\geq k + 1$; $v_k := $ number of $\tilde{e}_{i,k} = 1$ in the interior of intervals of level k of Φ; $\bar{\varepsilon}_k := $ number of sequences of 1's ending at some point (i, k) of Φ, but with the property that $(i, k + 1)$ is also inside Φ and that $S^{(k+1)}(x_i) \neq 0$; $\eta_k := $ number of $\tilde{e}_{i,k} = 1$ for which x_i is a zero of S of multiplicity $k + 1$; $d_k := $ number of 1's on the horizontal boundary of $\tilde{E}_\Phi$ of level k.

Let q be the degree of S. We note that $u_q = 0$, $\bar{\varepsilon}_q = 0$, $l_0 = 1$. We would like to find a lower bound μ_k for the number of continuous Rolle zeros of $S^{(k)}$ (they are disjoint from the zeros counted by u_k). We begin with $\mu_0 = 0$. We apply Lemma 7.10 to derive a lower estimate μ_{k+1} from μ_k. The derivative $S^{(k)}$ has $\mu_k + u_k$ continuous zeros inside the intervals of X_S of level k. To these we add $2l_k - d_k$ continuous zeros at the end points of the intervals. This will give $\mu_k + u_k + 2l_k - d_k$ continuous zeros of $S^{(k)}$ with

$$\mu_k + u_k + 2l_k - d_k - 1 \qquad (7.4.3)$$

intervals between them. The end points of the intervals cannot belong to the vertical boundary of Φ, for there $S^{(k)}$ is nonzero, piecewise constant; nor can they belong to the $l_{k+1} - 1$ complements $\tilde{I}$ of intervals of X_S of level $k + 1$. But intervals (7.4.3) may contain some $\tilde{I}$. If we omit all these, we end up with at least

$$\mu_k + u_k + 2l_k - l_{k+1} - d_k \qquad (7.4.4)$$

intervals I of level k, to which we can apply Lemma 7.10, for $S^{(k)}$, $S^{(k+1)}$ vanish only in isolated points on each of the I.

Application of Lemma 7.10 gives a point ξ inside each I. This ξ is even inside of the intervals of X_S of level $k+1$.

In particular, ξ may be a continuous zero of $S^{(k+1)}$ [special case of (A), Lemma 7.10]. Any other possibility would require $\xi = x_i$ for some x_i inside of the intervals of X_S of level $k+1$. To single out continuous zeros, we first discard all I that contain x_i where $S^{(k+1)}$ could be discontinuous. There are at most $(v_{k+1} - \eta_{k+1})$ such x_i. Then we discard x_i's at which $S^{(k)}$ is discontinuous, $S^{(k+1)}$ is continuous, but $S^{(k+1)}(x_i) \neq 0$. There are at most $\bar{\varepsilon}_k$ such x_i. Thus, the difference between (7.4.4) and $(v_{k+1} - \eta_{k+1}) + \bar{\varepsilon}_k$ will be a lower estimate for μ_{k+1}.

It is important that this estimate can be improved. We call a pair (i,k) *nonconfirming* (to Lemma 7.10) if x_i is counted by $(v_{k+1} - \eta_{k+1})$ or $\bar{\varepsilon}_k$, but cannot be obtained as a ξ of level k by Lemma 7.10. This will be the case if x_i does not belong to any I, or if $S^{(k)}$, $S^{(k+1)}$ do not alternate at x_i. For each $k = 0,\dots,q-1$ we examine all pairs $(i,k) \in \Phi$ and denote by N_k the number of nonconfirming (i,k) for this k. Since $\bar{\varepsilon}_k$ and $(v_{k+1} - \eta_{k+1})$ count disjoint pairs, $N = \sum_{k=0}^{q-1} N_k$ will be the total number of nonconfirming pairs in Φ.

At step k, we need to discard only

$$(v_{k+1} - \eta_{k+1}) + \bar{\varepsilon}_k - N_k$$

intervals I, the remaining will contain continuous Rolle zeros of $S^{(k+1)}$. Thus,

$$\mu_{k+1} \geq \mu_k + u_k + 2l_k - l_{k+1} - d_k - \bar{\varepsilon}_k - (v_{k+1} - \eta_{k+1}) + N_k. \qquad (7.4.5)$$

Since $l_0 = 1$, $2l_q = d_q$, we obtain by summation

$$0 = \mu_q \geq \sum_{k=0}^{q} (u_k + \eta_k) - \sum_{k=1}^{q} v_k - \sum_{k=0}^{q-1} \bar{\varepsilon}_k + N - \eta_0 - \sum_{k=0}^{q} d_k + \sum_{k=0}^{q} l_k + 1. \qquad (7.4.6)$$

The first sum is $Z(S)$, while $\sum_0^q (v_k + d_k) = |\tilde{E}_\Phi|$, $\sum_0^q l_k = L(\Phi)$. Hence

$$Z(S;(a,b)) \leq |\tilde{E}_\Phi| - L(\Phi) - 1 + \Delta - N, \qquad (7.4.7)$$

$$\Delta = \sum_{k=0}^{q-1} \bar{\varepsilon}_k + \eta_0 - v_0. \qquad (7.4.8)$$

We want to find an upper bound for $\Delta - N$. For a sequence (7.4.1) of $\tilde{E}_\Phi$, we study its contributions Δ_B, N_B to Δ, N. Here, N_B is the number of nonconfirming pairs in the sequence

$$(i, s-1) \quad (\text{if } s > 0), \quad (i,s),\dots,(i,s+t). \qquad (7.4.9)$$

Clearly, $N_B \geqslant 0$. The contribution of B to $\eta_0 - v_0$ is $\leqslant 0$, to $\Sigma_0^{q-1} \bar{\epsilon}_k$ the same as its contribution to $\bar{\epsilon}_{s+t}$, which may be 0 or 1. Hence

$$\Delta_B - N_B \leqslant 1 \qquad \text{for all } B. \qquad (7.4.10)$$

We shall discard many sequences B, for which the difference (7.4.10) is $\leqslant 0$.

In the first place, we can assume that none of the pairs (i, k) in (7.4.9) is nonconfirming. In particular, this would imply that x_i cannot be a continuous zero of $S^{(k)}$ for (i, k) in (7.4.9). Moreover, the sequence of derivatives $S^{(k)}$ will alternate at x_i.

We can further assume that the contribution of B to $\bar{\epsilon}_{s+t}$ is 1. Then $(i, s + t)$ will be an interior position in Φ; hence the whole sequence B will be interior to $\tilde{E}_\Phi$. Moreover, x_i will be an *odd point* of $S^{(s+t)}$. For if it were an even point, then alternation and $\bar{\epsilon}_{s+t+1} = 0$ would imply $S^{(s+t+1)}(x_i) = 0$, a contradiction.

Let B be one of the remaining even sequences (for which t is odd). First, let $s > 0$. By alternation, x_i is an odd point of $S^{(s-1)}$, hence a continuous zero of $S^{(s-1)}$ — a contradiction. If $s = 0$, then x_i is an even point of S, hence x_i is not a discontinuous zero of S. The contribution of B to η_0 is then 0, that to v_0 is 1, and we get $\Delta_B \leqslant 0$.

We need only count the odd interior sequences B of $\tilde{E}_\Phi$ and obtain

$$\Delta - N \leqslant \gamma(\tilde{E}_\Phi). \qquad \square$$

§7.5. COROLLARIES; BIRKHOFF–LORENTZ THEOREM

Theorem 7.11 provides an excellent approach to several results that count zeros of splines.

Theorem 7.12. *For an arbitrary Birkhoff spline $S \in \mathbb{S}_n^0(E, X)$ on $[-1, +1]$ we have*

$$Z(S;(-1,1)) \leqslant |\tilde{E}_\Phi| + \gamma(\tilde{E}_\Phi) - L(\Phi) - r \qquad (7.5.1)$$

where r is the number of intervals in the support of S.

Proof. We can write $S = S_1 + \cdots + S_r$, where the S_j are supported on disjoint intervals I_j. Adding relations (7.4.2) for $j = 1, \ldots, r$, we obtain (7.5.1). $\square$

The most important special case of Theorem 7.11 is the following result, which depends not on $\Phi(S)$, but only on the degree of the spline S. We formulate it in terms of the matrix E rather than $\tilde{E}$.

Theorem 7.13 (Birkhoff–Lorentz). *If $S \in \mathbb{S}_n^0(E, X)$ is a spline of degree q whose support is an interval, then*

$$Z(S;(-1,+1)) \leqslant |E_q| + \gamma(E) - q - 2. \qquad (7.5.2)$$

Here $\gamma(E)$ is the number of odd sequences in E supported on the left and E_q is the truncated matrix, consisting of the last $q + 1$ columns of E.

Proof. A spline $S \in \mathbb{S}_n^0(E, X)$ of degree q belongs also to $\mathbb{S}_q^0(E_q, X)$. Inequality (7.5.1), applied to $S \in \mathbb{S}_q^0(E_q, X)$, leads directly to (7.5.2).   □

We can replace $|E_q|$ by $|E|$ in (7.5.2). This gives an apparently weaker but actually equivalent formulation of Theorem 7.13.

If E satisfies the Pólya condition, then $|E| - |E_q| \geqslant n - q$. Hence:

Theorem 7.14 (Ferguson [40]). *If, in the situation of Theorem 7.13, the matrix E satisfies the Pólya condition, then*

$$Z(S;(-1,+1)) \leqslant |E| + \gamma(E) - n - 2. \qquad (7.5.3)$$

The history of Theorem 7.13 is interesting. For Birkhoff kernels K—which are a special case of splines of class $\mathbb{S}_n^0(E, X)$—Birkhoff [7] proved (7.5.2) with $Z(S)$ replaced by the number of *changes of sign* of K on $(-1, +1)$. (This also follows from Theorem 7.12.) Later, Ferguson [40] generalized this statement to Birkhoff splines S, assuming, however, that E satisfies the Birkhoff condition. Theorem 7.13 has been given in Lorentz [93]. Later Jetter [56] and Schumaker [157], who refer to [93], gave more general statements as far as the multiplicity of *interval zeros* is concerned (see Theorem 7.15, Corollary 7.16). In 1979, examining these proofs, the present authors found that they all contain the same error. The basic Lemma 7.10 was stated there incorrectly. They found that it is possible to correct the proof, keeping the same idea, but it becomes longer. It is interval zeros of S and of its derivatives that create the trouble. Our present approach via Theorem 7.11 gives a more general result and avoids this trouble.

To go further, we need the notion of the multiplicity μ of an interval zero $\xi = [c, d]$ of a spline S. A spline S on $[a, b]$ can have an interval zero $[c, d]$, $a < c < d < b$, only if $S(x) = 0$, $c < x < d$, and only if this is a maximal interval of this type, so that $S(c - h)S(d + h) \neq 0$ for all small $h > 0$. In this situation, $[c, d]$ is a *continuous zero* of S if S is continuous at c and d, and a *discontinuous zero* (of multiplicity 1) if S is discontinuous at one of the points and changes sign on $[c, d]$. The multiplicity μ of a zero $[c, d]$ is defined by means of the maximal sequence $S, \ldots, S^{(l)}$, each of these splines having $[c, d]$ as a zero, and each except perhaps $S^{(l)}$ having it as a continuous zero. Then $\mu = l + 1$. (This definition was given in [56, 157, 12].)

Clearly,

$$\mu \leqslant \min(q, q') + 1 \qquad (7.5.4)$$

where q, q' are the degrees of S on intervals adjoining $[c, d]$.

For kernels of certain operators, Jetter [56] has the following estimate, in which $Z^*(S;(-1, +1))$ is the number of all zeros of S in $(-1, +1)$, point zeros as well as interval zeros.

Theorem 7.15 (Jetter). *If $S \in \mathbb{S}_n^0(E, X)$, and if q is the degree of S,* then

$$Z^*(S; (-1, +1)) \leqslant |E| + \gamma(E) - r - q - 1 \qquad (7.5.5)$$

where r is the number of intervals in the support of S.

Proof. Let $S = \sum_{j=1}^r S_j$, where S_j are splines of degrees q_j, supported on disjoint intervals I_j, $j = 1, \ldots, r$. If J_j is the interval between I_j and I_{j+1}, and μ_j is its multiplicity as a zero of S, then $L(\Phi(S_j)) \geqslant q_j + 1$, and

$$\sum_{j=1}^{r-1} \mu_j \leqslant \sum_{1}^{r-1} \min(q_j, q_{j+1}) + r - 1 \leqslant \sum_{1}^r q_j - q + r - 1 \leqslant L(\Phi(S)) - q - 1.$$

Adding this to (7.5.1), we obtain (7.5.4).   □

In this proof, there is often a considerable loss of precision, and it is easy to prove (7.5.5) with more generous definitions of multiplicity of intervals than those given above; see [K]. Let T be a spline on $(-\infty, +\infty)$ of class $\mathbb{S}_n(E, X)$; it reduces to polynomials on $(-\infty, -1)$ and on $(1, +\infty)$.

Corollary 7.16. *Under the foregoing assumptions,*

$$Z^*(T; (-\infty, +\infty)) \leqslant |E| + \rho(E) + 2n - r - q + 1 \qquad (7.5.6)$$

where $\rho(E)$ is the number of odd sequences of E that do not begin in column 0.

Proof. Choose N so large that all the zeros of T are contained in $(-N, N)$. We construct a new pair $\bar{E}, \bar{X}$ by adjoining knots $x_0 = -N$, $x_{m+1} = N$ to X and corresponding first and last rows of 1's to E. Since $|\bar{E}| = |E| + 2(n+1)$ and since all sequences of E beginning in a column $s > 0$ are supported in $\bar{E}$, (7.5.6) follows from Theorem 7.15.   □

Our inequalities are similar to some given by Schumaker in [K, Theorems 8.37 and 8.39]. The first is identical with (7.5.6), except that it assumes that $S^{(n)}$ does not vanish on intervals (in which case $r = 1$). Theorem 8.39 is about Birkhoff splines, and is weaker than (7.5.5); it does not contain Birkhoff's Theorem 7.7.

In Lemma 7.10, instead of the hypothesis $S(\xi_1 +) = S(\xi_2 -) = 0$, we can assume that S has zeros at ξ_1 and ξ_2 in the wide sense (see §2.1). This provides an approach to a theorem of Jetter [58]:

Theorem 7.17 (Budan–Fourier theorem for splines). *Let S be a Birkhoff spline of degree n that corresponds to the $m \times (n+1)$ matrix E and a set of knots X. Then the number of zeros $Z(S; (x_1, x_m))$ of S in (x_1, x_m) with multiplicities counted satisfies*

$$\begin{aligned}
Z(S; (x_1, x_m)) \leqslant \;& \mathfrak{S}^- \{S(x_1 +), \ldots, S^n(x_1 +)\} \\
& - \mathfrak{S}^+ \{S(x_m -), \ldots, S^n(x_m -)\} \\
& + |\mathring{E}| + \gamma(E) - \varepsilon(x_1) - \varepsilon(x_2)
\end{aligned}$$

where $\mathfrak{S}^-, \mathfrak{S}^+$ are defined as in §2.1, $\mathring{E}$ is the matrix $[e_{ik}]_{i=2,\,k=0}^{m-1\,\,n}$ and $\varepsilon(y) = 1$ if S vanishes identically in a neighborhood of y, otherwise $\varepsilon(y) = 0$.

This is best obtained by replacing Lemma 7.10 by the generalized lemma, and following the proofs of Theorem 7.11 and 7.15. We omit the details.

§7.6. APPLICATIONS TO BIRKHOFF INTERPOLATION

It has been noticed by Lorentz [93] that the theorem of Atkinson and Sharma is a corollary of Birkhoff's Theorem 7.7:

Theorem 7.18. A Pólya matrix is order regular if it has no essential odd supported sequences.

Proof. The proof is by induction on $|E|$. The theorem is certainly true for $|E| = 2$. Let $|E| > 2$. If $E = E_1 \oplus E_2 \oplus \cdots \oplus E_\mu$ is the canonical decomposition with $\mu > 1$, then the theorem is proved by applying the induction hypothesis to the components E_ν; note that all E_ν are free from odd supported sequences. Therefore let us assume that E satisfies the Birkhoff condition. We can find an entry $e_{i,0} = 1$ in such a way that the reduced matrix $E_{i,0}$ which is obtained from E by putting $e_{i,0} = 0$ and deleting the last column satisfies the Pólya condition and does not contain odd supported sequences. (Take, e.g., $e_{i,0}$ as the first 1 in the first column.) Thus, $E_{i,0}$ is order regular according to the induction hypothesis. Now Birkhoff's kernel $K_E(X;t)$ has the strict degree $n-1$ as $D_{i,0}(X) = D(E_{i,0}, X) \neq 0$, and by Birkhoff's estimate the kernel does not change sign. The theorem follows from Corollary 7.6(a).   □

From Theorem 7.7 we can derive a little more:

Exercise. For the defect $d(E)$ of a Birkhoff matrix E we have the estimate $d(E) \leqslant \gamma(E)$ where $\gamma(E)$ is the number of odd supported sequences of E. (This is a weaker version of Theorem 2.1.)

With Jaffe [53] we call an interpolation matrix E *strongly order regular* if for each ordered set X: $x_1 < x_2 < \cdots < x_m$ of knots, Birkhoff's kernel $K_E(X;t)$ is of constant sign but does not vanish identically. For a strongly order regular matrix E the kernel is of the same sign for all X and t; this follows immediately from Theorem 7.1(c) and 7.6(a), since $D(E, X)$ does not change sign.

It is clear that the following implications hold for Birkhoff matrices:

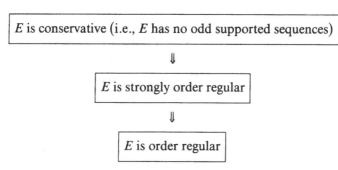

These implications cannot be inverted, as is shown by the following examples due to Jaffe [53]:

Example 1. The matrix

$$E = \begin{bmatrix} 1 & 1 & 0 & 0 & 0 & 0 \\ 0 & 1 & 0 & 0 & 1 & 0 \\ 1 & 1 & 0 & 0 & 0 & 0 \end{bmatrix} \tag{7.6.1}$$

is order regular but not strongly order regular.

Proof. The order regularity is known from §1.3. Let $x_1 = 0$, $x_2 = x$ $\in (0,1)$, and $x_3 = 1$. A straightforward computation shows that, for $t \in (x, 1)$,

$$K_E(X; t) = -\frac{1}{12} \left[\frac{1}{4} x (1-x)(1-t)^4 - x \left(\frac{1}{3} - \frac{x}{2} \right)(1-t)^3 \right].$$

It is clear that, for $x \neq 2/3$, $x < t$, and for t near to 1, the sign of $K_E(X; t)$ is given by that of $-x(1/3 - x/2)$, and the latter polynomial changes sign in the point $x = 2/3$. □

Example 2. The matrix

$$E = \begin{bmatrix} 1 & 1 & 1 & 1 & 0 & 0 & 0 \\ 0 & 1 & 0 & 1 & 0 & 0 & 0 \\ 1 & 0 & 0 & 0 & 0 & 0 & 0 \end{bmatrix} \tag{7.6.2}$$

is strongly order regular but not conservative.

Proof. This Birkhoff matrix has two odd supported sequences and is not conservative. We show that it is strongly order regular by explicit computation of its kernel $K_E(X, t)$. The proof that $K(X, t) \geq 0$ for all X, t is elementary, but not quite simple. Without loss of generality we can take $x_1 = 0$, $x_2 = x$, $x_3 = 1$. From the determinant expression for $K_E(X, t)$ we obtain

$$K_E(X,t) = \det \begin{bmatrix} \dfrac{1}{3!}x^3 & \dfrac{1}{4!}x^4 & \dfrac{1}{4!}(x-t)_+^4 \\[2mm] x & \dfrac{1}{2}x^2 & \dfrac{1}{2}(x-t)_+^2 \\[2mm] \dfrac{1}{4!} & \dfrac{1}{5!} & \dfrac{1}{5!}(1-t)_+^5 \end{bmatrix}.$$

Hence

$$K_E(X,t) = \begin{cases} \dfrac{1}{4!5!}(1-t)^5 x^5 & \text{for} \quad x \leqslant t \leqslant 1, \\[3mm] \dfrac{xt^2}{2 \cdot 4!5!} f(x,t) & \text{for} \quad 0 \leqslant t \leqslant x \end{cases}$$

where

$$f(x,t) = (20x^2 - 25x + 8)x^2 - 4tx(5x^3 - 5x + 2)$$
$$+ t^2(10x^4 - 5x + 2) - 2t^3 x^4.$$

Evidently $K_E(X,t) \geqslant 0$ for $x \leqslant t \leqslant 1$. To show that this inequality holds also for $0 \leqslant t \leqslant x$, we use critically the following equivalent expression for $f(x,t)$:

$$f(x,t) = 2t^2(1-x)^5 + 2t(x-t)(1-x)^4(x+4)$$
$$+ (x-t)^2(-2x^4 t + 20x^2 - 25x + 8).$$

If $0 \leqslant t \leqslant x/2$ we have

$$f(x,t) \geqslant (x-t)^2(-x^5 + 20x^2 - 25x + 8).$$

Let $g(x) = -x^5 + 20x^2 - 25x + 8$. In $0 \leqslant x \leqslant 1$ the function g attains its minimum inside the interval at a point x where $x^4 = 8x - 5$ and $0 < x < 1$. Hence at the minimum

$$g(x) = 12x^2 - 20x + 8 = 4(1-x)(2-3x).$$

But at the minimum $x = 5/8 + (1/8)x^4$. Hence $x < 5/8 + 1/8 = 3/4$, which in turn gives the better estimate $x < 5/8 + 1/8(3/4)^4 < 2/3$, showing that $g(x) \geqslant 0$ for $0 \leqslant x \leqslant 1$.

On the other hand if $x/2 \leqslant t \leqslant x$ we have $t \geqslant x - t$. Hence

$$f(x,t) \geqslant (x-t)^2 \left[2(1-x)^5 + 2(1-x)^4(x+4) - 2x^4 t + 20x^2 - 25x + 8 \right]$$

and it suffices to show that

$$h(x) = 10(1-x)^4 - 2x^5 + 20x^2 - 25x + 8$$
$$= 2(1-x)^5 - 20x^3 + 60x^2 - 55x + 16 \geqslant 0$$

for $0 \leqslant x \leqslant 1$. In turn it suffices to show that

$$k(x) = -20x^3 + 60x^2 - 55x + 16 \geqslant 0 \qquad \text{for} \quad 0 \leqslant x \leqslant 1.$$

The function k is at a minimum where $x^2 = 2x - 11/12$, at which point $k(x) = (10/3)(x - 0.7)$. The minimum occurs at $x = 1 - \sqrt{1/12} = 0.712$, which is greater than $7/10$.

This completes the proof that $K_E(X)$ does not change sign and hence that E is strongly order regular. $\qquad\qquad\qquad\qquad\qquad\square$

An interesting characterization of strongly order regular matrices has been given by Jaffe and Lorentz [53]:

Theorem 7.19. *An order regular Birkhoff matrix* $E = [e_{i,k}]_{i=1\,k=0}^{m\ \ n}$ *is strongly order regular if and only if each matrix*

$$E(p,q) = [\bar{e}_{i,k}]_{i=0\ \ k=0}^{m+1\ n+p+q}$$

$$= \begin{bmatrix}
0 & \cdots & 0 & 1 & \cdots & 1 & 0 & \cdots & 0 \\
e_{1,0} & \cdots & e_{1,n-1} & & & & & & \\
\vdots & & \vdots & & & & 0 & & \\
e_{m,0} & \cdots & e_{m,n-1} & & & & & & \\
0 & \cdots & 0 & 1 & \cdots & 1 & 0 & \cdots & 0
\end{bmatrix},$$

$p, q = 0, 1, 2, \ldots,$ *is order regular.*

Proof.

1. If E is strongly order regular, let $X: x_1 < \cdots < x_m$ and p, q be arbitrary. Let the polynomial $P \in \mathcal{P}_{n+p+q}$ be annihilated by $E(p,q)$ and $\tilde{X}$: $x_0 < x_1 < \cdots < x_m < x_{m+1}$. Since

$$P^{(k)}(x_i) = 0 \qquad \text{for all} \quad \bar{e}_{i,k} = 1,$$
$$i = 0, \ldots, m+1, \quad k = 0, \ldots, n+p+q,$$

we have, by Birkhoff's identity,

$$0 = \int_{x_1}^{x_m} P^{(n)}(t) K_E(X; t) \, dt.$$

This implies that either $P^{(n)} \equiv 0$, or that $P^{(n)}$ changes sign in (x_1, x_m). The latter case leads to a contradiction, since $P^{(n)} \in \mathcal{P}_{p+q}$ would have additional zeros in x_0 and x_{m+1}, of orders p and q, respectively. Thus $P^{(n)} \equiv 0$, and by the order regularity of E we would have $P \equiv 0$.

2. Now assume that all matrices $E(p, q)$, $p, q = 0, 1, \ldots$ are order regular. Without loss of generality we may assume that Birkhoff's kernel associated with $E(0, 0) = E$ and X: $x_1 < \cdots < x_m$ has a positive integral,

$$\int_{x_1}^{x_m} K_E(X; t) \, dt = D(E, X) > 0.$$

Let us put

$$F_{p, q}(x_0, x_{m+1}) = \int_{x_1}^{x_m} (t - x_0)^p (x_{m+1} - t)^q K_E(X; t) \, dt, \quad (7.6.3)$$

where $x_0 < x_1$, $x_{m+1} > x_m$, and $p, q = 0, 1, \ldots$ are arbitrary. We first prove that this expression is different from 0 for all values of the parameters.

Let us assume that $F_{p, q}(x_0, x_{m+1}) = 0$ for some choice of p, q, x_0, and x_{m+1}. We select a pair (i, k) in such a way that $D_{i, k}(X) \neq 0$ in (7.1.4); this is possible since $K_E(X; t)$ does not vanish identically. An application of Theorem 7.3 to the pair E, X, the function $g(t) = (t - x_0)^p (x_{m+1} - t)^q$, and constants $c_{i, k} = 0$ shows that there exists a polynomial $P \in \mathcal{P}_{n + p + q}$ such that

$$P^{(n)}(t) = (t - x_0)^p (x_{m+1} - t)^q$$

and

$$P^{(k)}(x_i) = 0 \quad \text{for} \quad e_{i, k} = 1, \quad i = 1, \ldots, m, \quad k = 0, \ldots, n.$$

This is a contradiction, since $E(p, q)$ is order regular.

Let us look at the sign of $F_{p, q}(x_0, x_{m+1})$. For $x_0 \to -\infty$ and $x_{m+1} \to +\infty$ we have

$$\left(\frac{1}{-x_0} \right)^p \left(\frac{1}{x_{m+1}} \right)^q F_{p, q}(x_0, x_{m+1})$$

$$= \int_{x_1}^{x_m} \left(1 - \frac{t}{x_0} \right)^p \left(1 - \frac{t}{x_{m+1}} \right)^q K_E(X; t) \, dt \to D(E, X).$$

It follows that

$$F_{p, q}(x_0, x_{m+1}) = \int_{x_1}^{x_m} (t - x_0)^p (x_{m+1} - t)^q K_E(X; t) \, dt > 0$$

for all choices of the parameters, and

$$\int_{x_1}^{x_m} \sum_{k=0}^{N} c_k (t - x_0)^k (x_{m+1} - t)^{N-k} K_E(X; t) \, dt \geq 0 \quad (7.6.4)$$

for all real numbers $c_k \geq 0$.

Now, if g is any positive continuous function on $[x_0, x_{m+1}]$, we may choose the numbers $c_k = c_{k, N}$ in such a way that

$$B_N(g; t) = \sum_{k=0}^{N} c_k (t - x_0)^k (x_{m+1} - t)^{N-k}$$

is the Nth Bernstein polynomial of g with respect to the interval $[x_0, x_{m+1}]$. The convergence property of the Bernstein polynomials, in connection with (7.6.4), shows that, necessarily,

$$\int_{x_1}^{x_m} g(t) K_E(X; t)\, dt \geq 0.$$

Since g was arbitrary, we must have $K_E(X; t) \geq 0$ for $t \in [x_1, x_m]$.   □

Exercise. Let u and v be the highest positions of 1's in rows 1 and m of E, respectively. Show that E cannot be strongly order regular if one of the matrices $E_{1, u}$ or $E_{m, v}$ is strongly order singular (i.e., if one of the determinants $D_{1, u}(X)$, $D_{m, v}(X)$ changes sign).

§7.7. NOTES

There is a simple formula for the Peano kernel. In general, let L be a linear functional that vanishes on all polynomials of degree $\leq n - 1$ and is defined on the space $A_n[a, b]$. Then for all f in this space,

$$L(f) = \int_a^b f^{(n)}(t) K_n(t)\, dt, \qquad K_n(t) = \frac{1}{(n-1)!} L\big((x - t)_+^{n-1}\big)$$

$$(7.7.1)$$

where the kernel K_n is obtained by applying L to $(x - t)_+^{n-1}$ as a function of x. This follows by applying L to the Taylor expansion of f, but requires the justification of the exchange of L and of the integral sign (this is not always easy).

Chapter 8

Almost-Hermitian Matrices; Special Three-Row Matrices

§8.1. INTRODUCTION

Several authors have tried to study classes of matrices, of modest generality, that contain the matrices (1.3.7) as special cases. The hope was that some insight into the general problem might be gained by deciding the regularity problem for this class. The failure to find a complete solution for even a relatively simple class of this type illustrates the complexity of the general problem, and to some extent justifies its probabilistic reformulation in §6.4.

A matrix E is *almost Hermitian* if all of its rows are Hermitian except for one interior row. A subclass consists of the three-row matrices $E(p, q; k_1, \ldots, k_s)$ where p and q are the lengths of the Hermitian first and third row while $0 < k_1 < \cdots < k_s$ are positions of 1's in the second row. We have then $p + q + s = n + 1$, and we shall always assume $1 \leqslant p \leqslant q$. The regularity problem is not completely solved even for matrices $E(p, q; k_1, k_2)$.

For almost-Hermitian matrices, many useful results follow from a representation of $D(E, X)$ given by Theorem 8.1. Most notable are a strong singularity criterion due to Lorentz [91] and DeVore, Meir, and Sharma [26] in §8.2 and the reciprocity theorem of Drols [29] in §8.3.

Sections 8.4–8.6 deal with the special matrices $E(p, q; k_1, k_2)$. In §8.4 we give the "chasing method" of Lorentz, Stangler, and Zeller [104] in a more geometric form. This solves the problem except for the two cases, (8.2.11) and (8.2.12). The second of them admits a complete solution. This is given in §8.5 using the methods of Lorentz, Stangler, and Zeller [104] and DeVore, Meir, and Sharma [26]. The other case is discussed in §8.6 where a certain region of regularity is defined. Some properties of this region are given, but the region is not explicitly known except for special cases. The geometric approach developed in this chapter is based on seminars of S. D. Riemenschneider at the University of Alberta.

§8.2. ALMOST-HERMITIAN MATRICES

An $m \times (n+1)$ Birkhoff matrix will be called *almost Hermitian* if all of its rows except the i_0th, $1 < i_0 < m$, are Hermitian. The i_0th row will have a Hermite sequence of length r and s additional 1's in positions

$$r < k_1 < \cdots < k_s < n.$$

The lengths of the other Hermitian sequences are denoted by r_i, $i \neq i_0$. The set of knots X: $x_1 < \cdots < x_m$, will often be taken with x_i, $i \neq i_0$, constant and with $x_{i_0} = x$ variable in the interval (a, b) where $a = x_{i_0 - 1}$ and $b = x_{i_0 + 1}$. Further, we set

$$H(t) = \prod_{i \neq i_0} (t - x_i)^{r_i}. \tag{8.2.1}$$

The importance of $H(t)$, which is a polynomial of degree $n - r - s + 1$, stems from the following result.

Theorem 8.1. *The almost-Hermitian matrix E is regular if and only if for each selection of x_i, $i \neq i_0$, the determinant*

$$\Delta(x) = \det\left[\frac{1}{(k_j - r)!} H^{(k_j - r)}(x), \dots, \right.$$

$$\left. \frac{1}{(k_j - r - s + 1)!} H^{(k_j - r - s + 1)}(x); j = 1, \dots, s \right] \tag{8.2.2}$$

does not vanish on (a, b). (The terms with $k_j - r - l < 0$ in $\Delta(x)$ are assumed to be 0.) Moreover,

$$D(E, X) = C(x)\Delta(x) \tag{8.2.3}$$

where $C(x)$ is nonzero in (a, b) and is a constant if $r = 0$.

Proof.   The polynomials (with $a_{l, n+l+1-s} = 1$)

$$P_l(t) = \frac{1}{(r+l)!}(t - x)^{r+l} H(t) = \sum_{\nu = 0}^{n + l + 1 - s} a_{l,\nu} \frac{t^\nu}{\nu!}, \qquad l = 0, \dots, s - 1,$$

are annihilated by the Hermitian rows $i \neq i_0$; that is, they satisfy $P_l^{(k)}(x_i) = 0$,

$0 \leqslant k < r_i$, $i \neq i_0$. We take $l = s - 1$ and obtain from this

$$P_{s-1}^{(k)}(x_i) = \sum_{\nu=0}^{n} a_{s-1,\nu} \frac{x_i^{\nu-k}}{(\nu-k)!} = 0, \qquad k < r_i, \quad i \neq i_0. \qquad (8.2.4)$$

We multiply the νth column, $\nu = 0,\dots,n-1$, of $D(E, X)$ with $a_{s-1,\nu}$, and add them to the last column. In view of (8.2.4), the elements of the last column for $i \neq i_0$, $k < r_i$ will become 0. In the identity

$$\frac{d^{k_j}}{dt^{k_j}}\left[(t-x)^{r+l}H(t)\right]\bigg|_{t=x} = \frac{k_j!}{(k_j - l - r)!} H^{(k_j - r - l)}(x),$$

$$l = 0,\dots,s-1,$$

the right-hand side is interpreted to be 0 if $k_j < l + r$. For $l = s - 1$ it follows that by the foregoing column operations the elements of the last column of $D(E, X)$ in row (i_0, k_j) will become $k_j!/(r+s-1)!$ times the last column in $\Delta(x)$. We repeat this operation, using the coefficients of the polynomial P_{s-2} to change the next to last column, and so on. Expanding the new determinant about the last s columns, we obtain

$$D(E, X) = CD(E', X)\Delta(x) \qquad (8.2.5)$$

where E' is the $m \times (n + 1 - s)$ matrix consisting of all Hermite sequences of E. Thus, $D(E', X) \neq 0$ for $a < x < b$ and does not depend on x if $r = 0$. $\quad\square$

Since the determinant $\Delta(x)$ contains only the differences $k_j - r$, $j = 1,\dots,s$, and not r and k_j separately, we obtain immediately

Corollary 8.2 (Drols [30]). *Let E' be the $m \times (n - r + 1)$ almost-Hermitian matrix formed from E by omitting the Hermite sequence in the row i_0, shifting the remaining 1's into positions $k_1 - r,\dots,k_s - r$, and omitting the last r columns. Then E is regular (weakly singular or strongly singular) if and only if E' is regular (weakly singular or strongly singular).*

From now on we shall always assume that $r = 0$ for an almost-Hermitian matrix. We shall treat the case $s = 2$ in some detail.

Theorem 8.3 ([26,91]). *A necessary condition for the regularity of the almost-Hermitian matrix E with $s = 2$ (and $r = 0$) is that for each selection of x_i, $i \neq i_0$, the polynomials $H^{(k_1-1)}$ and $H^{(k_2-1)}$ have no common zeros on (a, b) and that the zeros of $H^{(k_2-1)}$ in (a, b) strictly alternate with the zeros of $H^{(k_1-1)}$ in $[a, b]$. The matrix E is strongly singular if these conditions do not hold.*

Proof. The determinant $\Delta(x)$ is given in our case by

$$k_1!k_2!\Delta(x) = k_2 H^{(k_1)}(x)H^{(k_2-1)}(x) - k_1 H^{(k_2)}(x)H^{(k_1-1)}(x).$$

$$(8.2.6)$$

Next, $H(x)$ is a polynomial of degree $n - 1$ that is annihilated by the Hermite part E_1 of E with $n - 1$ ones. This means that the zeros of $H^{(k)}(x)$

are precisely those given by Rolle's theorem and the matrix E_1, and that the Rolle zeros that are not specified by E_1 are necessarily simple.

If some zero ξ of $H^{(k_1-1)}(x)$ coincides with a zero of $H^{(k_2-1)}(x)$ in (a, b), then $\Delta(\xi) = 0$ and E is singular by Theorem 8.1. Even more is true in this case. By the Taylor expansion about ξ, we have, for $h \to 0$,

$$k_1!k_2!\Delta(\xi + h) = (k_2 - k_1)H^{(k_2)}(\xi)H^{(k_1)}(\xi)h + o(h). \quad (8.2.7)$$

Since the zeros of $H^{(k_2-1)}(x)$ and $H^{(k_1-1)}(x)$ in (a, b) are simple, the first term on the right in (8.2.7) does not vanish. Thus, $\Delta(x)$ changes sign at $x = \xi$, and by (8.2.3), $D(E, X)$ changes sign.

We assume now that the zeros of $H^{(k_1-1)}$ and $H^{(k_2-1)}$ do not coincide on (a, b). Let ξ_1, ξ_2 be two adjacent zeros of $H^{(k_1-1)}$ on (a, b), and assume that $H^{(k_2-1)}$ does not vanish on $[\xi_1, \xi_2]$. For $x = \xi_1, \xi_2$, $k_1!k_2!\Delta(x) = k_2 H^{(k_1)}(x)H^{(k_2-1)}(x)$. Since $H^{(k_1)}$ has only a simple zero on (ξ_1, ξ_2), given by Rolle's theorem, $\Delta(x)$ changes sign on this interval. A similar argument applies for two adjacent zeros of $H^{(k_2-1)}(x)$ in (a, b).

If $\xi_1 = a$ and ξ_2 are zeros of $H^{(k_1-1)}$ in $[a, b]$ and $H^{(k_2-1)}$ does not vanish on $(a, \xi_2]$, then the sign of $\Delta(x)$ is determined by the quantity in square brackets in

$$k_1!k_2!\Delta(x) = H^{(k_1-1)}(x)H^{(k_2-1)}(x)\left[\frac{k_2 H^{(k_1)}(x)}{H^{(k_1-1)}(x)} - \frac{k_1 H^{(k_2)}(x)}{H^{(k_2-1)}(x)}\right].$$

In the Laurant expansion of this quantity, the dominant term for x near a is $[k_2 m_1 - k_1 m_2](x - a)^{-1}$ where m_1 and m_2 are the multiplicities of the zeros of $H^{(k_1-1)}$ and $H^{(k_2-1)}$ at a, respectively. This term is positive since $m_1 \geq m_2$ and $k_2 > k_1$; the first inequality holds since either the zeros of $H^{(k_1-1)}$ and $H^{(k_2-1)}$ are simple Rolle zeros or some of them are given by the Hermitian sequence of row $i_0 - 1$. If a is near ξ_2, then the dominant term is $k_2(x - \xi_2)^{-1}$, which is negative. The case in which $\xi_2 = b$ is similar.   □

Let $z(k)$, $Z(k)$ denote the numbers of zeros of $H^{(k)}$ in (a, b) and in $[a, b]$, respectively (we do not count multiplicities). Restricting ourselves to three-row matrices, we see that $z(k)$, $Z(k)$ are independent of the choice of a, b and that $Z(k) \geq 2$ for $k < n - 1$. Then the necessary condition of Theorem 8.3 implies that $Z(k_1 - 1) = z(k_2 - 1) + 1$. Hence:

Corollary 8.4. *A three-row matrix* $E(p, q; k_1, k_2)$ *is strongly singular if*

$$Z(k_1 - 1) \neq z(k_2 - 1) + 1. \quad (8.2.8)$$

It is easy to calculate these numbers; assuming that $1 \leq p \leq q$, we have from (5.1.4) [or directly from $H(t) = (t - a)^p(t - b)^q$]:

$$Z(k) = \begin{cases} k + 2 & \text{if } 0 \leq k < p - 1, \\ p + 1 & \text{if } p - 1 \leq k < q, \\ p + q - k & \text{if } q \leq k \leq p + q, \end{cases} \quad (8.2.9)$$

while

$$z(k) = \begin{cases} k & \text{if} \quad 0 \leqslant k < p, \\ p & \text{if} \quad p \leqslant k \leqslant q, \\ p+q-k & \text{if} \quad q < k \leqslant p+q. \end{cases} \tag{8.2.10}$$

By means of (8.2.9) and (8.2.10) it is easy to determine when we have equality $Z(k_1 - 1) = z(k_2 - 1) + 1$. Either both functions are at their maximum, equal to $p + 1$, in which case

$$p \leqslant k_1 < k_2 - 1 \leqslant q, \tag{8.2.11}$$

or their common value is less than $p + 1$. In the last case k_2 is to the right of the peak for $z(k-1)+1$ and k_1 is to the left of the peak for $Z(k-1)$. Then

$$q + 1 < k_2 \quad \text{and} \quad k_1 + k_2 = p + q + 1. \tag{8.2.12}$$

Together with Theorem 4.5 on coalescence, these facts lead to ([91, 104]):

Theorem 8.5. *If the almost-Hermitian matrix E with $r = 0$, $s = 2$, $k_1 < k_2 - 1$, has p ones in rows $1, \ldots, i_0 - 1$ and q ones in rows $i_0 + 1, \ldots, m$, and if $1 \leqslant p \leqslant q$, then E is strongly singular except possibly when (8.2.11) or (8.2.12) hold.*

These two exceptional cases will be treated in §8.5 and §8.6, respectively. We call them the *interior case* and the *symmetric exterior case*.

Functions $z(k)$, $Z(k)$ can be useful when dealing with matrices of more than three rows if the technique of independent knots of §5.4 is used. For example:

Proposition 8.6 [91]. *If the almost-Hermitian matrix of Theorem 8.5 has more than three rows, then E is also strongly singular when (8.2.12) holds.*

Proof. We can assume that there are at least two rows $i < i_0$, and by coalescing rows, if necessary, that $i_0 + 1 = m$. We set $x_{i_0-1} = a$, $x_{i_0+1} = b$ and place knots x_i, $i = 1, \ldots, i_0 - 2$, far away toward $-\infty$. It is possible to do this so that the Rolle zeros of $H^{(k)}$, $k = 0, 1, \ldots, n$, in $[a, b]$ are generated by zeros of $H^{(k-1)}$ in $[a, b]$. Therefore, the numbers $z(k)$ and $Z(k)$ are determined only by the three rows $i \geqslant i_0 - 1$.

If $p' < p$ is the number of 1's in row $i_0 - 1$, then we find that (8.2.9) and (8.2.10) hold with p replaced by p' and that there is equality in (8.2.8) if $q + 1 < k_2$ and $k_1 + k_2 = p' + q + 1$. But this contradicts (8.2.12). $\qquad \square$

Exercise. Prove Theorem 6.5 using Corollary 8.2 and Theorem 6.2. [*Hint:* First remove the Hermitian sequence of row i_0 by Corollary 8.2, then coalesce row $i_0 - 1$ to row i_0, and finally remove the resulting Hermitian sequence of row i_0, etc.]

§8.3. RECIPROCAL ALMOST-HERMITIAN MATRICES

A very useful idea is the concept of reciprocal almost-Hermitian matrices introduced by Drols [29]. In this section we assume that $r = 0$ and that the matrices have three rows. Two matrices of this type with the same p, q, and s, $E = E(p, q; k_1, \ldots, k_s)$, $E^* = E(p, q; k_1^*, \ldots, k_s^*)$ are *reciprocal* if the positions of the 1's in row 2 are related by the equations

$$k_j^* = n - k_{s-j+1}, \qquad j = 1, \ldots, s. \tag{8.3.1}$$

The usefulness of reciprocal matrices comes from an identity between derivatives of the polynomial $H(x)$ of (8.2.1). Let $X = \{-1, x, 1\}$; then this polynomial has the form

$$H_{p,q}(x) = (x+1)^p (x-1)^q, \qquad p \leqslant q. \tag{8.3.2}$$

Lemma 8.7 [104]. *If $l + s = p + q$, $0 \leqslant l \leqslant s$, $0 \leqslant p \leqslant q$, then the polynomial $H_{p,q}(x)$ satisfies the symmetry relation*

$$\frac{1}{s!} H_{p,q}^{(s)}(-x) = (-1)^l (x+1)^{q-s} (x-1)^{p-s} \frac{1}{l!} H_{p,q}^{(l)}(x), \qquad x \neq \pm 1. \tag{8.3.3}$$

Proof. We begin with the case $l \leqslant p \leqslant q \leqslant s$. Then from Leibnitz's formula we obtain

$$\frac{1}{l!} H_{p,q}^{(l)}(x) = \frac{1}{l!} \sum_{i=0}^{l} \binom{l}{i} i! \binom{p}{i} (x+1)^{p-i} (l-i)! \binom{q}{l-i} (x-1)^{q-l+i}$$

$$= \sum_{i=0}^{l} \binom{p}{i} \binom{q}{l-i} (x+1)^{p-i} (x-1)^{q-l+i} \tag{8.3.4}$$

$$= (x+1)^{s-q} (x-1)^{s-p} \sum_{i=0}^{l} \binom{p}{i} \binom{q}{l-i} (x+1)^{l-i} (x-1)^i,$$

since $p - l = s - q$ and $q - l = s - p$. Similarly, we may write

$$\frac{1}{s!} H_{p,q}^{(s)}(x) = \sum_{i=s-q}^{p} \binom{p}{i} \binom{q}{s-i} (x+1)^{p-i} (x-1)^{q-s+i}, \tag{8.3.5}$$

since the terms of this expansion are 0 for $i < s - q$ and $i > p$. Making the substitution $i = p - j$ and using $l + s = p + q$, we rewrite (8.3.5) as

$$\frac{1}{s!} H_{p,q}^{(s)}(-x) = (-1)^l \sum_{j=0}^{l} \binom{p}{j} \binom{q}{l-j} (x-1)^j (x+1)^{l-j}. \tag{8.3.6}$$

Comparing equations (8.3.4) and (8.3.6), we derive (8.3.3).

The case in which $p \leqslant l \leqslant s \leqslant q$ is similar. In this case the sum (8.3.4) is over the range $0 \leqslant i \leqslant p$; the sum (8.3.5) is also over $0 \leqslant i \leqslant p$, as is the sum in (8.3.6) after we substitute $j = p - i$. $\qquad \square$

The main result for reciprocal matrices is the following theorem due to Drols [29].

Theorem 8.8. *For reciprocal almost-Hermitian three-row matrices* E, E^* *and* $X = \{-1, x, 1\}$,

$$D(E, -x) = C(x)D(E^*, x)$$

where $C(x) \neq 0$, $x \neq \pm 1$. *In particular* E *is regular (weakly singular or singular) if and only if* E^* *is regular (weakly singular or singular).*

Proof. By Theorem 8.1,

$$D(E, -x) = C\Delta(-x), \qquad D(E^*, x) = C^*\Delta^*(x)$$

for some constants C, C^* where $\Delta(x)$ and $\Delta^*(x)$ are the determinants (8.2.2) with $r = 0$ and $H(x) = H_{p,q}(x)$. Since $k_j - l + (k^*_{s-j+1} - s + l + 1) = p + q$ by (8.3.1), we apply Lemma 8.7 to obtain

$$\frac{1}{(k_j - l)!} H^{(k_j - l)}_{p,q}(-x) = (-1)^{k^*_{s-j+1} - s + 1 + l}(x+1)^{q - k_j + l}(x-1)^{p - k_j + l}$$

$$\times \frac{1}{(k^*_{s-j+1} - s + 1 + l)!} H^{(k^*_{s-j+1} - s + 1 + l)}_{p,q}(x),$$

$$j = 1, \ldots, s, \quad l = 0, \ldots, s - 1. \qquad (8.3.7)$$

Substituting the expressions (8.3.7) into the determinant $\Delta(-x)$ and extracting the common factors from the rows and columns, we obtain

$$\Delta(-x) = \pm (x+1)^{\sigma_1}(x-1)^{\sigma_2}$$

$$\times \det\left[\frac{1}{(k^*_{s-j+1} - s + 1)!} H^{(k^*_{s-j+1} - s + 1)}_{p,q}(x), \ldots, \right.$$

$$\left. \frac{1}{k^*_{s-j+1}!} H^{(k^*_{s-j+1})}_{p,q}(x); j = 1, \ldots, s\right]$$

$$= C_1(x)\Delta^*(x)$$

for some integers σ_1, σ_2. In particular, $C_1(x) \neq 0$ if $x \neq \pm 1$. $\qquad \Box$

Additional theorems about the singularity of three-row almost-Hermitian matrices follow from results in Chapter 4. Recall that a three-row almost-Hermitian matrix is strongly order singular if the number $\gamma_2 - \alpha_{2,1} - \alpha_{2,3}$ is odd (Theorem 4.6). Here γ_2 is the coefficient of the maximal collision of row 2 in E.

Exercises

1 (Drols [29]). If the matrix $E(p, q; k_1, \ldots, k_s)$ is its own reciprocal and if either ps or qs is odd, then E is strongly singular. [*Hint*: The numbers $\gamma_2, \alpha_{2,1}, \alpha_{2,3}$ are easily computable (mod 2).]

2 (Drols [29]). If $E(p, q; k_1, \ldots, k_s)$ satisfies $p \leqslant k_1 < \cdots < k_s \leqslant q + s + 1$ and ps is odd, then E is strongly singular. [*Hint:* Compare the interchange numbers for the coalescences $((F_1 \cup F_2)^0 \cup F_3)^0$ and $(F_1 \cup (F_2 \cup F_3)^0)^0$.]

§8.4. NULL CURVES FOR THE MATRICES $E(p,q; k_1, k_2)$

The regularity problem of the matrices $E = E(p, q; k_1, k_2)$ can be studied from a geometric viewpoint by means of certain curves in the plane. As in the proof of Theorem 6.2, the curves are obtained by setting the partial derivatives of a certain determinant equal to 0. The matrix E will be singular if one of the curves crosses the diagonal $x = y$ inside the square $-1 \leqslant x, y \leqslant 1$. This technique can be viewed as a geometric interpretation of the approach of Lorentz, Stangler, and Zeller [104]. With this method we again prove Theorem 8.5 and, in the next two sections, treat the exceptional cases (8.2.11) and (8.2.12). We always assume that $0 < k_1 < k_2 - 1 < n - 1$, $1 \leqslant p \leqslant q$ and have $p + q + 1 = n$. Let E_1 be the $4 \times (n + 1)$ Hermite matrix with Hermite sequences of length $p, 1, 1, q$. It is obtained from E by moving the 1's, $e_{2, k_1} = e_{2, k_2} = 1$ to Hermite positions in separate rows. The knot set for E_1 will be $X_1 = \{-1, x, y, 1\}$; it is essential that x and y be allowed to take all real values. We define

$$P(x, y) = D(E_1, X_1). \tag{8.4.1}$$

For P we have the representation

$$P(x, y) = C(y - x)H(x)H(y) \tag{8.4.2}$$

where $H = H_{p,q}$ is the polynomial (8.3.2), and the constant C is different from 0. Indeed, for fixed (say) x, P is a polynomial in y of degree n, and since $p + q + 1 = n$, we can account for all of its zeros: of orders $p, 1, q$ at $x = -1, x = y, x = 1$, respectively. Hence P is divisible by $(y - x)H(y)$.

The significance of the polynomial $P(x, y)$ comes from the relationship between partial differentiation of $D(E_1, X_1)$ and shifts in E_1. We have

$$\pm D(E, X) = \frac{\partial^{k_1}}{\partial x^{k_1}} \frac{\partial^{k_2}}{\partial y^{k_2}} P(x, y) \bigg|_{x = y}. \tag{8.4.3}$$

We would like to find the null sets—that is, the set of points (x, y) in the plane where they vanish—for the polynomial

$$Q(x, y) = \frac{\partial^{k_2}}{\partial y^{k_2}} P(x, y) \tag{8.4.4}$$

and for R of (8.4.10). For Q we have

$$Q(x, y) = CH(x)\frac{\partial^{k_2}}{\partial y^{k_2}}[(y - x)H(y)]$$

$$= CH(x)[(y - x)H^{(k_2)}(y) + k_2 H^{(k_2-1)}(y)] \qquad (8.4.5)$$

$$= -CH(x)H^{(k_2)}(y)[x - \lambda_{k_2}(y)]$$

where we put

$$\lambda_k(y) = y + k\frac{H^{(k-1)}(y)}{H^{(k)}(y)}, \qquad k = 1,\dots,n. \qquad (8.4.6)$$

From (8.4.5) with $k = k_2$ we have that the null set of Q contains all points with $x = -1$ or $+1$, as well as all points with $y = -1$ (or $+1$) when $k_2 < p$ (or $< q$). We call this the trivial part of the null set and wish to find the remaining nontrivial part.

As a function of y, $(y - x)H(y)$ is annihilated by E_2, $X_2 = \{-1, x, 1\}$ where E_2 is the matrix obtained from E_1 by dropping its third row and last column. We can find the zeros of $[(y - x)H(y)]^{(k_2)}$ directly or by means of Rolle extensions of Chapter 5. Since E_2 is a $3 \times n$ Pólya matrix without odd supported sequences and $(y - x)H(y)$ is a polynomial of degree n in y, the Rolle extension of order k_2 is maximal and gives all zeros of the k_2nd derivative. By Lemma 5.3, the number of such zeros is at most

$$\mu_{k_2}(E_2) = \mu_{k_2-1}(E_2) - 1 + m_{k_2}(E_2)$$

(and equal to this number if there is no duplication). Here $m_{k_2}(E_2)$ counts the trivial zeros and by (5.1.5), we have at most

$$z_2 = M_{k_2-1}(E_2) - k_2 = M_{k_2-1}(E) - k_2 \geqslant 1 \qquad (8.4.7)$$

nontrivial zeros. As we shall see later, there are exactly z_2 zeros.

Similarly, $H(y)$ is annihilated by a two-row matrix with p ones in the first row and q ones in the second row and the knots $\{-1, 1\}$. Since there is one less entry 1 than in the matrix E_2, and no duplication, we obtain exactly $z_2 - 1$ zeros of $H^{(k_2)}(y)$ in $(-1, +1)$. We denote them by $y_1^* < \cdots < y_{z_2-1}^*$, and in addition, put $y_0^* = -\infty$, $y_{z_2}^* = +\infty$. For different null curves, obtained with computer aide, see Figures 8.1–8.3.

Lemma 8.10. *The nontrivial zeros of $Q(x, y)$ are given by z_2 curves defined for* $-\infty < x < +\infty$,

$$y_1(x) < \cdots < y_{z_2}(x). \qquad (8.4.8)$$

The curves do not enter the two regions $y > \max(x, 1)$ and $y < \min(x, -1)$. The functions y_j are continuous, strictly increasing, satisfy

$$y_{j-1}^* < y_j(x) < y_j^*, \qquad j = 1,\dots,z_2, \qquad (8.4.9)$$

and are asymptotic to the lines $y = y_{j-1}^*$, $y = y_j^*$.

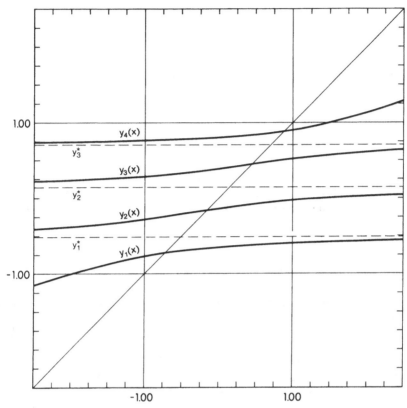

FIGURE 8.1 The auxiliary curves $y_j(x)$ for $E(4, 5; 2, 6)$.

Proof. We have $Q(x, y) = \pm D(\Lambda^{k_2} E_1, X_1)$ where the shift Λ^{k_2} moves the 1 of row 3 from $e_{3,0} = 1$ to $e_{3, k_2} = 1$. For knots satisfying $y > \max(x, 1)$ and $y < \min(x, -1)$, $\Lambda^{k_2} E_1$ has no odd supported sequences, hence $Q(x, y) \neq 0$.

The equation $Q = 0$ also has no solutions with $x, y \neq \pm 1$, $y = y_j^*$, $j = 1, \ldots, z_2 - 1$. For if $y \neq \pm 1$, then $H^{(k_2-1)}(y)$ and $H^{(k_2)}(y)$ cannot both vanish. The function $\lambda_{k_2}(y)$ of (8.4.6) has limits $+\infty$ $(-\infty)$ for $y \to +\infty$ $(-\infty)$ and is continuous except for infinite jumps at y_j^*, $j = 1, \ldots, z_2 - 1$, where it changes sign. It takes on all real values for $y \in (y_{j-1}^*, y_j^*)$. The function provides at least one solution of the equation $Q = 0$ in the strip. Hence there must be exactly one such solution, altogether z_2 of them. Moreover, $\lambda_{k_2}(y)$ is monotone for $y \in (y_{j-1}^*, y_j^*)$, and its inverses $y_j(x)$ are functions (8.4.8). $\square$

Our next theorem describes the null set of the polynomial

$$R(x, y) = \frac{\partial^{k_1}}{\partial x^{k_1}} Q(x, y) = \frac{\partial^{k_2}}{\partial y^{k_2}} Q_1(x, y) \qquad (8.4.10)$$

where

$$Q_1(x, y) = \frac{\partial^{k_1}}{\partial x^{k_1}} P(x, y).$$

Differentiating (8.4.5) we obtain the formula

$$\begin{aligned}
R(x, y) &= C\left[k_2 H^{(k_1)}(x) H^{(k_2-1)}(y) - k_1 H^{(k_1-1)}(x) H^{(k_2)}(y) \right. \\
&\quad \left. + (y-x) H^{(k_1)}(x) H^{(k_2)}(y)\right] \\
&= CH^{(k_1)}(x) H^{(k_2)}(y)\left[\lambda_{k_2}(y) - \lambda_{k_1}(x)\right]. \quad (8.4.11)
\end{aligned}$$

As we know from the proof of Lemma 8.10, each of the functions $\lambda_k(x)$ is monotone increasing with infinite jumps at the roots of $H^{(k)}(x)$ in $(-1, 1)$.

The possible zeros of R given by the roots of $H^{(k)}$, $k = k_1$ or k_2, at ± 1 are discarded as trivial. We would like to determine the nontrivial part of the null set of R.

Let $y \neq y_j^*$, $j = 1, \ldots, z_2 - 1$, be fixed. Then $Q(x, y)$, given by (8.4.5), is a polynomial of degree n in x with zeros of order $p, 1, q$ at $-1, \lambda_{k_2}(y), 1$. Hence, it is annihilated by the pair E_2 and $X_2 = \{-1, \lambda_{k_2}(y), 1\}$. We can apply once more the results about Rolle extensions to find that $R(x, y)$ has

$$z_1 = M_{k_1-1}(E) - k_1 + 1 \geqslant 2 \quad (8.4.12)$$

zeros for its k_1st derivative [which lie in $(-1, 1)$ if $\lambda_{k_2}(y)$ is in this interval]. Similarly, $H^{(k_1)}(x)$ has $z_1 - 1$ Rolle zeros in $(-1, +1)$ which we denote by $x_1^*, \ldots, x_{z_1-1}^*$, putting in addition $x_0^* = -\infty$, $x_{z_1}^* = +\infty$.

But we have also the possibility of using the second of the representations (8.4.10). Then we see that for each $x \neq x_i^*$, $i = 1, \ldots, z_1 - 1$, the polynomial $R(x, y)$, as a function of y, has z_2 nontrivial zeros.

Theorem 8.11. *The rectangles* $[x_{i-1}^*, x_i^*] \times [y_{j-1}^*, y_j^*]$, $i = 1, \ldots, z_1$, $j = 1, \ldots, z_2$ *subdivide the plane into* $z_1 z_2$ *regions, in each of which the nontrivial zeros of $R(x, y)$ form a continuous, strictly monotonic curve C connecting the corners* (x_{i-1}^*, y_{j-1}^*) *and* (x_i^*, y_j^*) *(If one of the corners is infinite, this is to be understood asymptotically.) The equation of the curve is*

$$\lambda_{k_1}(x) = \lambda_{k_2}(y), \qquad x_{i-1}^* < x < x_i^*, \quad y_{j-1}^* < y < y_j^*. \quad (8.4.13)$$

If a corner is finite, then the curve continues into the next rectangle. The curves C do not enter the four regions
(a) $1 < x \leqslant y$,
(b) $y \leqslant x < -1$,
(c) $x > 1$ *and* $y < -1$,
(d) $x < -1$ *and* $y > 1$.
There are $z_1 - 1$ *curves intersecting the open lower side of the square, z_2 curves intersecting its closed left side, altogether $z = z_1 + z_2 - 1$ curves. (See Figures 8.2 and 8.3.)*

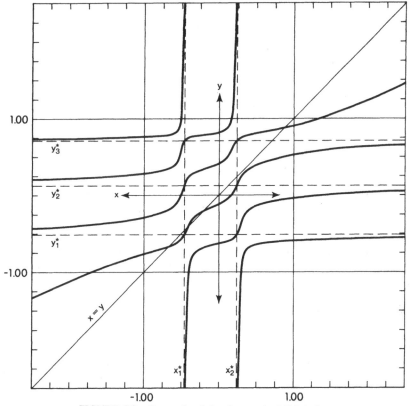

FIGURE 8.2 Curves $R = 0$ for the matrix $E(4, 5; 2, 6)$.

Proof. That each finite corner belongs to the null set of R follows from (8.4.11) since $H^{(k_1)}(x)$ and $H^{(k_2)}(y)$ vanish in each corner. Setting $R = 0$, we obtain the equations (8.4.13) for the curves. Since $\lambda_{k_1}(x), \lambda_{k_2}(y)$ are continuous and strictly monotone for $x^*_{i-1} < x < x^*_i$, $y^*_{j-1} < y < y^*_j$, it follows that the curves are also monotone.

The curves (8.4.13) do not enter the four regions (a)–(d), since in these regions

$$R(x, y) = \frac{\partial^{k_1}}{\partial x^{k_1}} Q(x, y) = \pm D(E^*, X^*)$$

where E^* is a permutation of the four-row matrix $\Lambda^{k_1}\Lambda^{k_2}E_1$ found by shifting the ones $e_{2,0} = e_{3,0} = 1$ in E_1 into positions $e_{2,k_1} = e_{3,k_2} = 1$, and X^* is the corresponding permutation of $X = \{-1, x, y, 1\}$ which orders the rows of the matrix. For the four regions, E^* will have no odd supported sequences.

Because of the empty region (b), the lowest curve coming from the left, which goes through (x^*_1, y^*_1), must intersect the left side of the square; it may pass through $(-1, -1)$. There are z_2 curves crossing this side. □

The curves of Theorem 8.11 divide the square $-1 < x, y < 1$ into regions where $D(E, X)$ has opposite signs. We have

Theorem 8.12. *The matrix $E(p, q; k_1, k_2)$ is singular if and only if one of the z curves of Theorem 8.11 has a point in common with the diagonal $x = y$ of the open square. It is strongly singular if and only if one of the curves crosses the diagonal.*

We would like to determine when some of the z curves of Theorem 8.11 pass through the corners $(-1, -1), (1, 1)$ of the square $-1 \leqslant x, y \leqslant 1$. This can happen only for the lowest curve entering the square from the left and for the highest curve leaving it on the right.

Proposition 8.13. *For the matrix $E = E(p, q; k_1, k_2)$ precisely $e_{1, k_2 - 1}$ curves of Theorem 8.11 pass through the point $(-1, -1)$ and precisely $e_{3, k_2 - 1}$ curves pass through the point $(1, 1)$.*

Proof. It is sufficient to treat, for example, the point $(-1, -1)$. This point is interior to the rectangle with corners $(-\infty, -\infty), (x_1^*, y_1^*)$. The equation of the curve is $\lambda_{k_2}(y) = \lambda_{k_1}(x)$, with the λ_k given by (8.4.6).

Now for $H(x) = (x + 1)^p (x - 1)^q$ we have $H^{(k)}(-1) = 0$ if $k < p$, $H^{(k)}(-1) \neq 0$ if $k \geqslant p$. If $e_{1, k_2 - 1} = 1$, or equivalently, $k_2 \leqslant p$, then also $k_1 \leqslant p$ and from the formula (8.4.6) we obtain $\lambda_{k_1}(x) \to -1, \lambda_{k_2}(y) \to -1$ if $x \to -1, y \to -1$. This means that the curve passes through $(-1, -1)$.

If $e_{1, k_2 - 1} = 0, e_{1, k_1 - 1} = 1$ we obtain in the same way $\lambda_{k_2}(y) \not\to -1$, $\lambda_{k_1}(x) \to -1$ for $x \to -1, y \to -1$, hence the curve does not pass through $(-1, -1)$. Finally, if $e_{1, k_1 - 1} = e_{1, k_2 - 1} = 0$, then

$$R(-1, -1) = \left. \frac{\partial^{k_2} \partial^{k_1}}{\partial y^{k_2} \partial x^{k_1}} D(E_1, X_1) \right|_{x, y = -1} = \pm D\left(\Lambda^{k_1} \Lambda^{k_2} E_1, X_1 \right)|_{x, y = -1}.$$

The last determinant, up to a sign, is the determinant of a regular two-row matrix. Thus, $R(-1, -1) \neq 0$, and no curve can pass through $(-1, -1)$. $\square$

From this we derive

Proposition 8.14. *The matrix $E(p, q; k_1, k_2)$ is strongly singular if*

$$z_2 \neq z_1 - 1 + e_{1, k_2 - 1} + e_{3, k_2 - 1} \tag{8.4.14}$$

or equivalently if

$$M_{k_2 - 1}(E) - M_{k_1 - 1}(E) \neq (k_2 - k_1) + e_{1, k_2 - 1} + e_{3, k_2 - 1}. \tag{8.4.15}$$

Proof. By (8.4.7) and (8.4.12), the two statements are equivalent. Since $M_{k-1}(E) - k$ is nondecreasing for $k_1 + 1 \leqslant k \leqslant k_2$ if $k_2 \leqslant q$, the only possibility for (8.4.14) in this case is

$$z_2 > z_1 - 1 + e_{1, k_2 - 1} + e_{3, k_2 - 1}. \tag{8.4.16}$$

Indeed, the right-hand side of (8.4.14) is estimated by

$$M_{k_1-1}(E) - k_1 + e_{1,k_2-1} + e_{3,k_2-1} = M_{k_1}(E) - k_1 - 1 + e_{1,k_2-1} - e_{1,k_1}$$

$$\leqslant M_{k_2-1}(E) - k_2 = z_2.$$

If $k_2 > q$, then $e_{1,k_2-1} = e_{3,k_2-1} = 0$ and (8.4.14) implies either $z_2 > z_1 - 1$, that means again (8.4.16) or

$$z_2 < z_1 - 1. \tag{8.4.17}$$

In the last case, at least one curve enters the square at the bottom and leaves on the top, crossing the diagonal.

In the case of (8.4.16), there are $z_2 - e_{3,k_2-1}$ curves leaving the square on the right below $(1,1)$ and $z_1 - 1 + e_{1,k_2-1}$ curves entering the square on the bottom, possibly through $(-1,-1)$. There is at least one curve that enters on the left above $(-1,-1)$ and leaves on the right below $(1,1)$. $\qquad\square$

We leave it to the reader to check that (8.4.15) happens exactly when (8.2.8) of Corollary 8.4 holds. We have thus obtained another proof of this corollary.

§8.5. THE EXTERIOR SYMMETRIC CASE

In this section we study the matrix $E(p, q; k_1, k_2)$ when $k_1 + k_2 = p + q + 1$, $k_2 > q + 1$. This means that the two points k_1, k_2 are outside of the interval $[p, q+1]$ and are symmetric with respect to its midpoint. For this case the problem of regularity can be solved completely. See [26, 91, 104] and Drols [29] for (iii).

Theorem 8.15. *If $E(p, q; k_1, k_2)$ is a matrix with*

$$k_1 + k_2 = p + q + 1, \qquad k_2 > q,$$

then (i) *E is regular if $p = q$,* (ii) *E is weakly singular if $q = p + 1$, and* (iii) *E is strongly singular if $q > p + 1$.*

We begin with lemmas about zeros of polynomials from [104].

Lemma 8.16. *If $P(x)$ is a polynomial of degree $n \geqslant 2$ with only real zeros, then the zeros of $P'(x)/P(x)$ will* (i) *strictly increase when one of the zeros of $P(x)$ increases;* (ii) *strictly move away from a zero of $P(x)$ when the multiplicity of that zero increases but the remaining zeros and their multiplicities are unchanged.*

Proof. Let $P(x) = \prod_{j=1}^{n}(x - x_j)^{r_j}$. We put

$$F(x) = \frac{P'(x)}{P(x)} = \sum_{j=1}^{n} \frac{r_j}{x - x_j}.$$

Between its poles, $F(x)$ is decreasing from $+\infty$ to $-\infty$. If x_{j_0} is replaced by $x_{j_0} + h$ to obtain $F_h(x)$, then at a zero ξ of $F(x)$, we have

$$F_h(\xi) = F_h(\xi) - F(\xi) = r_{j_0} h \left(\xi - x_{j_0} \right)^{-1} \left(\xi - x_{j_0} - h \right)^{-1} > 0$$

for small h. By the monotonicity of F_h, there will be a zero of F_h to the right of ξ.

If a multiplicity r_{j_0} is increased, then the value $F(x)$ will become larger for $x > x_{j_0}$ and smaller for $x < x_{j_0}$. The zeros of F will strictly move away from x_{j_0}.   □

The next lemma compares the zeros of two different polynomials $H_{p_1, q_1}, H_{p_2, q_2}$ of (8.3.2). Let $k \leqslant p_i$ and $p_i \leqslant q_i$, $i = 1, 2$. There are exactly k zeros of $H_{p_1, q_1}^{(k)}$ and of $H_{p_2, q_2}^{(k)}$ in $(-1, 1)$. We denote them by $-1 < x_1' < \cdots < x_k' < 1$ and $-1 < x_1'' < \cdots < x_k'' < 1$, respectively.

Lemma 8.17. (i) *Let $p_1 \leqslant p_2$ and $q_1 \geqslant q_2$. Then for $1 \leqslant k \leqslant p_1$,*

$$x_j' \leqslant x_j'', \qquad j = 1, \ldots, k, \tag{8.5.1}$$

with strict inequality unless $p_1 = p_2$ and $q_1 = q_2$. (ii) *If $k \leqslant p \leqslant q$, then the Rolle zeros x_j of $H_{p,q}^{(k)}(x)$ in $(-1, 1)$ satisfy*

$$x_j \leqslant -x_{k+1-j}, \qquad j = 1, \ldots, k, \tag{8.5.2}$$

with strict inequality unless $p = q$.

Proof. By Lemma 8.16, the Rolle zeros of $H_{p,q}'$ in the interval $(-1, 1)$ must shift to the right if p is increased or q is decreased. This strict increase of the Rolle zeros in turn causes a movement to the right of subsequent Rolle zeros of $H_{p,q}''$, again by Lemma 8.16. After a finite number of steps we arrive at (8.5.1) with strict inequality if either $p_1 \neq p_2$ or $q_1 \neq q_2$.

For (ii) we take $H_{p_1, q_1}(x) = H_{p, q}(x)$ and $H_{p_2, q_2}(x) = H_{q, p}(x) = \pm H_{p, q}(-x)$. The zeros of $H_{q, p}^{(k)}$ are the negatives of the zeros of $H_{p, q}^{(k)}$, that is, they are $-x_k < \cdots < -x_1$. Thus, (8.5.2) is a special case of (8.5.1).   □

Statement (iii) of Theorem 8.15 will follow from the following proposition about the curves C of Theorem 8.11. This method of proof of (iii) replaces the "Jagdmethode" of Lorentz, Stangler, and Zeller [104] and gives more, since even strong singularity is obtained. (Drols [29] also obtains the strong singularity by a different method.)

Proposition 8.18. *Let x_i', y_i' be the Rolle zeros of the derivatives $H_{p+1, q}^{(k_1)}$ and $H_{p+1, q}^{(k_2)}$, respectively. All points of the lattice (x_i', y_j'), $i, j = 1, \ldots, k_1$, lie on the curves C of Theorem 8.11. They satisfy the inequalities*

$$x_1^* < x_1' < x_2^* < \cdots < x_{k_1}^* < x_{k_1}' < 1, \tag{8.5.3}$$

$$-1 < y_1' < y_1^* < \cdots < y_{k_1-1}^* < y_{k_1}' < 1. \tag{8.5.4}$$

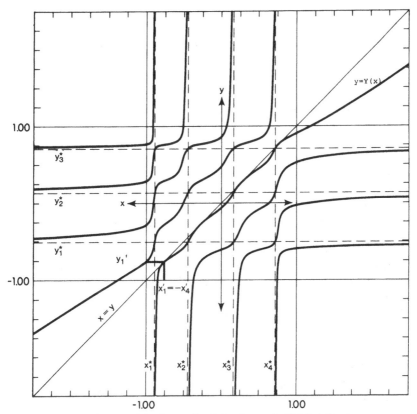

FIGURE 8.3 Curves $R = 0$ for the matrix $E(4,5;4,6)$.

Proof. Since

$$z_1 - 1 = M_{k_1-1}(E) - k_1 = k_1,$$

$$z_2 = M_{k_2-1}(E) - k_2 = p + q + 1 - k_2 = k_1, \qquad (8.5.5)$$

$$e_{1,k_2-1} = e_{3,k_2-1} = 0,$$

all k_1 of the curves entering at the bottom of the square exit on the right, and the first curve $y = Y(x)$ to enter on the bottom exits as the top curve on the right.

Let y'_j, $j = 1, \ldots, k_1$, be the ordinates of the points of intersection of the $z_2 = k_1$ curves of Lemma 8.10, which enter the square from the left, with the line $x = -1$. They satisfy (8.5.4). They are also the nontrivial zeros of Q

for $x = -1$. By (8.4.5),

$$\lim_{x \to -1} \frac{Q(x, y)}{(x+1)^p} = C'\left[(y+1)H_{p,q}^{(k_2)}(y) + k_2 H_{p,q}^{(k_2-1)}(y)\right]$$

$$= C'H_{p+1,q}^{(k_2)}(y),$$

hence y_i' are the zeros of $H_{p+1,q}^{(k_2)}$ in $(-1, +1)$.

Next, for each j, we study the zeros of $R = \partial^{k_1}Q/\partial x^{k_1}$ on the line $y = y_j'$. In each rectangle $[x_i^*, x_{i+1}^*] \times [y_{j-1}^*, y_j^*]$, $i = 1, \ldots, k_1$, there is exactly one point of intersection, $x = x_i'$, of the line $y = y_j'$ with the curve C in the rectangle. Since $\lambda_{k_2}(y_j') = -1$, by (8.4.5),

$$Q(x, y_j') = CH_{p,q}(x)(x+1)H_{p,q}^{(k_2)}(y_j') = C''H_{p+1,q}(x).$$

Hence the x_i' are the Rolle zeros of $H_{p+1,q}^{(k_1)}$; they do not depend on j. $\qquad\square$

To derive (iii) of Theorem 8.15, let C, with equation $y = Y(x)$, be the top curve of Theorem 8.11, coming into the square through the side $y = -1$. The curve passes through the rectangles $[x_j^*, x_{j+1}^*] \times [y_{j-1}^*, y_j^*]$, $j = 1, \ldots, k_1$, and through the points (x_j', y_j'). From the duality relation (8.3.3) we have $y_j' = -x_{k_1-j+1}'$, $j = 1, \ldots, k_1$; therefore by (8.5.2),

$$Y(x_j') = y_j' = -x_{k_1-j+1}' \geqslant x_j', \qquad j = 1, \ldots, k_1,$$

with strict inequality, unless $p + 1 = q$. The k_1 points (x_j', y_j') of C are above the diagonal if $p + 1 < q$, or on the diagonal if $p + 1 = q$. (Figure 8.3 illustrates the latter case.)

In order to show that a crossing does not occur when $p + 1 = q$ and also that none of the curves meet the diagonal $x = y$ when $p = q$, we compute the determinant directly, using Theorem 8.1 (see [26, 91]).

Lemma 8.19. *If $k_1 + k_2 = p + q + 1$, $q < k_2 < n$, then*

$$D(E, X) = \begin{cases} C_1(x^2-1)^{p-k_2}\left[\left(k_1 H_{p,p}^{(k_1-1)}(x)\right)^2 + (1-x^2)\left(H_{p,p}^{(k_1)}(x)\right)^2\right], \\ \qquad\qquad\qquad\qquad\qquad p = q, \\ C_2(x^2-1)^{p-k_2}(x-1)\left[H_{p+1,p+1}^{(k_1)}(x)\right]^2, \qquad p+1 = q. \end{cases}$$

$$(8.5.6)$$

Proof. From Theorem 8.1, $D = C\Delta$ with a constant $C \neq 0$; the determinant $\Delta = \Delta(x)$ of (8.2.2) is in this case

$$\Delta = \frac{1}{k_1!}H_{p,q}^{(k_1)}\frac{1}{(k_2-1)!}H_{p,q}^{(k_2-1)} - \frac{1}{(k_1-1)!}H_{p,q}^{(k_1-1)}\frac{1}{k_2!}H_{p,q}^{(k_2)}. \quad (8.5.7)$$

For $p = q$, $H_{p,p}(x) = H(x) = (x^2-1)^p$ is an even function, and $k_1 + k_2 =$

$2p + 1$. By (8.3.3),

$$\frac{1}{k_2!}H^{(k_2)}(x) = \frac{1}{k_2!}(-1)^{k_2}H^{(k_2)}(-x)$$

$$= (x^2 - 1)^{p-k_2}\frac{1}{(k_1 - 1)!}H^{(k_1-1)}(x),$$

$$\frac{1}{(k_2 - 1)!}H^{(k_2-1)}(x) = (x^2 - 1)^{p-k_2+1}\frac{1}{k_1!}H^{(k_1)}(x).$$

Substituting into (8.5.7), we obtain the case $p = q$ of the lemma. Part (i) of Theorem 8.15 is implied by this, because the roots of $H^{(k_1)}$ and those of $H^{(k_1-1)}$ alternate in $(-1, +1)$.

If $q = p + 1$, we make use of the identity

$$H^{(k)}_{p+1,p+1}(x) = (x+1)H^{(k)}_{p,p+1}(x) + kH^{(k-1)}_{p,p+1}(x)$$

and obtain

$$\Delta = \frac{1}{k_1!k_2!}\left[k_2 H^{(k_1)}_{p,p+1} H^{(k_2-1)}_{p,p+1} - k_1 H^{(k_2)}_{p,p+1} H^{(k_1-1)}_{p,p+1}\right]$$

$$= \frac{1}{k_1!k_2!}\left[H^{(k_1)}_{p,p+1} H^{(k_2)}_{p+1,p+1} - H^{(k_1)}_{p+1,p+1} H^{(k_2)}_{p,p+1}\right].$$

$(8.5.8)$

We again use (8.3.3) to eliminate derivatives of order k_2:

$$\frac{1}{k_2!}H^{(k_2)}_{p+1,p+1}(x) = (x^2 - 1)^{p+1-k_2}\frac{1}{k_1!}H^{(k_1)}_{p+1,p+1}(x)$$

and since $H_{p,p+1}(-x) = -H_{p+1,p}(x)$,

$$\frac{1}{k_2!}H^{(k_2)}_{p,p+1}(x) = (-1)^{k_2+1}\frac{1}{k_2!}H^{(k_2)}_{p+1,p}(-x)$$

$$= (x^2 - 1)^{p-k_2}(x-1)\frac{1}{(k_1 - 1)!}H^{(k_1-1)}_{p+1,p}(x).$$

Substituting this into (8.5.8), we have

$$\Delta(x) = \text{const } H^{(k_1)}_{p+1,p+1}(x)(x^2 - 1)^{p-k_2}(x - 1)$$

$$\times\left[(x+1)H^{(k_1)}_{p,p+1}(x) - k_1 H^{(k_1-1)}_{p+1,p}(x)\right].$$

$(8.5.9)$

From Leibnitz's rule and the observation that $H'_{p+1,p+1}(x) = (p+1)$ $[H_{p,p+1}(x) + H_{p+1,p}(x)]$, the expression in square brackets in (8.5.9) is equal to

$$H^{(k_1)}_{p+1,p+1} - k_1\left[H^{(k_1-1)}_{p,p+1} + H^{(k_1-1)}_{p+1,p}\right] = \text{const } H^{(k_1)}_{p+1,p+1}.$$

The second equation in (8.5.6) follows from this and (8.5.9).   □

§8.6. THE INTERIOR CASE

We have completely solved the regularity problem for the matrices $E = E(p, q; k_1, k_2)$, $1 \leqslant p \leqslant q$, $k_1 < k_2$ if $p = q$. In this case, E is regular if and only if the point (k_1, k_2) is on one of the lines $k_1 + k_2 = 2p + 1$ or $k_2 = k_1 + 1$. If $p < q$, then E is regular if $k_2 = k_1 + 1$, so we can assume that $k_2 \geqslant k_1 + 2$. Then we know that E is singular for pairs (k_1, k_2) outside of the triangle Λ defined by $k_1 \geqslant p$, $k_2 \leqslant q + 1$, $k_2 \geqslant k_1 + 1$. For fixed p, q with $p < q$, the *regularity domain* Ω of the matrices $E = E(p, q; k_1, k_2)$ is the set of all lattice points $(k_1, k_2) \in \Lambda$ for which E is regular. Some properties of Ω follow from a theorem given by Lorentz, Stangler, and Zeller [104], and refined by Drols [29]:

Theorem 8.20. *If the matrix $E(p, q; k_1, k_2)$ is singular and if*

$$p \leqslant k_1' \leqslant k_1 < k_2 - 1 \leqslant k_2' - 1 \leqslant q, \qquad (8.6.1)$$

and either $k_1' \neq k_1$ or $k_2' \neq k_2$, then the matrix $E(p, q; k_1', k_2')$ is strongly singular.

Proof. First we compute the constants z_1, z_2 of §8.4:

$$
\begin{aligned}
z_1 &= p + 1, & e_{1, k_2 - 1} &= 0 & &\text{always,} \\
z_2 &= p + 1, & e_{3, k_2 - 1} &= 1 & &\text{if } k_2 < q + 1, \qquad (8.6.2) \\
z_2 &= p, & e_{3, k_2 - 1} &= 0 & &\text{if } k_2 = q + 1.
\end{aligned}
$$

Let $C_1, \ldots, C_{z_1 + z_2 - 1}$ be the curves of Theorem 8.11, enumerated according to their intersection with the bottom side of the square from right to left, then according to their intersection with its left side, from bottom up. We find that $C_1, \ldots, C_{z_1 - 1}$ enter the square at bottom, leave at right. The curve C_{z_1} enters on the left, and either leaves at the right, passing through $(1, 1)$ if $k_2 < q + 1$, or crosses the top side if $k_2 = q + 1$. The rest of the curves enter on the left, leave at the top.

For each y, $-1 < y < 1$, there are exactly $p + 1$ intersections of the corresponding horizontal line with the curves C_ν. Their positions are given by the nontrivial Rolle zeros $x_j(y)$ of the k_1th derivative of $Q = \text{const}(x - \lambda_{k_2}(y))(x + 1)^p(x - 1)^q$. Only one of these zeros can be outside of the square, so that we have one of the two possibilities

$$
\begin{aligned}
x_1(y) &< \cdots < x_{p+1}(y) < 1, \\
x_1(y) &< \cdots < x_p(y) < 1 \leqslant x_{p+1}(y).
\end{aligned}
\qquad (8.6.3)
$$

If we increase k_1 by one, the new Rolle zeros $x_j'(y)$ will be derived from the $x_j(y)$ and 1, but without the participation of -1. Thus, *inside the square they will move to the right*, $x_j'(y) > x_j(y)$. This yields: *If k_1 increases, but k_2, p, q remain constant, then each of the curves C_ν moves to the right. In* particular: The lowest left curve C_{z_1} cannot intersect the diagonal $y = x$. For if it were to do so, then by increasing k_1 we would obtain the singularity of $E(p, q; k_2 - 1, k_2)$.

Now let $E(p, q; k_1, k_2)$ be singular, then C_{z_1-1} must have a point in common with the diagonal. Then C_{z_1-1} strictly intersects it if k_1 is replaced by k_1', $p \leqslant k_1' < k_1$, so that $E(p, q; k_1', k_2)$ is strongly singular. This proves the theorem if $p \leqslant k_1' < k_1$, $k_2' = k_2$. It remains to treat the case $k_1' = k_1$, $k_2 < k_2' \leqslant q + 1$. This is done by means of the reciprocal matrices and Theorem 8.8. ☐

The regularity set $\Omega \subset \Lambda$ can now be partially described. Each line entering Ω from the left or from above will be contained in Ω until its intersection with $k_2 = k_1 + 1$. Each of these lines contains at most one point of weak singularity. By Drols' theorem (Theorem 8.8), Ω is symmetric about the line $k_1 + k_2 = p + q + 1$.

Up to now, we have been able to determine the regularity or singularity of a matrix $E(p, q; k_1, k_2)$ simply in terms of the parameters. Apparently, the only way to describe the shape of Ω is by going back to the determinant $D(E, X)$. All we have is a corollary of Theorem 8.2:

For given k_1, the point $k_2 = \phi(k_1)$ on the upper boundary of Ω is the largest k_2 for which

$$k_1! k_2! \Delta(x) = k_2 H_{p,q}^{(k_1)}(x) H_{p,q}^{(k_2-1)}(x) - k_1 H_{p,q}^{(k_2)}(x) H_{p,q}^{(k_1-1)}(x) \qquad (8.6.4)$$

does not change sign for $-1 < x < 1$.

Sometimes it is possible to obtain simple conditions from this. For $p = 1$, the region Ω is the intersection of the triangle and the interior of an ellipse:

Theorem 8.21 (DeVore, Meir, Sharma [26]). *Let*

$$\rho(k_1, k_2) = (q + 2)(k_1 + k_2 - 1)^2 - 4(q + 1)k_1 k_2. \qquad (8.6.5)$$

The matrix $E(1, q; k_1, k_2)$ is regular when $\rho(k_1, k_2) < 0$, weakly singular when $\rho(k_1, k_2) = 0$, and strongly singular when $\rho(k_1, k_2) > 0$. Thus

$$\Omega = \{(k_1, k_2): (k_1, k_2) \in \Lambda, \rho(k_1, k_2) < 0\}. \qquad (8.6.6)$$

Proof. In (8.6.4) we take $p = 1$ and make the change of variable $u = \frac{1}{2}(x - 1)$. Then $H_{1,q}(x)$ is replaced by a constant multiple of $Q(u) = (u + 1)u^q$. By Leibnitz's formula, we see that (8.6.4) is a constant multiple of

$$u^{2q-k_1-k_2+1}\{k_2(q - k_1 + 2)[(q + 1)u + q + 1 - k_1][(q + 1)u + q + 2 - k_2]$$

$$- k_1(q - k_2 + 2)[(q + 1)u + q + 1 - k_2][(q + 1)u + q + 2 - k_1]\}.$$

If we make the substitution $v = (q + 1)u + (q + 1)$, then the expression in curved brackets becomes

$$(k_2 - k_1)\big[(q+2)v^2 - (k_2 + k_1 - 1)(q+2)v + (q+1)k_1k_2\big].$$

The discriminant of this expression is a positive multiple of $\rho(k_1, k_2)$.   □

§8.7. NOTES

It is possible to compare regularity properties of matrices $E(p, q; k_1, k_2)$ when p and q change. For example (Drols [29]): *If $E(p, q; k_1, k_2)$ is regular or weakly singular and $p \leqslant k_1 < k_2 - 1 \leqslant q + 1$, then $E(p, q+1; k_1 + 1, k_2)$ is regular.* For changing p, Drols [28] shows: *If $E(p, q; k_1, k_2)$, $p < k_1 < k_2 - 1 \leqslant q$, is singular, then so are $E(p+1, q; k_1, k_2)$ and $E(p+1, q; k_1 + 1, k_2 + 1)$.*

From these statements we can gain more information about the set Ω—for example, by carrying over results concerning the case $p = 1$ (see Drols [28]).

Another statement about the region Ω given in [104, Theorem 1(b)], is incorrect (at least for $k_1 + k_2 \geqslant p + q + 1$), as is the version of Markov's theorem on which it was based.

Chapter 9

Applications

§9.1. INTRODUCTION

There is a large literature on uniform approximation by polynomials whose values, or those of their derivatives, are constrained in some way. The simplest case is the problem of monotone approximation: Approximate a given function $f \in C[a, b]$ by increasing (or decreasing) algebraic polynomials of degree at most n. More generally, let $0 < k_1 < \cdots < k_p \leqslant n$ be given integers, and let $\varepsilon_j = \pm 1$, $j = 1, \ldots, p$, be given signs. Let $\mathscr{P}_n^* = \mathscr{P}_n(k_1, \ldots, k_p; \varepsilon_1, \ldots, \varepsilon_p)$ consist of all polynomials P of degree at most n that satisfy

$$\varepsilon_j P^{(k_j)}(x) \geqslant 0, \qquad a \leqslant x \leqslant b, \quad j = 1, \ldots, p. \qquad (9.1.1)$$

The problem of approximating a given continuous function f in $C[a, b]$ by a polynomial $P \in \mathscr{P}_n^*$ is also called the problem of monotone approximation. More generally, the derivatives of P may be required to satisfy the constraints

$$l_j(x) \leqslant P^{(k_j)}(x) \leqslant u_j(x), \qquad a \leqslant x \leqslant b, \quad j = 1, \ldots, p, \qquad (9.1.2)$$

for some suitable functions l_j, u_j (see §9.3).

There are two groups of questions that have been studied in this connection. One is the degree of approximation of f by polynomials $P \in \mathscr{P}_n^*$. Here it is essential to assume that the approximated function should have properties matching those of $P \in \mathscr{P}_n^*$. These studies do not generally require Birkhoff interpolation (see notes).

The other group of questions—questions of Chebyshev type—concern the existence, characterization, and uniqueness of a polynomial $P \in \mathcal{P}_n^*$ of best approximation to f. We shall see that the theorems of this group of problems are usually independent of whether the approximated function also satisfies the constraints. The existence question for an element of best approximation always follows from the usual compactness arguments provided that the constraints are not so severe that the class $\mathcal{P}_n^*$ is empty. The uniqueness question depends in an essential way on Birkhoff interpolation; this is seen in §9.4.

In §9.2, we first discuss monotone approximation by polynomials $\mathcal{P}_n^*$ satisfying (9.1.1). There is more specific information available for this problem than for the general constrained problem (9.1.2), treated in §9.3; but the essential features remain the same.

In §9.5, the uniqueness result of §9.4 is applied in order to study the simultaneous approximation of a function and its derivatives by a polynomial and its derivatives.

Finally, in §9.6 we give an application of a completely different nature. We show how Birkhoff interpolation can be used to determine when a system of powers forms a Chebyshev system on the real line.

§9.2. MONOTONE APPROXIMATION

Most of the problems for monotone approximation would be irrelevant if we knew that the polynomial of best approximation of a monotone function were itself monotone. That this assumption is false is seen in the exercise at the end of this section.

Let k_j be a selection of integers, ε_j be a selection of signs, and $\mathcal{P}_n^* = \mathcal{P}_n(k_1,\ldots,k_p; \varepsilon_1,\ldots,\varepsilon_p)$ be defined by (9.1.1). Each set $\mathcal{P}_n^*$ is nonempty in the following strong sense. There exists a polynomial q of degree $\leqslant n$ for which $\varepsilon_j q^{(k_j)}(x) > 0$, $a \leqslant x \leqslant b$. For the proof we can take, with some $\alpha < a$,

$$q(x) = c_p(x-\alpha)^{k_p} + c_{p-1}(x-\alpha)^{k_{p-1}} + \cdots + c_1(x-\alpha)^{k_1} \quad (9.2.1)$$

where $c_p = \varepsilon_p$ and the sign of c_j is ε_j, $j = 1,\ldots,p-1$. We may choose $|c_{p-1}|,\ldots,|c_1|$ inductively to be sufficiently large so that the corresponding derivatives have the desired sign.

The purpose of this section is to obtain a characterization of polynomials of best approximation from $\mathcal{P}_n^*$. The following subsets of $[a,b]$, associated with $f \in C[a,b]$ and $P \in \mathcal{P}_n^*$, will be needed:

$$A = A(f,P) = \{x : |f(x) - P(x)| = \|f - P\|\}, \quad (9.2.2)$$

$$B_j = B_j(P) = \{x : P^{(k_j)}(x) = 0\}, \quad j = 1,\ldots,p. \quad (9.2.3)$$

These sets are extremal for P; on the set A, $|f(x) - P(x)|$ attains its maximum, on the set B_j the function $\varepsilon_j P^{(k_j)}(x)$ attains its minimum. The sets A, B_j may be finite or infinite; if a set B_j is infinite, then it is identical with $[a, b]$. In addition, we have

$$P^{(k_j+1)}(y) = 0 \qquad \text{if} \quad a < y < b, \quad y \in B_j. \qquad (9.2.4)$$

In fact, $\varepsilon_j P^{(k_j)}$ has a local minimum at y. Another useful remark is as follows:

If $k_{j+1} = k_j + 1$ and if the degree of $P \in \mathscr{P}_n^*$ is at least k_j, then $B_j(P)$ contains at most one point, which is either a or b. (9.2.5)

Indeed, since $\varepsilon_{j+1} P^{(k_j+1)} \geq 0$, $P^{(k_j)}$ is monotone, and not identically 0. Thus, $P^{(k_j)}$ can have at most one zero in $[a, b]$, and since the derivative has a constant sign, the zero must be a or b.

For unrestricted approximation of a function $f \in C[a, b]$ by polynomials of degree $\leq n$, there exist two characterizations of polynomials P of best approximation (see, e.g., [I, pp. 18 and 30]), due to Chebyshev and to Kolmogorov. The theorem of Chebyshev gives a very concrete necessary and sufficient condition for P, namely, that $f(x) - P(x)$ should attain the maximum of its modulus with alternating sign in some $n + 2$ points. The conditions of Kolmogorov's theorem are much less concrete. What is then the justification of Kolmogorov's theorem? It lies in its adaptability to different situations. Thus, for the monotone approximation, we begin with a theorem of Kolmogorov's type.

Theorem 9.1 [144, 106]. *Let* $f \in C[a, b]$, *let* $P \in \mathscr{P}_n^* = \mathscr{P}_n(k_1, \ldots, k_p; \varepsilon_1, \ldots, \varepsilon_p)$ *and* $f \neq P$. *Then* P *is a polynomial of best approximation for* f *in* $\mathscr{P}_n^*$ *if and only if there is no polynomial* Q *of degree* $\leq n$ *satisfying*

$$[f(x) - P(x)]Q(x) < 0, \qquad x \in A(f, P), \qquad (9.2.6)$$

$$\varepsilon_j Q^{(k_j)}(y) \leq 0, \qquad y \in B_j(P), \quad j = 1, \ldots, p, \qquad (9.2.7)$$

and, equivalently, if and only if there is no Q *satisfying* (9.2.6) *and*

$$\varepsilon_j Q^{(k_j)}(y) < 0, \qquad y \in B_j(P), \quad j = 1, \ldots, p. \qquad (9.2.8)$$

Proof. First, the two sets of conditions are equivalent. We have only to show that if there is a polynomial Q satisfying (9.2.6) and (9.2.7), then there exists also a polynomial Q_1 of degree $\leq n$ that satisfies (9.2.6) and (9.2.8). But it is clear that a possible choice is $Q_1 = Q - \delta q$, where $\delta > 0$ is sufficiently small and q is given by (9.2.1).

The conditions of the theorem are necessary. Assume that there exists a polynomial Q satisfying (9.2.6), (9.2.8). We show that P is not a polynomial of best approximation for f. In fact, if $P_1 = P - \lambda Q$, where $\lambda > 0$ is sufficiently small, then the usual argument [I, p. 19] shows that $\|f - P\| > \|f - P_1\|$. It remains to show that $P_1 \in \mathscr{P}_n^*$ for all small $\lambda > 0$.

We can find open neighborhoods $G_j \supset B_j$ of the sets B_j, $j = 1, \ldots, p$, for which

$$\varepsilon_j Q^{(k_j)}(x) < 0, \qquad x \in G_j.$$

From the definition of the sets B_j follows the existence of an $\alpha > 0$ such that

$$\varepsilon_j P^{(k_j)}(x) \geqslant \alpha, \qquad x \notin G_j, \quad j = 1, \ldots, p.$$

Thus, if $x \in [a, b] \backslash G_j$, then

$$\varepsilon_j P_1^{(k_j)}(x) \geqslant \alpha - \lambda |Q^{(k_j)}(x)| \geqslant \alpha - \lambda M$$

where $M = \max_j \|Q^{(k_j)}\|$. On the other hand, for $x \in G_j$,

$$\varepsilon_j P_1^{(k_j)}(x) \geqslant -\varepsilon_j \lambda Q^{(k_j)}(x).$$

If $\lambda < \alpha M^{-1}$, the last two inequalities yield the desired result.

The conditions are sufficient. Assume that P is not a polynomial of best approximation. We shall show that there is a Q of degree $\leqslant n$ that satisfies both (9.2.6) and (9.2.7). Let $P_1 \in \mathcal{P}_n^*$ be a polynomial of best approximation for f. For $x \in A(f, P)$, $P_1(x)$ is closer to $f(x)$ than is $P(x)$; the signs of the differences $P(x) - f(x)$ and $P(x) - P_1(x)$ are the same. Thus, for $Q_1 = P - P_1$ we have

$$[f(x) - P(x)]Q_1(x) < 0, \qquad x \in A(f, P),$$

and moreover,

$$\varepsilon_j Q_1^{(k_j)}(y) = -\varepsilon_j P_1^{(k_j)}(y) \leqslant 0, \qquad y \in B_j(P), \quad j = 1, \ldots, p \qquad \square$$

Theorem 9.1 and Theorem 9.2 below will be used to prove the uniqueness theorem in §9.4. There are other forms of these theorems. In particular, it may be desirable to reduce the number of points x, y that appear in conditions (9.2.6), (9.2.7). Using the well-known technique of separation of convex sets (see, e.g., Cheney [C]), we can replace these conditions by certain equations, which contain only $m \leqslant n + 2$ points of $[a, b]$ taken from the sets A, B_j, $j = 1, \ldots, p$. Unlike the Chebyshev case, inequality $m < n + 2$ is sometimes possible. See Lorentz and Zeller [106] for details and some examples.

Exercise. The function $f(x) = x^{2n+1}$, $n = 0, 1, 2, \ldots$, and each of its even-order derivatives is monotone increasing on $[-1, +1]$. The polynomial of best (unrestricted) approximation to f of degree $\leqslant 2n$ is

$$P_{2n}(x) = x^{2n+1} - 2^{-2n} C_{2n+1}(x)$$

where C_k is the Chebyshev polynomial of degree k. Show that all of the derivatives $P_{2n}^{(k)}$ change sign in $[-1, +1]$ if $k \leqslant n/\pi$. [*Hint:* Setting $x = \cos t$,

$C_k(x) = \cos kt$, we have

$$P'_{2n}(x) = (2n+1)\left[\cos^{2n}t - 2^{-2n}\frac{\sin(2n+1)t}{\sin t}\right].$$

The second term in the square brackets is larger than the first at $t = (1 + 2k/2n + 1)\pi/2$ if $k\pi/(2n+1) \leqslant 1/2$.]

§9.3. APPROXIMATION WITH RESTRICTIONS FOR DERIVATIVES

The ideas introduced in the study of monotone approximation have wider application, as was noted by Roulier and Taylor [145] and Chalmers [15]. Let $0 \leqslant k_1 < \cdots < k_p \leqslant n$ be integers as before except that now we allow the possibility $k_1 = 0$. We wish to approximate a given function f by polynomials having restricted ranges of their derivatives given by (9.1.2). For the bounding functions l_j and u_j, it is necessary to make some assumptions. We assume that either $u_j \equiv +\infty$ or that $u_j(x)$ is finite on $[a, b]$ and differentiable for $a < x < b$; similarly, that either $l_j \equiv -\infty$ or l_j is finite and differentiable on (a, b); further that $l_j(x) < u_j(x)$ on $[a, b]$. We exclude the case when $l_j \equiv -\infty$, $u_j \equiv +\infty$. Sometimes we shall also require that

$$l_1(x) \leqslant f(x) \leqslant u_1(x) \qquad \text{if} \quad k_1 = 0. \tag{9.3.1}$$

Let $\mathcal{P}_n^* = \mathcal{P}_n(k_j, u_j, l_j;\ j = 1, \ldots, p)$ consist of all polynomials of degree at most n that satisfy (9.1.2). This time it is necessary to assume that $\mathcal{P}_n^*$ is nonempty (this is always true for monotone approximation). With given f and $P \in \mathcal{P}_n^*$, we associate the set $A(f, P)$ of (9.2.2) and the sets

$$B_u^j = B_u^j(P) = \left\{x:\ P^{(k_j)}(x) = u_j(x)\right\}, \tag{9.3.2}$$

$$B_l^j = B_l^j(P) = \left\{x:\ P^{(k_j)}(x) = l_j(x)\right\}, \qquad j = 1, \ldots, p. \tag{9.3.3}$$

A difference with the simpler monotone case is that a direct analogue of Theorem 9.1 is not valid without additional assumptions (see the exercise at the end of this section). However, we can prove

Theorem 9.2 [145]. *Let $f \in C[a, b]$ and $P \in \mathcal{P}_n^*$. Then P is a best approximation for f from $\mathcal{P}_n^*$ if and only if*

$$\max_{x \in A}\left[f(x) - P(x)\right]Q(x) \geqslant 0 \tag{9.3.4}$$

for each polynomial Q of degree $\leqslant n$ that satisfies

$$l_j(x) \leqslant P^{(k_j)}(x) - Q^{(k_j)}(x) \leqslant u_j(x), \qquad j = 1, \ldots, p, \quad x \in [a, b]. \tag{9.3.5}$$

Proof. Since $P^{(k_j)}(x) - \lambda Q^{(k_j)}(x)$, $0 < \lambda < 1$, is always between $P^{(k_j)}(x)$ and $P^{(k_j)}(x) - Q^{(k_j)}(x)$, $j = 1, \ldots, p$, $P_\lambda = P - \lambda Q$ belongs to $\mathcal{P}_n^*$ if

(9.3.5) holds. If (9.3.4) does not hold for Q, then as in Theorem 9.1, for small λ, we have $\|f - P\| > \|f - P_\lambda\|$. On the other hand, if $P_1 \in \mathcal{P}_n^*$ is a polynomial of best approximation and $\|f - P\| > \|f - P_1\|$, then for $Q = P - P_1$, the conditions (9.3.5) hold but inequality (9.3.4) does not. $\square$

At first we must allow the possibility of several polynomials of best approximation in $\mathcal{P}_n^*$. Let $\mathcal{B}$ be the set of all $P \in \mathcal{P}_n^*$ of best approximation to a given continuous function $f \in C[a, b]$. Since uniformly convergent sequences of polynomials of degree $\leqslant n$ can be termwise differentiated, $\mathcal{B}$ is closed in $C[a, b]$. Clearly, $\mathcal{B}$ is convex and the polynomials in $\mathcal{B}$ have uniformly bounded norms.

An essential step toward the uniqueness theorem is the existence of "minimal polynomials" [106]. They exist in every compact convex subset $\mathcal{B}$ of $\mathcal{P}_n^*$.

We consider the sets $A(f, P), B_u^j(P), B_l^j(P)$ for a fixed f and different $P \in \mathcal{B}$. A polynomial $P_0 \in \mathcal{B}$ will be called *minimal* if for each $P \in \mathcal{B}$

$$A(f, P_0) \subset A(f, P), \tag{9.3.6}$$

$$B_u^j(P_0) \subset B_u^j(P), \quad B_l^j(P_0) \subset B_l^j(P), \quad j = 1, \dots, p, \tag{9.3.7}$$

$$P_0(x) = P(x) \quad \text{for} \quad x \in A(f, P_0), \tag{9.3.8}$$

and the degree of P does not exceed the degree of P_0.

Theorem 9.3. *Let $\mathcal{B}$ be a compact convex subset of $\mathcal{P}_n^*$. For each $f \in C[a, b]$ satisfying (9.3.1), there exists a minimal polynomial in $\mathcal{B}$.*

Proof. Let

$$A = \bigcap_{P \in \mathcal{B}} A(f, P); \quad B_u^j = \bigcap_{P \in \mathcal{B}} B_u^j(P), \quad B_l^j = \bigcap_{P \in \mathcal{B}} B_l^j(P), \quad j = 1, \dots, p.$$
$$\tag{9.3.9}$$

Each open covering of the separable metric space $[a, b]$ contains a countable subcovering. Dually, each intersection of a family of closed sets is an intersection of a countable subfamily. Consequently, each of the intersections (9.3.9) can be written as a countable intersection. One can do this even with the same sequence of polynomials $P_r \in \mathcal{B}$, $r = 1, 2, \dots$, for each of them:

$$A = \bigcap_{r=1}^{\infty} A(f, P_r), \quad B_u^j = \bigcap_{r=1}^{\infty} B_u^j(P_r), \quad B_l^j = \bigcap_{r=1}^{\infty} B_l^j(P_r)$$
$$\tag{9.3.10}$$

for $j = 1, \dots, p$. Furthermore, if ν is the maximal degree of the polynomials $P \in \mathcal{B}$, we can assume that P_1 is of degree ν.

The polynomial

$$P_0(x) = \sum_{r=1}^{\infty} a_r P_r(x), \quad a_r > 0, \quad \sum_{r=1}^{\infty} a_r = 1$$

belongs to $\mathcal{B}$ since $\mathcal{B}$ is compact and convex. Let $x \in A(f, P_0)$; then

$$E = \|f - P_0\| = |f(x) - P_0(x)| \leqslant \sum_{r=1}^{\infty} a_r |f(x) - P_r(x)| \leqslant E.$$

It follows that we have $|f(x) - P_r(x)| = E$ for each r. Thus, $A(f, P_0) \subset A$. Then necessarily $A(f, P_0) = A$. Similarly, $B_u^j(P_0) = B_u^j$, $B_l^j(P_0) = B_l^j$, $j = 1, \ldots, p$. Moreover, the a_r may be selected so that P_0 is of exact degree ν.

It remains to show that P_0 satisfies (9.3.8). If $P \in \mathcal{B}$ were not equal to P_0 on A, then the polynomial $\frac{1}{2}(P + P_0)$ would belong to $\mathcal{B}$ and would have a strictly smaller A set than P_0; this is impossible. $\qquad\square$

Exercise [145]. Under the assumption that there exists a $P \in \mathcal{P}_n^*$ for which strict inequalities hold in (9.1.2), prove the analogue of Theorem 9.1 for $\mathcal{P}_n(k_j, u_j, l_j; j = 1, \ldots, p)$ where (9.2.7) is replaced by

$$Q^{(k_j)}(x) \geqslant 0, \quad x \in B_u^j(P); \quad Q^{(k_j)}(x) \leqslant 0, \quad x \in B_l^j(P), \quad j = 1, \ldots, p,$$
$$(9.3.11)$$

and (9.2.8) is replaced by (9.3.11) with strict inequalities.

§9.4. THE UNIQUENESS THEOREM

The proof of uniqueness for restricted derivative approximation requires the Atkinson–Sharma theorem from the theory of Birkhoff interpolation. This was first used by G. G. Lorentz and Zeller [106] for monotone approximation with one derivative restricted, and in a more sophisticated way by R. A. Lorentz [109] for the general case of monotone approximation. The proof of the last paper is slightly modified here, as suggested by Roulier and Taylor [145] and Kimchi and Richter-Dyn [79], to include the general case.

Theorem 9.4 (R. A. Lorentz). *Let $f \in C[a, b]$ satisfy* (9.3.1). *Then the polynomial P of best approximation from $\mathcal{P}_n(k_j, u_j, l_j; j = 1, \ldots, p)$ to f is unique.*

The main technical device of the proof is to set up a Birkhoff interpolation matrix that satisfies the Atkinson–Sharma theorem. Let P_0 be a minimal polynomial in $\mathcal{B}$ (the set of all polynomials of best approximation to f), let $P_0 \neq f$, and let A, B_u^j, B_l^j, $j = 1, \ldots, p$, be the sets (9.2.2), (9.3.2), (9.3.3) for P_0. We assume that there exists another polynomial $P \in \mathcal{B}$, $P \neq P_0$. Let $\tilde{\nu}$, $1 \leqslant \tilde{\nu} \leqslant \nu$, be the degree of $\tilde{P} = P_0 - P$. The non-zero polynomials $\tilde{P}$ and $\tilde{P}^{(k_j)}$, $k_j \leqslant \tilde{\nu}$, vanish, respectively, on the sets A and $B^j = B_u^j \cup B_l^j$, $k_j \leqslant \tilde{\nu}$. Consequently, each of the sets has finitely many points, which we denote by $x_i, y_{i,j}$ and we have

$$\tilde{P}(x_i) = 0, \qquad x_i \in A, \tag{9.4.1}$$

$$\tilde{P}^{(k_j)}(y_{i,j}) = 0, \qquad y_{i,j} \in B^j, \quad k_j \leqslant \tilde{\nu}. \tag{9.4.2}$$

In addition [compare (9.2.4)],

$$\tilde{P}^{(k_j+1)}(y_{i,j}) = 0, \qquad y_{i,j} \in B^j \cap (a,b), \quad k_j < \tilde{\nu}. \tag{9.4.3}$$

Indeed, if, for example, u_j is finite at $x = y_{i,j} \in B_u^j$, then the differences $P_0^{(k_j)} - u_j$, $P^{(k_j)} - u_j$ vanish at x, and do not change sign at this point.

The equations obtained are distinct (for an arbitrary polynomial $\tilde{P}$) with the following exceptions. An equation in (9.4.2) may be identical with some equation in (9.4.3) if $k_{j+1} = k_j + 1$. Similarly, an equation in (9.4.1) may also be an equation in (9.4.2) if $k_1 = 0$. (It can be shown that none of these duplications are possible for the special case of monotone approximation.)

The *distinct* equations from (9.4.1)–(9.4.3) define a homogeneous Birkhoff interpolation problem for $\tilde{P}$, with an interpolation matrix E. Because of the possible overlap of (9.4.2) and (9.4.3), E may have odd interior sequences. We form a new matrix E^* from E by omitting the last 1 in any odd non-Hermitian sequence of an interior row of E. Since equations (9.4.2) and (9.4.3) always occur in pairs for interior knots, this can be accomplished by dropping some of the equations (9.4.3) while all equations (9.4.2) remain. We denote by $y_{i,j}^*$ the points in $B^j \cap (a,b)$ corresponding to the equations (9.4.3) which remain after this operation and are not duplicated in (9.4.2). If $|E^*| = N^* + 1$ and $\tilde{\nu} < N^*$, we add to E^* columns of 0's numbered $k = \tilde{\nu} + 1, \ldots, N^*$. Thus, E^* will be a conservative matrix that annihilates the polynomial $\tilde{P} = P_0 - P$.

By the Atkinson–Sharma theorem we will have $\tilde{P} \equiv 0$, and Theorem 9.4 will follow, if we can establish

Lemma 9.5. *The matrix E^* satisfies the Pólya conditions.*

Proof. Since $|f(x) - P_0(x)| = \|f - P_0\|$ for at least one point, the set A is nonempty and $m_0(E^*) \geq 1$. Suppose that the Pólya condition fails for E^* and let $k = k^*$ be the first $k \geq 1$ for which it fails. Note that $k^* \leq \tilde{\nu}$. Then

$$M_{k^*}(E^*) \leq k^*, \qquad M_{k^*-1}(E^*) = k^*$$

and there are only 0's in the k^*th column of E^*.

We consider the matrix E_1^* consisting of columns $k = 0, \ldots, k^* - 1$ of E^*. No maximal sequence of E^* can cross the k^*th column, hence E_1^* is also conservative. The matrix E_1^*, which has k^* ones and satisfies the Pólya conditions, is regular.

This means in particular that there is a polynomial Q_1 of degree $\leq k^* - 1$ that solves the equations

$$Q_1(x_i) = -\operatorname{sign}[f(x_i) - P_0(x_i)], \qquad x_i \in A, \tag{9.4.4}$$

$$Q_1^{(k_j)}(y_{i,j}) = \begin{cases} 1 & \text{if} \quad y_{i,j} \in B_u^j, \quad k_j < k^*, \\ -1 & \text{if} \quad y_{i,j} \in B_l^j, \quad k_j < k^*, \end{cases} \tag{9.4.5}$$

and

$$Q_1^{(k_j+1)}(y_{i,j}^*) = 0, \qquad y_{i,j}^* \in B^j \cap (a,b), \quad k_j + 1 < k^*. \qquad (9.4.6)$$

Here (9.3.1) is needed to show that (9.4.4) and (9.4.5) give the same value for $Q_1(x)$ if $k_1 = 0$ and $x \in A \cap B^1$.

We also have

$$Q_1^{(k)}(x) \equiv 0, \qquad k \geqslant k^*.$$

For each $y_{i,j} \in B^j$, $k_j < k^*$, it is possible to find an open (with respect to $[a,b]$) interval $G_{i,j}$ containing $y_{i,j}$ which is so small that on $G_{i,j}$, $Q_1^{(k)}(y)$ has the sign of $Q_1^{(k_j)}(y_{i,j})$.

Once the $G_{i,j}$ have been selected, we take $\varepsilon > 0$ so small that

$$l_j(x) + \varepsilon \leqslant P_0^{(k_j)}(x) \leqslant u_j(x) - \varepsilon, \qquad x \notin \cup_i G_{i,j}; \qquad (9.4.7)$$

the lower inequality holds on $G_{i,j}$ if $y_{i,j} \in B_u^j$, and the upper inequality if $y_{i,j} \in B_l^j$.

After these preparations, we apply Theorem 9.2, taking $P = P_0$, $Q = \lambda Q_1$. Then we have (9.3.5) for all x, provided that $\lambda \max_j \|Q_1^{(k_j)}\| < \varepsilon$. Indeed, if $x \notin \cup_i G_{i,j}$, we use (9.4.7). If $x \in G_{i,j}$, and for example $y_{i,j} \in B_u^j$, then $P_0^{(k_j)}(x) - \lambda Q_1^{(k_j)}(x) \leqslant P_0^{(k_j)}(x) \leqslant u_j(x)$, and also $P_0^{(k_j)}(x) - \lambda Q_1^{(k_j)}(x) \geqslant l_j(x) + \varepsilon - \varepsilon = l_j(x)$. In addition, (9.3.4) fails because of (9.4.4). By Theorem 9.2, P_0 is not a polynomial of best approximation to f; a contradiction. □

Remark. When a characterization theorem of the type of Theorem 9.1 is available (see the exercise at the end of §9.3) and $k_1 > 0$, then (9.4.5) may be replaced by homogeneous conditions. This simplifies the last steps of the proof.

Exercise (Schmidt [149]). For $f \in C[a,b]$, let $T_n(f)$ denote the polynomial of best (monotone) approximation from $\mathscr{P}(k_1, \ldots, k_p; \varepsilon_1, \ldots, \varepsilon_p)$. The operator T_n is continuous at each f; more explicitly, $\|g_m - f\| \to 0$ implies $\|T_n(g_m) - T_n(f)\| \to 0$. [*Hint:* Establish the two inequalities $\|T_n(g_m)\| \leqslant 2\|g_m\|$ and $\|\|g_m - T_n(g_m)\| - \|f - T_n(f)\|\| \leqslant \|g_m - f\|$.]

§9.5. SIMULTANEOUS APPROXIMATION

Many authors (see, e.g., Chalmers [16] and R. A. Lorentz [110]) have studied simultaneous approximation of a function and of its derivatives by a polynomial P and by its derivatives. This is equivalent to the approximation of a function $f \in C^k[a,b]$ by polynomials of degree $\leqslant n$ in the norm

$$\||g|\| = \max_{i=1,\ldots,p} \max_{a \leqslant x \leqslant b} |g^{(k_i)}(x)|, \qquad (9.5.1)$$

where the integers $0 = k_1 < k_2 < \cdots < k_p \leqslant k$ are given. That a best approximation to f in $\|\|\cdot\|\|$ norm exists among polynomials of degree n follows from compactness arguments. Let $E(f, \|\|\cdot\|\|) = E^*(f) \geqslant 0$ be the best approximation of a given function $f \in C^k[a, b]$ in this norm. Let $\mathscr{B}^* = \mathscr{B}^*(f)$ be the set of all polynomials of degree at most n of best approximation to f in the norm $\|\|\cdot\|\|$.

Another description of the set $\mathscr{B}^*(f)$ is the following: It is the collection of all polynomials P of degree $\leqslant n$ that satisfy the constraints

$$f^{(k_j)}(x) - E^*(f) \leqslant P^{(k_j)}(x) \leqslant f^{(k_j)}(x) + E^*(f), \qquad j = 1, \ldots, p. \quad (9.5.2)$$

We shall assume in this section that $f \in C^{k_p+1}[a, b]$. Then

$$u_j(x) = f^{(k_j)}(x) + E^*(f), \quad l_j(x) = f^{(k_j)}(x) - E^*(f), \qquad j = 1, \ldots, p \quad (9.5.3)$$

are admissible bounding functions of §9.3. They also satisfy (9.3.1). The sets $B^j = B^j_u \cup B^j_l$ now become

$$B^j(f, P) = \{x : |f^{(k_j)}(x) - P^{(k_j)}(x)| = E^*(f)\}, \qquad j = 1, \ldots, p. \quad (9.5.4)$$

It is easy to find examples when the set $\mathscr{B}^*(f)$ consists of more than one polynomial. Then there arises the question about the dimension of this set. We shall use the Theorem 9.4 and the notion of minimal polynomials to obtain the following result of R. A. Lorentz [110].

Theorem 9.6. *If $f \in C^{(k_p+1)}[a, b]$ is not a polynomial of degree $\leqslant k_p$, then the dimension q of $\mathscr{B}^*(f)$ is the largest k_s with the property that there exist polynomials $P \in \mathscr{B}^*(f)$ for which*

$$\|f^{(k_j)} - P^{(k_j)}\| < E^*(f), \qquad j = 1, \ldots, s - 1. \quad (9.5.5)$$

Proof. We shall call $P_0 \in \mathscr{B}^*(f)$ a *minimal polynomial* of $\mathscr{B}^*(f)$ if for any $P \in \mathscr{B}^*(f)$

$$B^j(f, P_0) \subset B^j(f, P), \qquad j = 1, \ldots, p, \quad (9.5.6)$$

and the degree of P does not exceed the degree of P_0. Theorem 9.3 holds for $\mathscr{B}^*(f)$ and this definition of minimal polynomials.

For any $P \in \mathscr{B}^*(f)$, we have $B^j(f, P) = \varnothing$ exactly when $\|f^{(k_j)} - P^{(k_j)}\| < E^*(f)$. For minimal polynomials P_0, the sets $B^j(f, P_0)$ are uniquely defined. Therefore (9.5.5) is equivalent to the statements

$$B^j(f, P_0) = \varnothing, \quad j = 1, \ldots, s - 1, \qquad B^s(f, P_0) \neq \varnothing. \quad (9.5.7)$$

Assume first that (9.5.5) is valid; we estimate the dimension of $\mathscr{B}^*(f)$ from above. Let Q_1 be the polynomial of best uniform approximation to $f^{(k_s)}$ by polynomials Q of degree $\leqslant n - k_s$ that satisfy the constraints

$$f^{(k_{s+j})}(x) - E^*(f) \leqslant Q^{(k_{s+j}-k_s)}(x) \leqslant f^{(k_{s+j})}(x) + E^*(f), \qquad j = 0, \ldots, p - s \quad (9.5.8)$$

By Theorem 9.4, Q_1 is unique.

Let $P \in \mathscr{B}^*(f)$. We wish to prove that $P^{(k_s)} = Q_1$, from which it would follow that $\dim \mathscr{B}^*(f) \leqslant k_s$. For each $P \in \mathscr{B}^*(f)$, by (9.5.6) and (9.5.7),

$$\|f^{(k_s)} - P^{(k_s)}\| = E^*(f).$$

If $P^{(k_s)}$ is not equal to Q_1, then $P^{(k_s)}$ is not a polynomial of best approximation to $f^{(k_s)}$ for the problem (9.5.8); consequently

$$\|f^{(k_s)} - Q_1\| < E^*(f). \tag{9.5.9}$$

Now let $\tilde{P}$ be any polynomial with $\tilde{P}^{(k_s)} = Q_1$. Because of (9.5.8), $\tilde{P}$ satisfies inequality (9.5.2) for $j \geqslant s$. On the other hand, a minimal polynomial P_0 satisfies (9.5.2) for $j = 1, \ldots, p$, and with *strict* inequalities (9.5.2) for $j < s$. Therefore, if $\varepsilon > 0$ is sufficiently small, the polynomial

$$P_\varepsilon = P_0 + \varepsilon(\tilde{P} - P_0) = (1 - \varepsilon)P_0 + \varepsilon\tilde{P}$$

will satisfy (9.5.2) for $j = 1, \ldots, p$. Hence $P_\varepsilon \in \mathscr{B}^*(f)$. It follows that

$$E^*(f) = \|f^{(k_s)} - P_\varepsilon^{(k_s)}\| \leqslant (1 - \varepsilon)\|f^{(k_s)} - P_0^{(k_s)}\| + \varepsilon\|f^{(k_s)} - Q_1\| < E^*(f).$$

This contradiction proves our assertion.

On the other hand, if (9.5.5) holds for some polynomial P_0 of degree $\leqslant k_p$ and if Q is an arbitrary polynomial of degree $\leqslant k_s - 1$, then for $P = P_0 + \varepsilon Q$, and all sufficiently small ε, P satisfies $\|f^{(k_j)} - P^{(k_j)}\| \leqslant E^*(f)$, $j = 1, \ldots, p$. Hence $\dim \mathscr{B}^*(f) \geqslant k_s$. □

§9.6. CHEBYSHEV SYSTEMS ON $[-1, +1]$

As a simple application of Theorem 6.5, we can decide when a system of powers

$$\mathcal{G} = \{x^{k_0}, x^{k_1}, \ldots, x^{k_p}\}, \qquad 0 \leqslant k_0 < \cdots < k_p, \tag{9.6.1}$$

k_j integers, forms a Chebyshev system on **R** (or, equivalently, on $[-1, +1]$). The following two theorems were proved independently by Passow [129] and Lorentz [91].

Theorem 9.7. *The system (9.6.1) is a Chebyshev system on the set $A = (-\infty, 0) \cup (0, +\infty)$ if and only if all differences $k_j - k_{j-1}, j = 1, \ldots, p$, are odd.*

Proof. Let $n = k_p$ and let q be a fixed integer $1 \leqslant q \leqslant p$. We construct a $(p + 2) \times (n + 1)$ Pólya matrix E with 1's only in column 0 or in row $q + 1$ as follows: Column 0 will be all 1's except possibly in the $(q + 1)$st row; the $(q + 1)$st row has 0's exactly in positions $k_0, k_1, \ldots, k_p$.

A polynomial P of degree $\leqslant n$ satisfies the homogeneous conditions of row $q + 1$, with $x_{q+1} = 0$, if and only if it is of the form

$$P(x) = \sum_{j=0}^{p} a_j x^{k_j}. \tag{9.6.2}$$

This P is annihilated by E and some set of knots X with $x_{q+1} = 0$ if and only if it has q distinct zeros to the left of 0 and $p + 1 - q$ zeros to the right of 0.

Each such P is identically 0 if and only if the matrix E is regular. According to Theorem 6.5, this is the case if and only if all differences $k_j - k_{j-1}$ are odd.

The condition obtained in this way is the same for all q, $1 \leqslant q \leqslant p - 1$. But if $q = 0$, or $q = p + 1$, there is no condition, the corresponding matrices being regular. This implies the theorem. $\square$

For the whole line **R**, we have the additional necessary condition that $k_0 = 0$, for otherwise all polynomials (9.6.2) will vanish at 0.

Theorem 9.8. *The system* (9.6.1) *is a Chebyshev system on* **R** (*or on* $[-1, +1]$) *if and only if* $k_0 = 0$ *and all differences* $k_j - k_{j-1}, j = 1, \ldots, p$, *are odd.*

Proof. We have already observed the necessity of the condition. Suppose that the conditions hold and P has $p + 1$ distinct zeros on **R**. If $a_0 \neq 0$ in (9.6.2), then all zeros are in A and $P \equiv 0$ by Theorem 9.7. If $a_0 = 0$, then P is a polynomial in $x^{k_1}, \ldots, x^{k_p}$ that has at least p zeros in A and Theorem 9.7 applies again. $\square$

The same technique can be used to characterize the Chebyshev systems of order r among all systems (9.6.1). A system $\mathcal{G}$ is a Chebyshev system of order r if the only polynomial P in $\mathcal{G}$ that has $p + 1$ zeros counting multiplicities up to order r is $P \equiv 0$. For the case in which $r = n + 1$, the following result is due to Passow [132].

Corollary 9.9. *The system* (9.6.1) *is a Chebyshev system of order r on* **R** (*or on* $[-1, +1]$) *if and only if* $k_j = j$, $j = 0, \ldots, r - 1$, *and all differences* $k_j - k_{j-1}, j = 0, \ldots, p$, *are odd.*

The proof is left as an exercise.

§9.7. NOTES

9.7.1. The material in §§9.1–9.5 is only a small part of the vast literature on uniform approximation with constraints. This whole area has been the subject of a recent survey article by Chalmers and Taylor [19], to which we refer the reader for more details concerning specific results and as a good source for the literature.

9.7.2. Questions concerning the degree of approximation $E_n^*(f)$ to a monotone function f by monotone polynomials were discussed in the papers of Lorentz and Zeller [105, 106], Lorentz [89], Beatson [6], and Švedov [175].

They prove theorems of Jackson type for $E_n^*(f)$. A strong result is that of DeVore [25],

$$E_n^*(f) = \mathcal{O}\left(n^{-m}\omega\left(f^{(m)}, 1/n\right)\right)$$

for an increasing function $f \in C^m$ and increasing polynomials. On the other hand [105],

$$\limsup_{n \to \infty} \frac{E_n^*(f)}{E_n(f)} = +\infty$$

for some $f \in C^m$, $f^{(m)} \geq 0$, and for approximation by polynomials with $P^{(m)} \geq 0$.

9.7.3. Chalmers [15] has abstracted the essential features of the uniqueness results in 9.2–9.4 to produce a general framework that includes these as examples. Let V be an n-dimensional subspace of the space $C(E)$ of continuous real-valued functions on a compact subset E of the real line. Let L_α, $\alpha \in A$, be a set of linear functionals defined on V, and set

$$V_0 = \{P \in V: l_\alpha \leq L_\alpha P \leq u_\alpha, \alpha \in A\}$$

where l_α, u_α are extended real-valued functions on A. If the set of functionals L_α has what Chalmers calls a "maximal extremal extension," he is able to prove the uniqueness of best approximation from V_0. Chalmers [17] has also developed a Remez exchange algorithm for finding the best approximation, applicable in many cases when his abstract method works.

9.7.4. Although the polynomial of best monotone approximation is unique, Fletcher and Roulier [43] have shown that it is not necessarily strongly unique; that is, there does not exist a constant $\gamma = \gamma(f) > 0$ such that

$$\|f - P\| \geq \|f - T_n(f)\| + \gamma\|T_n(f) - P\|^\alpha, \quad \alpha = 1, \qquad (9.7.1)$$

for all $P \in \mathscr{P}_n^*$, where $T_n(f)$ is the polynomial of best monotone approximation. Schmidt [149] has shown that (9.7.1) is valid with $\alpha = 2$ for polynomials $P \in \mathscr{P}_n(k_1, \ldots, k_p; \varepsilon_1, \ldots, \varepsilon_p)$ of bounded norm. From this he obtains that the best approximation operator satisfies a Lipschitz condition of order $\frac{1}{2}$: There is a constant $\lambda(K)$ with the property that $f, g \in C[a, b]$, $\|g\| \leq K$, imply, whenever the degree of $T_n(f)$ is $\geq k_p$,

$$\|T_n(f) - T_n(g)\| \leq \lambda(K)\|g - f\|^{1/2}.$$

Examples of Fletcher and Roulier show that Schmidt's results are best possible.

9.7.5. There have been several papers dealing with mixed interpolation and approximation. For the case of Birkhoff interpolatory conditions, we mention the papers of Kimchi and Richter–Dyn [77–79]. They treat questions of the Chebyshev–Kolmogorov type: characterization, uniqueness, and alternation theorems. The interpolation matrix involved is assumed to remain regular even if a certain number of 1's in Lagrange or Hermite positions are added. In particular, matrices with only even non-Hermitian sequences satisfy this requirement.

9.7.6. Keener [75] and R. A. Lorentz [199] have considered the uniqueness (or nonuniqueness) of best simultaneous approximation from $(n + 1)$-dimensional subspaces other than algebraic polynomials. For example [199], Theorem 9.6 remains true if the approximation is by polynomials from a Birkhoff system $\mathcal{G} = \{g_0, \ldots, g_n\}$ for which the span $\{g_0^{(k_p)}, \ldots, g_n^{(k_p)}\}$ has dimension $n + 1 - k_p$.

Chapter 10

Birkhoff Quadrature Formulas

§10.1. DEFINITIONS; q-REGULAR PAIRS E, X

Quadrature formulas that give the value of an integral $\int_a^b f\,dg$ in terms of values of the function f and of its derivatives will be treated for the general case of the Birkhoff data $f^{(k)}(x_i)$. We want to find formulas that are exact for functions f belonging to a fixed finite system $\mathcal{G}$. In this chapter the system $\mathcal{G}$ will be the polynomials $\mathcal{P}_n$. The exposition in the first three sections of the chapter will follow the paper of Lorentz and Riemenschneider [101] (see also Stieglitz [171]); the next sections are based on work of Dyn [32], Micchelli and Rivlin [126], and others.

Without loss of generality we shall assume in what follows that $a = 0$, $b = 1$. The measure dg will always be a *positive regular Borel measure*, so that $\int_0^1 f\,dg$ is a Stieltjes integral with monotone increasing function $g(x)$.

Let $E = [e_{i,k}]$ be an $m \times (n+1)$ matrix with zeros and ones, with $|E| = N+1$ ones. For a given measure dg and a given set of knots X: $0 \leqslant x_1 < \cdots < x_m \leqslant 1$, we would like to answer the following question: do there exist constants $c_{i,k}$ defined if $e_{i,k} = 1$, *with the property that*

$$\int_0^1 f\,dg = \sum_{e_{i,k}=1} c_{i,k} f^{(k)}(x_i) \tag{10.1.1}$$

holds for all $f \in \mathcal{P}_n$? We then call (10.1.1) a *Birkhoff quadrature formula*.

The pair E, X will be called *quadrature regular* (or *q-regular*) with respect to dg, $\mathcal{P}_n$, if a formula of this type exists.

A standard way to obtain a quadrature formula is to integrate a corresponding interpolation formula. Let E, X be a regular pair for Birkhoff interpolation. This interpolation reproduces each polynomial $P \in \mathcal{P}_n$. If $p_{i,k}$ (defined for pairs i, k with $e_{i,k} = 1$) are the fundamental interpolation polynomials, then $P = \sum_{e_{i,k}=1} P^{(k)}(x_i) p_{i,k}$. Integrating this and putting

$$c_{i,k} = \int_0^1 p_{i,k} \, dg, \qquad e_{i,k} = 1, \tag{10.1.2}$$

we get the quadrature formula with $N = n$,

$$\int_0^1 P \, dg = \sum_{e_{i,k}=1} c_{i,k} P^{(k)}(x_i), \qquad P \in \mathcal{P}_n.$$

This is the *interpolatory quadrature formula* for the pair E, X.

Exercise (Dyn). If in an interpolatory formula, $x_0 = 0$, $x_m = 1$, all interior Hermite sequences are odd, and all interior non-Hermite sequences are even, then $c_{i,k} > 0$ whenever $e_{i,k} = 1, k$ even, belongs to an interior Hermite sequence.

Among the quadrature formulas not of this type, the most important are the *Gaussian formulas*, for which $|E|$ is smaller than $n + 1$. As Gauss found in 1821, for each integer $N \geqslant 1$ there exist points $0 < x_1 < \cdots < x_N < 1$ and constants $c_1, \ldots, c_N$, $c_k > 0$, $k = 1, \ldots, N$, so that the formula

$$\int_0^1 f \, dx = c_1 f(x_1) + \cdots + c_N f(x_N) \tag{10.1.3}$$

is exact for all polynomials f of degree $\leqslant 2N - 1$. For a given odd n, we can take N as small as $N = (n+1)/2$. Later M. G. Krein in his famous paper [84] extended this to arbitrary positive measures dg and to polynomials in a Chebyshev system. Let $u_0, \ldots, u_n$ be a Chebyshev system on $[0,1]$. We want to establish

$$\int_0^1 f \, dg = \sum_{k=1}^p c_k f(x_k), \qquad c_k > 0, \quad k = 1, \ldots, p, \tag{10.1.4}$$

valid for all f of the form $f = \sum_0^n a_i u_i$.

We define the *index I* of formula (10.1.4) or of the set of points $x_1, \ldots, x_p$ to be the sum of weights of all points x_k appearing in it. To a point x_k we assign weight 1 if $0 < x_k < 1$, and weight $\frac{1}{2}$ if x_k is 0 or 1. Thus the index I of (10.1.4) could be p, $p - \frac{1}{2}$, or $p - 1$.

Theorem 10.1 (M. G. Krein). *For each measure dg there exist quadrature formulas with $I \leqslant \frac{1}{2}(n+1)$. For all nonexceptional measures, $I \geqslant \frac{1}{2}(n+1)$ for each such formula. Moreover, for nonexceptional measures there exist exactly two formulas with $I = \frac{1}{2}(n+1)$, one of them with $x_1 = 0$, the other with $x_1 > 0$.*

We explain this in more detail, using the dimension $n_1 = n + 1$ of the space spanned by the Chebyshev system. Measures that are exceptional in this theorem are discrete measures supported on a set of points of index $I < \frac{1}{2}n_1$. For odd n_1 there exist two minimal formulas (10.1.4) with $p = \frac{1}{2}(n_1 + 1)$ and with points $0 = x_1 < \cdots < x_p < 1$ (Case I) and with $0 < x_1^* < \cdots < x_p^* = 1$ (Case II), respectively. For even n_1 there exist two formulas with minimal I, one of them with knots $0 = x_1 < x_2 < \cdots < x_{p+1} = 1$ (Case III), the other with $0 < x_1^* < \cdots < x_p^* < 1$ (Case IV), $p = \frac{1}{2}n_1$. The knots x_k, x_k^* interlace for both groups of formulas. For a nice exposition of this theorem see [H, pp. 37–51].

Remark. From the unicity in Theorem 10.1 it follows for nonexceptional measures that the quadrature formulas mentioned there are *continuous*. This means the following. Let $\mathcal{G}_j$: $u_{0, j}, \ldots, u_{n, j}, j = 1, 2, \ldots$, be Chebyshev systems on $[0, 1]$, which converge to the Chebyshev system $\mathcal{G}$: $u_0, \ldots, u_n$, in the sense that $\|u_{k, j} - u_k\| \to 0$ for $j \to \infty$, $k = 0, \ldots, n$. Then the quadrature formulas of one of the types mentioned in the theorem

$$\int_0^1 f \, dg = \sum_{j=1}^p c_{k, j} f(x_{k, j}), \qquad c_{k, j} > 0, \tag{10.1.5}$$

converge to the formula (10.1.4) of the same type: $c_{k, j} \to c_k$, $x_{k, j} \to x_k$ for $j \to \infty$, $k = 0, \ldots, n$. For the proof, we first show that the $c_{k, j}$ must be uniformly bounded. This follows from the fact that the $c_{k, j}$ are positive and that the span of $\mathcal{G}$ contains a strictly positive function [H, p. 51]. Then, if the desired convergences do not take place, we would have two different formulas (10.1.4).

We return to quadrature formulas for $\mathcal{P}_n$. Then (10.1.1) is valid for all $f(x) = x^j/j!$, $j = 0, \ldots, n$. This leads to a system of linear equations for the coefficients $c_{i, k}$. Let

$$\mu_j = \int_0^1 \frac{x^j}{j!} \, dg, \qquad j = 0, 1, \ldots,$$

denote the *moments* of the measure dg. The equations are

$$\sum_{e_{i, k} = 1} c_{i, k} \frac{x_i^{j-k}}{(j-k)!} = \mu_j, \qquad j = 0, \ldots, n. \tag{10.1.6}$$

We compare this with the equations for the coefficients a_j of the polynomial $P(x) = \sum_{j=0}^n a_j x^j/j!$, which solves the Birkhoff problem $P^{(k)}(x_i) = \gamma_{i, k}$:

$$\sum_{j=0}^n a_j \frac{x_i^{j-k}}{(j-k)!} = \gamma_{i, k}, \qquad e_{i, k} = 1. \tag{10.1.7}$$

It follows that the matrix of coefficients $c_{i, k}$ in (10.1.6) is the transpose of the $(N+1) \times (n+1)$ matrix $A(E, X)$ of (10.1.7) (see also §1.3). The augmented matrix of the system (10.1.6) is the transpose of

$$B(E, X; dg) = \begin{bmatrix} A(E, X) \\ \mu_0, \mu_1, \ldots, \mu_n \end{bmatrix}. \tag{10.1.8}$$

Hence

Theorem 10.2. (i) *The pair* E, X *is q-regular for* dg *and* $\mathcal{P}_n$ *if and only if*

$$\text{rank } A(E, X) = \text{rank } B(E, X; dg). \tag{10.1.9}$$

(ii) *If the pair* E, X *is regular, it is also q-regular. In this case* $N = n$, *and the quadrature formula* (10.1.1) *is unique. It is the interpolatory formula of the pair.*

Let E_r be the matrix consisting of the first $r + 1$ columns of E. The first $r + 1$ columns of (10.1.8), if 0's corresponding to $e_{i,k} = 1$ with $k > r$ are removed from them, form $B(E_r, X; dg)$. Hence

Corollary 10.3. *If* E, X *is q-regular for* dg *and* $\mathcal{P}_n$, *then* E_r, X *is q-regular for* dg *and* $\mathcal{P}_r$.

Part (ii) of Theorem 10.2 has an inverse: *If the pair* E, X *is q-regular for all measures* dg, *then* E, X *is regular.*

Indeed, for any measure the row of its moments belongs to the span of the rows of $A(E, X)$. Taking dg to be the point measure of mass 1 at t, $0 < t < 1$, we see that all rows $1, t, \ldots, t^n/n!$ belong to the span. Thus, the rows of $A(E, X)$ span $\mathbf{R}^{n+1}$.

Exercise. This inverse remains true if instead of all measures $dg \geqslant 0$ we take only absolutely continuous measures, that is, measures $dg = \omega(x) dx$, $\omega \in L_1$, $\omega(x) \geqslant 0$.

If $dg = dx$, we can replace (10.1.9) by a condition involving only matrices of type $A(E, X)$.

Let X^* be derived from X by adding knots 0 and 1 (if necessary); let E^* be derived from E by adding a new 0th column of 0's and two 1's for the knots 0 and 1 (and possibly one or two new rows). For example, if X is $x_1, \ldots, x_m$ with $0 = x_1 < \cdots < x_m < 1$, then X^* consists of knots $x_1, \ldots, x_m, 1$, and

$$E^* = \begin{bmatrix} 1 & & \\ 0 & & \\ \vdots & & E \\ 0 & & \\ \hline 1 & 0 & \cdots & 0 \end{bmatrix}. \tag{10.1.10}$$

Thus, E^* has m, $m + 1$ or $m + 2$ rows, $n + 2$ columns and $|E^*| = N + 3$ ones.

Corollary 10.4. *For* $dg = dx$ *condition* (10.1.9) *reads*

$$\text{rank } A(E^*, X^*) = \text{rank } A(E, X). \tag{10.1.11}$$

For example, in the case of (10.1.10), we subtract the first row of the $(N+3) \times (n+2)$ matrix $A(E^*, X^*)$ from the last, and obtain

$$
\begin{bmatrix}
1 & 0 & \cdots & & 0 \\
\hline
0 & & & & \\
\vdots & & A(E, X) & & \\
\vdots & & & & \\
0 & & & & \\
\hline
0 & 1/1! & \cdots & & 1/(n+1)!
\end{bmatrix}
=
\begin{bmatrix}
1 & 0 & \cdots & & 0 \\
\hline
0 & & & & \\
\vdots & & & & \\
& & B(E, X; dx) & & \\
0 & & & &
\end{bmatrix}.
$$

The following observations serve to clarify the relations between the different notions of regularity and q-regularity.

Lemma 10.5. *Let* $\lambda_j \neq 0$, $j = 0, \ldots, n$. *Then for any distinct numbers* y_j, $j = 0, \ldots, n$, *the set of* $n+1$ *rows* $(\lambda_0, \lambda_1 y_j, \ldots, \lambda_n y_j^n)$ *spans* $\mathbf{R}^{n+1}$.

Indeed, the determinant of the rows is $\lambda_0 \cdots \lambda_n V$, where V is the Vandermonde determinant of the y_j. Using this, we obtain

Proposition 10.6. *Let* dg *be a measure with moments* $\mu_j \neq 0$, $j = 0, \ldots, n$. *If the pair* E, X *is* q-*regular with respect to* dg *for* (i) *all* X *with* $X \subset (0, 1)$, *or* (ii) *all* $X \subset [0, 1)$ *with* $x_1 = 0$, *or* (iii) *all* $X \subset (0, 1]$ *with* $x_m = 1$, *then* rank $A(E, X) = n+1$ *for all* X. *In particular, if* $N = n$, *then* E *is regular.*

Proof. We consider only the case (ii). Let $X \subset [0, 1)$ with $x_1 = 0$ be arbitrary. Then also the set X' given by $x_i' = \alpha x_i$, $i = 1, \ldots, m$, $0 < \alpha < 1$, belongs to the class (ii). By the assumption and Theorem 10.2 (i), the row $(\mu_0, \ldots, \mu_n)$ is a linear combination of the rows of $A(E, X')$:

$$
\left(\frac{x_i'^{-k}}{(-k)!}, \ldots, \frac{x_i'^{n-k}}{(n-k)!} \right) = \alpha^{-k} \left(\frac{x_i^{-k}}{(-k)!}, \ldots, \alpha^n \frac{x_i^{n-k}}{(n-k)!} \right).
$$

Hence the row $(\mu_0, \alpha^{-1}\mu_1, \ldots, \alpha^{-n}\mu_n)$ is spanned by the rows of $A(E, X)$. In view of Lemma 10.5, $A(E, X)$ spans $\mathbf{R}^{n+1}$ for all X of the class (ii), as required.   □

Here is another case when q-regularity implies regularity:

Lemma 10.7. *Let* E, X *be* q-*regular with respect to the positive measure* dg *which is not supported on any set of three points* $0, x_i, 1$. *If a submatrix* E' *of* E *has only even sequences in each row with* $0 < x_i < 1$, *and if its Pólya function satisfies* $M_k(E') \geq k$, $k = 0, \ldots, n$, *then* E *satisfies the Pólya conditions and* E, X *is complete for* $\mathcal{P}_n$.

(See the definition in §1.5.)

Proof. If E, X were not complete for $\mathcal{P}_n$, then there would exist a nontrivial polynomial $P \in \mathcal{P}_n$ that is annihilated by E, X. By (10.1.1) and the q-regularity of E, X, we have $\int_0^1 P \, dg = 0$. Since the submatrix E'

stipulates an even number of zeros of P at points x_i, $0 < x_i < 1$, $e_{i,0} = 1$, there is a zero or multiplicity of a zero for P not specified by E'. Extending E' to account for this zero, we obtain a conservative Pólya matrix that annihilates P for some knots—a contradiction.   □

§10.2. q-REGULAR MATRICES

The q-regularity of a pair E, X has been defined in §10.1. In view of Proposition 10.6, in the corresponding definition for a matrix we must restrict the set of knots X: An $m \times (n+1)$ matrix E of 0's and 1's is *quadrature regular* (or *q-regular*) for dg and $\mathcal{P}_n$ if each pair E, X with $x_1 = 0$, $x_m = 1$ is regular for dg and $\mathcal{P}_n$; otherwise E is *q-singular*.

Here we allow the first and last row of E to contain only zero entries. There exist q-regular matrices that are not regular.

Example. We consider the matrices

$$\begin{bmatrix} 0 & 1 & 0 & 0 \\ 1 & 0 & 1 & 0 \\ 0 & 1 & 0 & 0 \end{bmatrix}, \quad \begin{bmatrix} 1 & 0 & 1 & 0 & 0 & 0 \\ 1 & 0 & 1 & 0 & 0 & 0 \\ 1 & 0 & 1 & 0 & 0 & 0 \end{bmatrix}, \quad (10.2.1)$$

the measure $dg = dx$, and the set of knots $X = \langle 0, x, 1 \rangle$, $0 < x < 1$. Simple calculations show that E, X is singular only if $x = \frac{1}{2}$, and that (10.1.11) holds for both matrices for $x = \frac{1}{2}$. Thus, pairs E, X are q-regular for all $X = \langle 0, x, 1 \rangle$, $0 < x < 1$.

Corollary 10.3 implies: If E is q-regular for dg and $\mathcal{P}_n$, then E_r is q-regular for dg and $\mathcal{P}_r$ for all $1 \le r \le n$.

It is more difficult to find q-singular than singular matrices among all Pólya matrices, but we have nevertheless:

Proposition 10.8. *If E is an $m \times (n+1)$ Pólya matrix with one odd supported sequence, all other sequences of its interior rows being even, and if dg is not supported on any m points of $[0,1]$, then E is q-singular for dg and $\mathcal{P}_n$.*

Proof. By Theorem 6.2, E, X is singular for at least one X, and by Lemma 10.7, for this X the pair E, X is q-singular.   □

We shall discuss the influence of decomposition $E = E_1 \oplus E_2$ on q-regularity. We obtain only an incomplete analogue of Theorem 1.4.

Lemma 10.9. *Let A be a square matrix of the form*

$$A = \left[\begin{array}{c|c} A_1 & * \\ \hline 0 & A_2 \end{array}\right] \begin{array}{l} \updownarrow r+1 \\ \updownarrow n-r \end{array}$$
$$\underset{r+1}{\longleftrightarrow} \ \underset{n-r}{\longleftrightarrow}$$

where $*$ *stands for arbitrary elements. If the square submatrix* A_1 *has full rank* $r + 1$, *then*

$$\operatorname{rank} A = \operatorname{rank} A_1 + \operatorname{rank} A_2. \tag{10.2.2}$$

Theorem 10.10. *Let* $E = E_1 \oplus E_2$ *be the decomposition of the* $m \times (n + 1)$ *matrix* E, *and let* E_2 *be regular. Then* E *is* q-*regular if and only if* E_1 *is* q-*regular.*

Proof. Necessity follows from Corollary 10.3. To prove sufficiency, let $X: x_1 < \cdots < x_m$ be an arbitrary set of knots for which $x_1 = 0$, $x_m = 1$. Since E_2 is regular, rank $A(E_2, X) = n - r$ if we assume that E_1 has $r + 1$ columns, E_2 has $n - r$ columns. Modulo row interchanges we have the following matrix equations

$$A(E, X) = \begin{bmatrix} A(E_1, X) & \vert & * \\ - - - - - & \llcorner & - - - - \\ 0 & \vert & A(E_2, X) \end{bmatrix},$$

$$B(E, X; dg) = \begin{bmatrix} A(E_1, X) & \vert & * \\ - - - - - & \llcorner & - - - - - \\ 0 & \vert & A(E_2, X) \\ - - - - - & \llcorner & - - - - - \\ \mu_0 \cdots \mu_r & \vert & \mu_{r+1} \cdots \mu_n \end{bmatrix} = \begin{bmatrix} B(E_1, X; dg) & \vert & * \\ - - - - - & \llcorner & - - - - - \\ 0 & \vert & A(E_2, X) \end{bmatrix}.$$

Then E is q-regular because rank $A(E, X) = \operatorname{rank} A(E_1, X) + \operatorname{rank} A(E_2, X) = \operatorname{rank} B(E_1, X; dg) + n - r \geq \operatorname{rank} B(E, X; dg)$. $\square$

If, conversely, $E = E_1 \oplus E_2$, and the matrix E_1 is regular, E_2 is q-regular, it does not necessarily follow that E is q-regular. To obtain an example, we take for E_1 the single column $0, 1, 0$, for E_2 either of the matrices (10.2.1) and apply Proposition 10.8.

§10.3. PÓLYA CONDITIONS; GAUSSIAN FORMULAS

Our main result of this section is

Theorem 10.11. *If the measure* dg *is not supported on the two points* $0, 1$, *then each matrix that is* q-*regular with respect to* dg *and* $\mathcal{P}_n$ *must satisfy the Pólya conditions.*

This does not follow if merely a pair E, X is q-regular for some X because of the existence of Gaussian formulas. For the Lebesgue measure the theorem is due to Stieglitz [171], but the general case is more difficult.

The moment space $\mathfrak{M}_n$ is the set of moment sequences $\mu_0, \ldots, \mu_n$ generated by all positive measures dg on $[0, 1]$. It is the properties of the moments rather than of the measure itself that determine whether a matrix is q-regular or q-singular with respect to dg. Every moment sequence in $\mathfrak{M}_n$

admits a representation

$$\mu_i = \frac{1}{i!} \sum_{j=1}^{n+2} \alpha_j t_j^i, \qquad i = 0, \ldots, n, \qquad (10.3.1)$$

where the t_j are distinct points of $[0, 1]$ and $\alpha_j \geqslant 0$. (See [H, p.39].) Let $\mathfrak{M}_n^0$ denote the subset of $\mathfrak{M}_n$ consisting of moment sequences admitting a representation (10.3.1) with $\alpha_j > 0$ for some $t_j \in (0, 1)$.

Note that for *any* measure $dh \geqslant 0$ satisfying $i!\mu_i = \int_0^1 t^i \, dh(t)$, $i = 0, \ldots, n$, the sequence $\{i!\mu_i\}_{i=0}^n$ is constant exactly when dh has support on the two-point set $\{0, 1\}$. In particular, $\{\mu_i\}_{i=0}^n \notin \mathfrak{M}_n^0$ exactly when dh has this support.

For brevity, we shall say that a measure dg on $[0, 1]$ has the Pólya property of order m if each $m \times (r + 1)$, $r = 0, \ldots, n$, matrix that is q-regular with respect to dg satisfies the Pólya conditions.

Lemma 10.12. *If a measure dg has the Pólya property of order $m = 2$, then it has this property for all m.*

Proof. Let μ_i be the moments of the measure dg. Suppose that E is an $m \times (n + 1)$ matrix, q-regular with respect to dg, which fails to satisfy the Pólya conditions. Let E_r be the first $r + 1$ columns of E such that the Pólya conditions are satisfied for E_{r-1} but not for E_r. By Corollary 10.3, E_r is q-regular. Both E_r and E_{r-1} have r ones. For an arbitrary set X, $0 = x_1 < \cdots < x_m = 1$, the matrix $A(E_r, X)$ has r rows and relation (10.1.9) implies

$$\det \begin{bmatrix} A(E_r, X) \\ \mu_0, \mu_1, \ldots, \mu_r \end{bmatrix} = 0. \qquad (10.3.2)$$

Let $X^* = \{0, 1\}$, and let E^* be a row coalescence of E_r to two rows. Using the formula (3.5.2) for the coalesced determinant, which applies also to determinants with some rows of constants, we get

$$\det \begin{bmatrix} A(E^*, X^*) \\ \mu_0, \mu_1, \ldots, \mu_r \end{bmatrix} = 0. \qquad (10.3.3)$$

On the other hand, the coalescence of the matrix E_{r-1} to two rows is regular; hence rank $A(E^*, X^*) = r$. Since E^* does not satisfy the Pólya conditions, it is not q-regular, and (10.1.9) implies that the determinant (10.3.3) is not 0, a contradiction. $\square$

Proposition 10.13. *The following statements are equivalent for a nonzero measure $dg \geqslant 0$ and $n \geqslant 2$:*

(a) *dg has the Pólya property of order 2;*
(b) *the moment set of dg belongs to $\mathfrak{M}_n^0$;*
(c) *dg is not supported on the two-point set $\{0, 1\}$.*

Proof. Statement (b) implies (a). Assume that the moment set $\{\mu_i\}$ of dg belongs to $\mathfrak{M}_n^0$. Since $dg \neq 0$, $\mu_0 > 0$. Consequently, the Pólya property of order 2 holds for dg and $r = 0$. Let r be the first integer $\leqslant n$ for which the Pólya property of order 2 fails for dg. There exists a $2 \times (r + 1)$ matrix E that is q-regular with respect to dg but does not satisfy the Pólya conditions. By our choice of r and Corollary 10.3, E_{r-1} must contain precisely r ones. With $X = \{0, 1\}$, $A(E, X)$ is an $r \times (r + 1)$ matrix, hence by the q-regularity of E,

$$\det \begin{bmatrix} A(E, X) \\ \mu_0, \ldots, \mu_r \end{bmatrix} = 0. \tag{10.3.4}$$

Let $X_t = \{0, t, 1\}$ and let E' be formed from E by adding an interior row $(1, 0, \ldots, 0)$. Since E' satisfies the Pólya conditions and has no odd supported sequences,

$$\det \begin{bmatrix} A(E, X) \\ 1, \ldots, t^r/r! \end{bmatrix} = (-1)^\sigma \det A(E', X_t) \neq 0, \qquad 0 < t < 1, \tag{10.3.5}$$

where σ is the number of 1's in the second row of E. The right-hand side of (10.3.5) is a continuous function $\phi(t)$ defined for $0 \leqslant t \leqslant 1$; hence it is of constant sign.

Using the representation (10.3.1) of the μ_i, we obtain

$$\det \begin{bmatrix} A(E, X) \\ \mu_0, \ldots, \mu_r \end{bmatrix} = (-1)^\sigma \sum_{j=1}^{n+2} \alpha_j \phi(t_j), \tag{10.3.6}$$

with at least one j for which $\alpha_j > 0$, $0 < t_j < 1$. Thus the determinant (10.3.6) is not 0, and E is not q-regular. Therefore, dg has the Pólya property of order 2.

Not (c) implies not (a). Since in this case the measure dg is supported on $\{0, 1\}$, it is sufficient to note that the $2 \times (n + 1)$ matrix, $n \geqslant 2$,

$$\begin{bmatrix} 1 & 0 & \cdots & 0 \\ 1 & 0 & \cdots & 0 \end{bmatrix}$$

does not satisfy the Pólya conditions but is q-regular with respect to dg.

That (b) is equivalent to (c), we have already seen. $\qquad\square$

Theorem 10.11 follows at once from Lemma 10.12 and Proposition 10.13.

We will discuss different types of matrices and of associated quadrature formulas (10.1.1). In view of Lemma 10.7 and Theorem 10.11, it is natural to assume that the measure dg is supported on more than m points of $[0, 1]$. We can classify Pólya $m \times (n + 1)$ matrices with $|E| = n + 1$ as follows:

   I. *Regular matrices E.* Here for each $X \subset [0, 1]$ there is a unique quadrature formula, which is the interpolatory formula.
   II. *Singular, q-regular matrices E.* If the pair E, X is regular, there is again a unique (interpolatory) formula. If E, X is singular and

$x_1 = 0$, $x_m = 1$, then the pair is q-regular. In this case rank $A(E, X) = n + 1 - d$ with $d > 0$ and the condition (10.1.9) is satisfied. We can omit d entries $e_{i,k} = 1$ from E, obtaining a matrix E_1 for which rank $A(E_1, X) = $ rank $B(E_1, X; dg) = n + 1 - d$. *There is then a Gaussian formula [i.e., a formula (10.1.1) with $N < n$] for $\mathcal{P}_n$, associated with E_1 and X.*

III. *Pólya matrices E that are not q-regular.* For almost all X for which E, X is regular, we still have a unique interpolatory quadrature formula. For some X with $x_1 = 0$, $x_m = 1$, the pair E, X is not q-regular, and there is no corresponding quadrature formula for $\mathcal{P}_n$.

In all three categories, there may appear Gaussian formulas, either because some of the $c_{i,k}$ are 0 in (10.1.1), or because formula (10.1.1) happens to be valid for all polynomials in $\mathcal{P}_{n_1}$, with some $n_1 > n$.

Exercise. Let

$$E = \begin{bmatrix} 1 & 0 & 0 \\ 0 & 1 & 0 \\ 1 & 0 & 0 \end{bmatrix}, \qquad X = \{0, x, 1\}, \qquad 0 < x < 1, \quad dg = dx,$$

(10.3.7)

then the quadrature formula is

$$\int_0^1 f \, dx = Af(0) + Bf'(x) + Cf(1).$$ (10.3.8)

There are values of x for which (a) E, X is not q-regular; (b) (10.3.8) is a Gaussian formula exact for $\mathcal{P}_3$; and (c) for a submatrix of E, (10.3.8) is a Gaussian formula exact for $\mathcal{P}_2$.

A quadrature formula (10.1.1), associated with the pair E, X, valid for polynomials of degree $\leqslant n$, is said to have the *drop*

$$\delta = \delta(E, X) = n - N$$ (10.3.9)

if $N + 1$ (with $N \leqslant n$) is the number of its nonzero terms (i.e., terms with $c_{i,k} \neq 0$).

For example, the drops of the formulas of Theorem 10.1 are $\delta = \frac{1}{2}n$ if n is even, $\delta = \frac{1}{2}(n - 1)$ and $\delta = \frac{1}{2}(n + 1)$ if n is odd. If d is the defect of a singular, q-regular matrix E, then for some X, $\delta(E, X) \geqslant d$ (see II).

We can estimate $\delta(E, X)$ from below for sets of knots X that satisfy the conditions $x_1 = 0$, $x_m = 1$, or satisfy any fixed one of the two conditions, or are not restricted at all. We get several similar statements. For the first case we have

Proposition 10.14. *Assume that for a pair E, X with $x_1 = 0$, $x_m = 1$ there is a quadrature formula for dg and $\mathcal{P}_n$. Let ρ be the smallest number of*

1's that must be added to E to obtain a matrix with only even sequences in its interior rows, then

$$\delta = \delta(E, X) \leqslant \rho. \tag{10.3.10}$$

Moreover, we even have the "delayed Pólya conditions" for E:

$$M_k \geqslant k + 1 - \rho, \qquad k = 0, \ldots, n. \tag{10.3.11}$$

Proof. (a) First let $\rho = 0$. If the Pólya conditions are not satisfied by E, there is an $r > 0$, so that E_{r-1} satisfies these conditions, and E_r does not. Then Corollary 10.3 and Lemma 10.7 provide a contradiction.

(b) For $\rho > 0$, we can add ρ ones to E, obtaining a matrix E^* with no odd sequences in interior rows. Then by (a), E^* satisfies the Pólya conditions. $\qquad\square$

§10.4. KERNELS OF QUADRATURE FORMULAS

Some interesting results for quadrature formulas have been obtained by Schoenberg by a study of their kernels. Certain properties of kernels follow from our theorems of Chapter 7. Later we shall see that results about spline interpolation from Chapters 13 and 14 also help with quadrature formulas (see §14.5). Let (10.1.1) be the quadrature formula associated with E, X, dg; let

$$R(f) = \int_0^1 f\, dg - \sum_{e_{i,k}=1} c_{i,k} f^{(k)}(x_i) \tag{10.4.1}$$

be its remainder. The kernel $K(t)$ of (10.4.1) can be described as follows.

Theorem 10.15. *The function*

$$K(t) = \int_0^1 \frac{(x-t)_+^n}{n!}\, dg(x) - \sum_{e_{i,k}=1} c_{i,k} \frac{(x_i - t)_+^{n-k}}{(n-k)!} \tag{10.4.2}$$

has its support contained in $[0, 1]$ if and only if (10.1.1) is exact for polynomials of degree $\leqslant n$. In this case, K is the kernel of the remainder $R(f)$:

$$R(f) = \int_0^1 K(t) f^{(n+1)}(t)\, dt \tag{10.4.3}$$

for each function $f \in C^n[0, 1]$ with absolutely continuous derivative $f^{(n)}$.

Proof. Obviously, $K(t) = 0$ for $t > 1$. It is 0 for $t < 0$ precisely when the formula (10.1.1) is valid for all functions $(x - c)^n/n!$, $c < 0$, and they span $\mathcal{P}_n$. To establish (10.4.3), we apply the functional R, which is orthogonal to all polynomials in $\mathcal{P}_n$, to the Taylor formula of $f(x)$,

$$f(x) = \sum_{k=0}^n \frac{f^{(k)}(0)}{k!} x^k + \int_0^1 \frac{(x-t)_+^n}{n!} f^{(n+1)}(t)\, dt. \tag{10.4.4}$$

The integral part of $R(f)$ commutes with the integral in (10.4.4) by Fubini's theorem; the sum in $R(f)$ commutes with the integral because of the rules of differentiation. We thus obtain (10.4.3). $\qquad\qquad\qquad\qquad\qquad$ □

An application of this theorem tells how the q-regularity of E_{r+1}, X depends on the q-regularity of E_r, X for submatrices E_r of E (see §10.1).

Exercise. Suppose E_r, X is q-regular with respect to dg. The pair E_{r+1}, X is q-regular for dg if and only if either: (a) E_{r+1} has an entry $e_{i,r+1} = 1$, or (b) there is a Birkhoff quadrature formula for E_r, X, dg with a kernel K_r of mean value 0,

$$\int_0^1 K_r(t)\, dt = 0.$$

We now discuss when the kernel is a Birkhoff spline. If $dg = dx$, the kernel (10.4.2) takes the form

$$K(t) = \frac{(1-t)_+^{n+1}}{(n+1)!} - \frac{(0-t)_+^{n+1}}{(n+1)!} - \sum_{e_{i,k}=1} c_{i,k} \frac{(x_i - t)_+^{n-k}}{(n-k)!}, \quad (10.4.5)$$

and we have: *If (10.1.1) is a Birkhoff quadrature formula for E, X, dx, then $K(t)$ is a Birkhoff spline for the $m \times (n+2)$ matrix E^*, which is obtained by adding the column $[1,0,\ldots,0,1]$ in front of E.*

Theorem 7.13 can serve to estimate the number of zeros of $K(t)$. This gives an inequality obtained, in different degrees of generality, by Johnson [63], Karlin and Micchelli [71], Micchelli [123], Jetter [58,60]:

Theorem 10.16. *For the kernel $K(t)$ of the Birkhoff quadrature formula associated with E, X, dx, we have*

$$Z(K,(0,1)) \leqslant |E| + \rho(E) - (n+1) \qquad\qquad (10.4.6)$$

where $\rho(E)$ is the number of odd sequences in the interior rows of E.

Proof. From Theorem 7.13 we obtain

$$Z(K,(0,1)) \leqslant |E^*| + \gamma(E^*) - (n+3).$$

Inequality (10.4.6) follows, since $|E^*| = |E| + 2$ and $\gamma(E^*) = \rho(E)$. $\qquad$ □

Exercise. When E is a normal matrix and (10.1.1) is the unique interpolatory quadrature formula for E, X, dx, then

$$K(t) = \pm D(E, X) K_{E_1^*}(X, t)$$

where E_1^* is obtained from E^* by adding a last column of 0's and $K_{E_1^*}(X, t)$ is its Birkhoff kernel. [*Hint:* Apply Theorem 7.2 to $F(x) = \int_0^x f(t)\, dt$.]

§10.5. DYN'S THEOREMS

There is an interesting approach to Gaussian formulas for Birkhoff interpolation due to N. Dyn [31,32]. She obtains some new theorems and many old ones. The only disadvantage of her theory is that it does not yield proofs of uniqueness of the knots X in cases when this uniqueness is known to exist.

We begin with a general lemma. We say that a matrix E satisfies the *delayed Pólya conditions with constant* $\sigma (\sigma = 0, 1, \ldots)$ if for this matrix

$$M_k \geqslant k + 1 - \sigma, \qquad k = 0, \ldots, n, \qquad M_n = |E| = n + 1 - \sigma. \quad (10.5.1)$$

This is equivalent to the assumption that the matrix E' obtained from E by adding σ ones to column 0 is a Pólya matrix.

In this section we assume that dg is a fixed positive measure on $[0, 1]$ not supported on any m points.

We shall need simplices of various kinds in $\mathbf{R}^m$. Let S_m be the m-dimensional open simplex consisting of points $(x_1, \ldots, x_m)$ for which $0 < x_1 < \cdots < x_m < 1$; let S_m^* be similarly defined by the inequalities $0 = x_1 < x_2 < \cdots < x_m < 1$ (or, alternatively, by $0 < x_1 < \cdots < x_m = 1$); and let S_m^{**} be the "open" simplex $0 = x_1 < x_2 < \cdots < x_m = 1$. Let $\bar{S}_m, \bar{S}_m^*, \bar{S}_m^{**}$ be their closures.

Lemma 10.17. (i) *Let E^* be an $m \times (n + 1)$ matrix that satisfies the delayed Pólya conditions with constant σ, has 0's in its 0th column, and has no odd sequences. For $X \in \bar{S}_m$ let $\mathcal{G} = \mathcal{G}(E^*, X)$ be the set of all polynomials in $\mathcal{P}_n$ annihilated by E^*, X; if some of the coordinates of X coincide, E^* is replaced by its corresponding coalescence. Then: (a) $\mathcal{G}$ is a space of dimension σ; (b) $\mathcal{G}$ is a Chebyshev space; and (c) $\mathcal{G}$ has a basis that depends continuously on X.*

(ii) *Statements (a)–(c) hold for a matrix E^* even if we also allow odd sequences in rows 1 and m, provided $X \in \bar{S}_m^{**}$.*

Proof. Consider case (i). To prove (a), let $1 < y_1 < \cdots < y_\sigma$ be arbitrary points taken so far to the right that the set $Y = X \cup \{y_1, \ldots, y_\sigma\}$ is independent in the sense that for all $X \subset [0, 1]$, the maximal Rolle extension of E, Y exists [this is necessary in the case of (ii)]. Let E be obtained from E^* by adding to it σ Lagrangian rows. The pair E, Y is regular. Let $p_1(t), \ldots, p_\sigma(t)$ be the fundamental polynomials of interpolation for this pair, which have values 1 at the points $y_1, \ldots, y_\sigma$, respectively. They span $\mathcal{G}$, hence we have (a). From the regularity of another pair E', Z we derive: A polynomial $P \in \mathcal{G}$ that vanishes at any σ points $0 \leqslant z_1 < \cdots < z_\sigma \leqslant 1$ is identically 0. This yields (b). Finally, each polynomial $p_j(X; t)$, for fixed $y_1, \ldots, y_\sigma$, is a continuous function of $X \in \bar{S}_m$. Indeed, let $f_j, j = 1, \ldots, \sigma$, be fixed functions in $C^n(-\infty, +\infty)$ that vanish on $[0, 1]$ and satisfy $f_j(y_j) = 1$, $f_j(y_k) = 0, j \neq k$. Then p_j interpolates f_j at the pair E, Y and is continuous in X by Theorem 5.22.

A similar proof applies for (ii); the matrix E' needed in the proof is still conservative because the extremal rows of E^* are parts of the extremal rows of E'. $\qquad \square$

Theorem 10.18. *Let E be an $m \times (n + 1)$ matrix all of whose Hermitian sequences are odd, their number being p, and all the rest of whose* *sequences are even. If E satisfies the delayed Pólya conditions* (10.5.1) *with constant $\sigma = p$, then there is a Gaussian quadrature formula, valid for all polynomials $f \in \mathcal{P}_n$,*

$$\int_0^1 f \, dg = \sum_{e_{i,k}=1} c_{i,k} f^{(k)}(x_i), \qquad 0 < x_1 < \cdots < x_m < 1, \quad (10.5.2)$$

with drop p.

Proof. If we add a 1 after each of the p Hermitian sequences of E, we obtain a matrix $E^+ = [e_{i,k}^+]$ with $n+1$ ones which satisfies the Pólya conditions and is conservative. For E^+ and an arbitrary $X \in S_m$ we write the interpolatory quadrature formula (see §10.1)

$$\int_0^1 f \, dg = \sum_{e_{i,k}^+=1} c_{i,k} f^{(k)}(x_i), \qquad f \in \mathcal{P}_n, \quad (10.5.3)$$

with $n+1$ terms, where

$$c_{i,k} = \int_0^1 p_{i,k}(x) \, dg, \qquad e_{i,k}^+ = 1. \quad (10.5.4)$$

Then we have to show that for a proper choice of X in S_m we get $c_{i,k} = 0$ for the additionally introduced 1's.

Let $1 \leqslant i_1 < \cdots < i_p \leqslant m$ be the positions of the p odd Hermitian sequences of E. From E we derive its submatrix E^- by replacing all ones $e_{i_j,0} = 1$, $j = 1,\ldots,p$, by 0's. To E^- we can apply Lemma 10.17 (i) with $\sigma = 2p$ and obtain a Chebyshev subspace $\mathcal{G} = \mathcal{G}(E^-, X)$ of $\mathcal{P}_n$ of dimension $2p$, which depends continuously on $X \in \bar{S}_m$. By Theorem 10.1, Case IV, for each X there is a Z: $0 < z_1 < \cdots < z_p < 1$, $Z \in S_p$, with the formula

$$\int_0^1 f \, dg = \sum_{j=1}^p c_j f(z_j), \qquad f \in \mathcal{G}. \quad (10.5.5)$$

By the Remark to Theorem 10.1 and Lemma 10.17 (i,c), this defines a continuous map $Z = T_2 X$ of $X \in \bar{S}_m$ into S_p.

We would like to find an $X \in S_m$ for which

$$x_{i_j} = z_j, \qquad j = 1,\ldots,p. \quad (10.5.6)$$

This can be achieved by means of a fixed-point theorem.

For $Z \in \bar{S}_p$, let $X = T_1 Z$ be a continuous map of $\bar{S}_p$ into $\bar{S}_m$ that satisfies (10.5.6) and also has the property that for $Z \in S_p$ and the corresponding $X \in S_m$, inequalities $0 < z_1 < \cdots < z_p < 1$ imply $0 < x_1 < \cdots < x_m < 1$. For example, we can require the x_i not yet defined by (10.5.6) to be equally spaced in the intervals between $0, x_{i_1},\ldots,x_{i_p}, 1$.

Now $T = T_1 T_2$ is a continuous map of $\bar{S}_m$ into itself, and by Brouwer's theorem it has a fixed point X. The corresponding Z belongs to the open simplex S_p, and therefore the fixed point $X = T_1 Z$ belongs to S_m.

For this X we have

$$\int_0^1 f\,dg = \sum_{j=1}^p c_j f(x_j), \qquad f \in \mathcal{G}. \tag{10.5.7}$$

In particular, the polynomials $p_{i,k}$ of (10.5.3) and (10.5.4), which correspond to i, k with $e_{i,k}^+ = 1$, $e_{i,k} = 0$, are in $\mathcal{G}$ and vanish at the points x_{i_j}, hence from (10.5.7) we derive $c_{i,k} = 0$ for these pairs i, k. Formula (10.5.3) reduces to (10.5.2). □

A special case of Theorem 10.18 is a result of Turán and Popoviciu [138, 178], which is as follows. For a $m \times (n+1)$ Hermite matrix E with all odd sequences and $n + 1 - m$ ones there is an X: $0 < x_1 < \cdots < x_m < 1$ and a quadrature formula (10.5.2) for $f \in \mathcal{P}_n$ with the drop $p = m$. It can also be shown (see [72, p. 119]) that X is unique.

At the cost of decreasing somewhat the number p, we can relax the conditions for the exterior rows of E. In addition, we can then enforce $x_1 = 0$ or $x_m = 1$, or both. In this way Dyn obtains four theorems in all, one of which is Theorem 10.18. We give another of them:

Theorem 10.19. *Let E be an $m \times (n+1)$ matrix all of whose Hermitian sequences in interior rows are odd, and all of whose other interior sequences are even. If the number of Hermitian sequences in the interior rows $1 < i < m$ is p, and if E satisfies (10.5.1) with the constant $\sigma = p$, then there is a Gaussian quadrature formula for all $f \in \mathcal{P}_n$,*

$$\int_0^1 f\,dg = \sum_{e_{i,k}=1} c_{i,k} f^{(k)}(x_i), \qquad 0 = x_1 < \cdots < x_m = 1, \tag{10.5.8}$$

with drop p.

Proof. This is a variation of the proof of Theorem 10.18. The matrix E^+ now differs from E by 1's added after each *interior* Hermite sequence of E. The matrix E^- is obtained by dropping all 1's in the 0th column of E. Their number is $p + \tau$, where τ is the number of 1's among $e_{1,0}, e_{m,0}$. They are located in rows $i_1 < \cdots < i_{p+\tau}$.

First let $\tau = 0$. We take an arbitrary point $X \in \bar{S}_m^{**}$ and with the help of Lemma 10.17 (ii) obtain a Chebyshev subspace $\mathcal{G}$ of $\mathcal{P}_n$ of dimension $2p$. We apply Theorem 10.1, Case IV, obtain the formula (10.5.5), and by means of Brouwer's fixed-point theorem applied to a map of $\bar{S}_m^{**}$ into itself via $\bar{S}_p$ find a point $X \in S_m^{**}$ for which $z_j = x_{i_j}$, $j = 1, \ldots, p$. Then we proceed as before.

If $\tau = 2$, we have $0 = x_{i_1} < \cdots < x_{i_{p+2}} = 1$. For an arbitrary $X \in \bar{S}_m^{**}$, we obtain $\mathcal{G}$ of dimension $2p + 2$ by means of Lemma 10.17 (ii). Case III of Theorem 10.1 yields a quadrature formula for $f \in \mathcal{G}$,

$$\int_0^1 f\,dg = \sum_{j=1}^{p+2} c_j f(z_j), \qquad 0 = z_1 < \cdots < z_{p+2} = 1. \tag{10.5.9}$$

By means of the fixed-point theorem applied to a map of $\bar{S}_m^{**}$ into itself via the simplex $\bar{S}_p^{**}$ we obtain an X in the simplex $\bar{S}_m^{**}$ with the property $z_j = x_{i_j}$, $j = 1, \ldots, p + 2$. For this X, the interpolatory formula (10.5.3) reduces to (10.5.8). Finally, if $\tau = 1$, we use Case I of Theorem 10.1 with a similar proof.   □

The special case of Theorem 10.19, when all rows of E are Lagrangian, was proved earlier by Micchelli and Rivlin [126] by a different method. If all rows of E are Lagrangian, the formula (10.5.9) is often called a Lobatto formula. We leave it to the reader to prove Dyn's theorems concerning sets of knots $0 = x_1 < \cdots < x_m < 1$ or $0 < x_1 < \cdots < x_m = 1$.

Finally, we may ask how a matrix E of Theorem 10.19 can be derived from a Pólya matrix.

Theorem 10.20. *Let E^* be an $m \times (n + 1)$ Pólya matrix. If it contains λ interior even Hermitian and μ interior odd Hermitian sequences, and q interior odd non-Hermitian sequences with $q \leqslant \mu$, then E^* has a submatrix E with the Gaussian formula of type (10.5.8) for polynomials of degree $\leqslant n^* = n + (\mu - q)$ and drop $\lambda + \mu$ (hence for polynomials of degree $\leqslant n$ with drop $q + \mu$).*

Proof. We omit q ones from odd non-Hermitian sequences of E^*, making them even, and λ ones from even Hermitian sequences, making them odd Hermitian sequences. The matrix E obtained, considered as an $m \times (n^* + 1)$ matrix, satisfies the Pólya conditions with delay $\lambda + \mu$. Then we apply Theorem 10.19.   □

§10.6. NOTES

10.6.1. In [153], Schoenberg pointed out an interesting relation between monosplines and quadrature formulas (for $dg(x) = dx$). This has been often imitated (see [170, 65]). Schoenberg showed that if dg is absolutely continuous, then the weight $c_{i,k}$ ($e_{i,k} = 1$) of the formula is the jump of $K^{(k)}$ at the point x_i except for a sign.

10.6.2. Theorem 10.16 remains true for kernels of Birkhoff quadrature formulas associated with E, X, and an absolutely continuous, positive measure dg. Theorem 10.16 can be applied to the derivatives of K in order to improve the estimates (10.3.12) slightly (Jetter [60]).

10.6.3. The quadrature formula (10.5.2) is unique for $dg(x) = dx$ if E is Hermitian or quasi-Lagrangian. This has been proved by Popoviciu [138], Micchelli and Rivlin [126], Karlin and Pinkus [72], and Barrow [4].

10.6.4. In [198] Jetter, using a new method, proves the existence of quadrature formulas

$$\int_0^1 f \, dg = \sum_{i=1}^n a_i f^{(k_i)}(x_i), \qquad f \in \mathscr{P}_{2n-1},$$

where $0 < x_1 < \cdots < x_n < 1$, and the orders of the derivatives are selected in a "pyramidal way":

$$k_i \leqslant k_{i-1} \leqslant k_{i+1}, \quad i = 2,\ldots,I; \qquad k_i = 0, \quad i = I,\ldots,J;$$
$$k_i \leqslant k_{i+1} \leqslant k_i + 1, \quad i = J,\ldots,n-1.$$

It would be interesting to have a uniqueness result here.

Chapter 11

Interpolation at the Roots of Unity

§11.1. LACUNARY INTERPOLATION

The idea of Turán and his associates was to couple a singular matrix E that has identical rows $i = 1, \ldots, m$ with a specially chosen set of knots to obtain a regular pair E, X. If the rows have 1's in positions $m_0 = 0 < m_1 < \cdots < m_q$, we have the $(m_0, \ldots, m_q)$ *lacunary interpolation*. For example, the $(0,2)$ interpolation corresponds to the matrix of Example 2, §2.3, and to prescribed values $P(x_i)$, $P''(x_i)$, $i = 1, \ldots, m$. We assume the Pólya conditions, which take the form

$$m_k \leqslant km, \qquad k = 1, \ldots, q. \tag{11.1.1}$$

The main purpose was to obtain for the $(0,2)$ interpolation results known for the $(0,1)$ case (this corresponds to the Hermite–Fejér interpolation, see [J, Vol. III, p. 57]).

Let x_i, $-1 < x_1 < \cdots < x_n < +1$, be the roots of the nth Chebyshev polynomial. The Lagrange interpolation polynomial of a continuous function $f(x)$ at the knots x_i in general does not converge to $f(x)$ for $n \to \infty$. Fejér noticed that the situation changes if we consider the Hermite interpolation polynomial $P_{2n-1}(f, x)$ with prescribed values $P_n(x_i) = f(x_i)$, $P_n'(x_i) = \beta_i = \beta_{i,n}$. We can take $\beta_{i,n} = f'(x_i)$ if f' is continuous, or $\beta_{i,n} = 0$, or we can even take $\beta_{i,n}$ arbitrary, subject only to the condition $\max_i |\beta_{i,n}| = o(n)$. With this assumption we have the uniform convergence $P_{2n-1}(f, x) \to f(x)$ for $f \in C[-1, +1]$.

For the $(0, 2)$ interpolation, Turán and his students select as knots the roots of the derivative P_n' of the Legendre polynomial; more precisely, they take the $n + 1$ knots

$$X: -1 < x_1 < \cdots < x_{n-1} < +1 \qquad (11.1.2)$$

where x_k are these roots. (The pair E, X is regular if n is even.) As an analogue to Fejér's theorem, they establish the convergence $P_n(f, x) \to f(x)$ for functions f of the Zygmund class. This investigation (presented in Chapter 12) requires an astonishing amount of information about the properties of Legendre polynomials.

Since then, several authors have continued investigations of lacunary interpolation for the knots (11.1.2)—for example, Saxena and Sharma [146, 147] for the $(0, 1, 3)$ interpolation, or Varma [182] for the $(0, 1, 2, 4)$ interpolation. Lacunary interpolation at the roots of other classical polynomials has also been studied (see §12.6).

In this chapter, we will begin with lacunary interpolation by polynomials at the roots of unity on the circle $|z| = 1$. This case is easier, but perhaps not less important, than Turán's case. The investigation for $(0, 2)$ and $(0, 2, 3)$ interpolation was started by Kis and Sharma. Now it is possible to treat arbitrary $(m_0, m_1, \ldots, m_q)$ interpolation. Regularity was established by Cavaretta, Sharma, and Varga [14], and convergence for a proper class of functions by Riemenschneider and Sharma [140].

Another related problem is that of trigonometric interpolation at equidistant points. In §11.5 we treat the case of an odd number of points $2\pi k / (2n + 1)$, $k = 0, 1, \ldots, 2n$, and an odd number of derivatives, of orders $0 = m_0 < \cdots < m_q$. The interpolating system is then

$$1, \cos x, \sin x, \ldots, \cos Nx, \sin Nx \qquad (11.1.3)$$

with $2N + 1 = (2n + 1)(q + 1)$. If the number of knots is even, or for $(0, 2)$ interpolation, we have to omit $\cos Nx$ or $\sin Nx$ from the system (11.1.3). This has led to many separate cases (see §11.6), often quite different in character, which we cannot treat in this book.

§11.2. REGULARITY OF
$(m_0, m_1, \ldots, m_q)$ INTERPOLATION

The regularity of polynomial interpolation at complex knots $x_1, \ldots, x_m$ has been completely described by Ferguson's Theorem 4.10. However, if the knots are restricted to the unit circle in the complex plane, then the conditions of Theorem 4.10 are no longer necessary for the regularity of interpolation. The regularity problem for interpolation on the circle is essentially different from the problems of real or order regularity. The common ground is that the Pólya conditions are necessary for polynomial interpolation at any particular set of knots.

Examples

1. Let E be a normal matrix consisting of a Hermite part and an additional non-Hermite one, $e_{i,k} = 1$. In general, E is not order regular, by Theorem 6.2. However, E is regular on the circle. Indeed, let the nontrivial polynomial P be annihilated by E, Z with the knots $Z = \{z_i\}$ on the circle. The Hermite portion of E gives all n zeros of P_n. The zeros of P', different from the z_i, are strictly in $|z| < 1$. The same applies to derivatives of higher orders. But $P^{(k)}$ has a zero on the circle, a contradiction.

2. The $3 \times (n+1)$ matrix

$$E = \begin{bmatrix} 1 & 0 & 0 & \cdots & 0 & 0 \\ 0 & 1 & 1 & \cdots & 1 & 0 \\ 1 & 0 & 0 & \cdots & 0 & 0 \end{bmatrix}, \qquad n > 2,$$

is not regular on the circle; it is order regular exactly when n is odd. The first statement is true if n is odd, for $P(z) = (z-1)^n - (e^{i\pi/n} - 1)^n$ is annihilated by E and the knots $e^{i\pi/n}, 1, e^{-i\pi/n}$. If n is even, we take instead $P(z) = (z-1)^n - (e^{2i\pi/n} - 1)^n$ and the knots $e^{2\pi i/n}, 1, e^{-2\pi i/n}$.

The problem of interpolation on the unit circle has been studied only for the special knots $Z_n = \{z_k = e^{2\pi i k/n}: k = 1, \ldots, n\}$, the nth roots of unity. The early results were patterned on the work of Turán's school (Chapter 12) and dealt with regularity of interpolation and convergence in certain special cases. Kis [80] basically considered $(0,2)$ and $(0,1,3)$ interpolation at Z_n. Sharma [158, 159] generalized these results to $(0, m)$ and $(0, 2, 3)$ interpolation [his methods suffice to prove the regularity of $(0, m_1, m_2)$ interpolation for Z_n]. Recently, Cavaretta, Sharma, and Varga [14] have shown the regularity of any $(m_0, m_1, \ldots, m_q)$ Pólya matrix at the roots of unity. We give the proof of Riemenschneider and Sharma [140], which also provides a basis for the convergence theorem. For this purpose, we have to develop properties of certain systems of linear differential equations.

For natural numbers $m_0 = 0 < m_1 < \cdots < m_q$, let $\bar{m} = (m_0, \ldots, m_q)$. Let $E_{\bar{m}}$ denote the $n \times (q+1)n$ interpolation matrix that has 1's in columns $m_j, j = 0, \ldots, q$, and 0's elsewhere.

Theorem 11.1 [14, 140]. *The pair $E_{\bar{m}}, Z_n$ is regular if and only if $E_{\bar{m}}$ is a Pólya matrix, that is, if $m_j \leqslant jn, j = 0, \ldots, q$.*

Proof. Since the Pólya conditions are necessary for polynomial interpolation, we can assume that $E_{\bar{m}}$ is a Pólya matrix. Suppose that $P(z)$ is a polynomial of degree $(q+1)n - 1$ that is annihilated by $E_{\bar{m}}, Z_n$. Then $P(z)$ may be written in the form

$$P(z) = P_0(z) + z^n P_1(z) + \cdots + z^{qn} P_q(z) \qquad (11.2.1)$$

where each polynomial $P_\lambda(z) = \sum_{\nu=0}^{n-1} a_{\lambda,\nu} z^\nu$ is of degree $n-1$.

From Leibnitz' formula for differentiation of a product, we have

$$\left(z^s P_\lambda(z)\right)^{(m_j)}\Big|_{z=z_k} = \sum_{l=0}^{m_j} \binom{m_j}{l}(s)_l z_k^{s-l} D^{m_j-l} P_\lambda(z_k)$$

$$= z_k^{s-m_j} \sum_{l=0}^{m_j} \binom{m_j}{l}(s)_l z_k^{m_j-l} D^{m_j-l} P_\lambda(z_k)$$

$$= z_k^{s-m_j} G_{j,s}(D) P_\lambda(z)\big|_{z=z_k}$$

where $G_{j,s}(D)$ is the Euler differential operator

$$G_{j,s}(D) = \sum_{l=0}^{m_j} \binom{m_j}{l}(s)_l z^{m_j-l} D^{m_j-l}, \qquad 0 \leqslant j \leqslant q. \qquad (11.2.2)$$

[Recall that $(a)_l = a(a-1)\cdots(a-l+1)$, $(a)_0 = 1$.] With the help of (11.2.2) we can write the interpolating conditions $P^{(m_j)}(z_k) = 0$, $0 \leqslant j \leqslant q$, $1 \leqslant k \leqslant n$, for the polynomial (11.2.1) in the form

$$P_0^{(m_j)}(z_k) + z_k^{-m_j} \sum_{\lambda=1}^{q} G_{j,\lambda n}(D) P_\lambda(z)\bigg|_{z=z_k} = 0, \qquad (11.2.3)$$

$k = 1, \ldots, n$, $j = 0, \ldots, q$. Since $G_{j,\lambda n}(D) P_\lambda(z)$ is again a polynomial of degree $n-1$, the equations (11.2.3) yield that the polynomials $P_\lambda(z)$, $\lambda = 0, 1, \ldots, q$, satisfy the system of differential equations

$$\sum_{\lambda=0}^{q} G_{j,\lambda n}(D) P_\lambda(z) = 0, \qquad j = 0, \ldots, q. \qquad (11.2.4)$$

Since differential operators of the type (11.2.2) commute, the system of differential equations (11.2.4) can be easily handled. Let $H_{\bar{m}}(D)$ be the determinant

$$H_{\bar{m}}(D) = \det\left[G_{l,0}(D), G_{l,n}(D), \ldots, G_{l,qn}(D); l = 0, \ldots, q\right]$$

and let $H_{j,\lambda n}(D)$ be the algebraic complement of the element $G_{j,\lambda n}(D)$ in $H_{\bar{m}}(D)$. Multiplication of (11.2.4) with $H_{j,\lambda}(D)$ and summation yield

$$H_{\bar{m}}(D) P_\lambda(z) = 0, \qquad \lambda = 0, 1, \ldots, q. \qquad (11.2.5)$$

From (11.2.2) we have

$$G_{j,\lambda n}(D) z^\nu = z^\nu \sum_{l=0}^{m_j} \binom{m_j}{l}(\lambda n)_l (\nu)_{m_j-l}$$

$$= z^\nu (\nu + \lambda n)_{m_j}, \qquad 0 \leqslant j, \lambda \leqslant q. \qquad (11.2.6)$$

Therefore, by the linearity of $H_{\bar{m}}(D)$, for any polynomial $P(z) = \sum_{\nu=0}^{n-1} a_\nu z^\nu$,

$$H_{\bar{m}}(D) P(z) = \sum_{\nu=0}^{n-1} a_\nu M(\nu) z^\nu$$

where $M(\nu)$ are the determinants

$$M(\nu) = \det\left[(\nu)_{m_j}, (\nu+n)_{m_j}, \ldots, (\nu+qn)_{m_j}; j = 0, \ldots, q\right],$$

$$\nu = 0, \ldots, n-1. \quad (11.2.7)$$

In particular for P_λ,

$$\sum_{\nu=0}^{n-1} a_{\lambda,\nu} M(\nu) z^\nu = 0.$$

By Corollary 4.18, the determinants (11.2.7) are all positive. Hence, $a_{\lambda,\nu} = 0$ for $\nu = 0, 1, \ldots, n-1$ and $\lambda = 0, 1, \ldots, q$. Therefore by (11.2.1) $P(z) \equiv 0$. This shows that $E_{\bar{m}}, Z_n$ is regular.   □

We note also the formula

$$H_{j,\lambda}(D)\left(\sum_{\nu=0}^{n-1} a_\nu z^\nu\right) = \sum_{\nu=0}^{n-1} a_\nu M_{j,\lambda}(\nu) z^\nu$$

where $M_{j,\lambda}(\nu)$ is the algebraic complement of the j, λ element of $M(\nu)$.

The proof of the convergence theorem for the $E_{\bar{m}}, Z_n$ interpolation is more difficult. It requires properties and estimates of the fundamental polynomials. They can be obtained by the technique that has just been developed. We shall need the fundamental polynomials for Lagrange interpolation at the roots of unity. These are the unique polynomials of degree $n-1$ that satisfy $l_k(z_k) = 1$, $l_k(z_l) = 0$, $k \neq l$, $1 \leqslant k, l \leqslant n$. They are given by

$$l_k(z) = \frac{z_k}{n} \frac{z^n - 1}{z - z_k} = \frac{1}{n} \sum_{\nu=0}^{n-1} z_k^{-\nu} z^\nu; \qquad k = 1, \ldots, n. \quad (11.2.8)$$

The *fundamental polynomials* $\alpha_{\bar{m}, k, j}(z)$ for the $E_{\bar{m}}, Z_k$ interpolation are the unique polynomials of degree $(q+1)n - 1$ satisfying

$$\alpha_{\bar{m}, k, j}^{(m_r)}(z_l) = \begin{cases} 1, & r = j, l = k, \\ 0 & \text{otherwise,} \end{cases} \quad 0 \leqslant r, j \leqslant q; \quad 1 \leqslant k, l \leqslant n$$

$$(11.2.9)$$

Let k and j be fixed. If we ask for $\alpha_{\bar{m}, k, j}(z)$ of the form (11.2.1), then exactly as was the case for (11.2.3), the conditions (11.2.9) give rise to the system of equations

$$\left. P_0^{(m_r)}(z_l) + z_l^{-m_r} \sum_{\lambda=1}^{q} G_{r,\lambda n}(D) P_\lambda(z) \right|_{z=z_l}$$

$$= \begin{cases} l_k(z_l), & r = j, \\ 0 & \text{otherwise,} \end{cases} \qquad 1 \leqslant l \leqslant n.$$

It follows that the polynomials $P_\lambda(z)$ must satisfy the system of differential equations

$$\sum_{\lambda=0}^{q} G_{r,\lambda n}(D)P_\lambda(z) = \begin{cases} z_k^{m_j} l_k(z), & r=j, \\ 0, & r \ne j, \end{cases} \qquad \text{for} \quad 0 \leqslant r \leqslant q.$$

With the help of the operators $H_{\bar{m}}$, $H_{j,\lambda}$ we obtain

$$H_{\bar{m}}(D)P_\lambda(z) = H_{j,\lambda}(D)(z_k^{m_j} l_k(z)), \qquad \lambda = 0,\dots,q. \quad (11.2.10)$$

Using (11.2.7) and (11.2.8), equations (11.2.10) can be rewritten as

$$\sum_{\nu=0}^{n-1} a_{\lambda,\nu} M(\nu) z^\nu = \frac{1}{n} \sum_{\nu=0}^{n-1} z_k^{m_j-\nu} M_{j,\lambda}(\nu) z^\nu.$$

Hence

$$z^{\lambda n} P_\lambda(z) = \frac{1}{n} \sum_{\nu=0}^{n-1} z_k^{m_j-\nu} z^{\nu+\lambda n} \frac{M_{j,\lambda}(\nu)}{M(\nu)}.$$

Substituting this expression into (11.2.1) and writing the resulting terms as determinants, we readily obtain the following representation for the fundamental polynomials.

Theorem 11.2 [14]. *The fundamental polynomials $\alpha_{\bar{m},k,j}(z)$ of the $E_{\bar{m}}$, Z_n interpolation have the representation*

$$\alpha_{\bar{m},k,j}(z) = \frac{1}{n} \sum_{\nu=0}^{n-1} z_k^{m_j-\nu} \frac{M^j(z)}{M(\nu)}; \qquad 1 \leqslant k \leqslant n, \quad 0 \leqslant j \leqslant q, \quad (11.2.11)$$

where $M(\nu)$ is given by (11.2.7) and $M^j(z)$ is obtained from $M(\nu)$ by replacing the jth row by $(z^\nu, z^{\nu+n},\dots,z^{\nu+qn})$.

Example. $\bar{m} = (0, m)$, with $m \leqslant n$. From equation (11.2.9) we obtain

$$\alpha_{\bar{m},k,0}(z) = \frac{1}{n} \sum_{\nu=0}^{n-1} z_k^{-\nu} z^\nu + (1-z^n)\frac{1}{n} \sum_{\nu=0}^{n-1} \frac{z_k^{-\nu}(\nu)_m z^\nu}{(\nu+n)_m - (\nu)_m}$$

$$(11.2.12)$$

and

$$\alpha_{\bar{m},k,1}(z) = \frac{(z^n-1)}{n} \sum_{\nu=0}^{n-1} \frac{z_k^{m-\nu} z^\nu}{(\nu+n)_m - (\nu)_m}. \qquad (11.2.13)$$

It is also immediately clear that these formulas give the fundamental polynomials of the $(0, m)$ interpolation.

§11.3. ESTIMATES OF THE FUNDAMENTAL POLYNOMIALS

The fundamental polynomials given in Theorem 11.2 can be estimated by using the representation

$$\alpha_{\bar{m},k,j}(z) = \sum_{\lambda=0}^{q} z^{\lambda n} P_{\lambda,k,j}(z) \qquad (11.3.1)$$

where the polynomials $P_{\lambda,k,j}(z)$ satisfy the differential equations (11.2.10). This requires a closer look at the differential operators in (11.2.10).

The estimates obtained below will be valid for fixed $m_1,\ldots,m_q$ and for large n. We always assume that $m_j \leqslant jn, j = 1,\ldots,q$.

An inequality of M. Riesz [I, p. 40] states: *If $P(z)$ is a polynomial of degree n, then*

$$|P^{(m)}(z)| \leqslant n^m \max_{|w|\leqslant 1} |P(w)|, \qquad |z|\leqslant 1. \qquad (11.3.2)$$

We shall extend this inequality to operators $H_{j,\lambda}(D)$. Let $\sigma_j = \Sigma_{l\neq j} m_l$.

Lemma 11.3. *There is a constant $C(\overline{m})$, independent of n, so that*

$$|H_{j,\lambda}(D)P(z)| \leqslant C(\overline{m})n^{\sigma_j} \max_{|w|\leqslant 1} |P(w)| \qquad (11.3.3)$$

for $|z| \leqslant 1, 0 \leqslant j, \lambda \leqslant q$ and for any polynomial $P(z)$ of degree n.

Proof. By (11.2.2) and (11.3.2) it follows easily that

$$|G_{l,\lambda n}(D)P(z)| \leqslant C(m_l,\lambda)n^{m_l} \max_{|w|\leqslant 1} |P(w)|, \qquad |z|\leqslant 1, \quad (11.3.4)$$

for $0 \leqslant l, \lambda \leqslant q$ and any polynomial of degree n. The estimate (11.3.3) now follows from the definition of $H_{j,\lambda}(D)$ by repeated application of (11.3.4). $\square$

The differential operator $H_{\overline{m}}(D)$ defined in (11.2.5) and appearing on the left-hand side of (11.2.10) is invertible as an operator on polynomials of degree $n-1$. In fact, if $P(z) = \Sigma_{\nu=0}^{n-1} a_\nu z^\nu$, then

$$H_{\overline{m}}(D)P(z) = \sum_{\nu=0}^{n-1} a_\nu M(\nu)z^\nu$$

where the $M(\nu)$ are the determinants (11.2.7). Therefore, we may formally define

$$H_{\overline{m}}(D)^{-1}P(z) = \sum_{\nu=0}^{n-1} \left(\frac{a_\nu}{M(\nu)}\right)z^\nu. \qquad (11.3.5)$$

The analysis of $H_{\overline{m}}(D)^{-1}$ requires some additional knowledge of the $M(\nu)$. It is not quite correct that $M(\nu)$ is increasing in ν. The following is an approximation to this.

Lemma 11.4. *Let $M(\nu)$ be the determinants (11.2.7), then*

$$\frac{n^{\Sigma_{l=1}^{q} m_l}}{M(\nu)} = g(\nu) + \mathcal{O}\left(\frac{1}{n}\right) \qquad (11.3.6)$$

where $g(\nu)$ is bounded and decreasing for $0 \leqslant \nu \leqslant n$.

Proof. We observe that

$$(\nu + \lambda n)_{m_l} = n^{m_l}\left(\frac{\nu}{n} + \lambda\right)\left(\frac{\nu}{n} + \lambda - \frac{1}{n}\right)\cdots\left(\frac{\nu}{n} + \lambda - \frac{m_l-1}{n}\right)$$

$$= n^{m_l}\left[\left(\frac{\nu}{n} + \lambda\right)^{m_l} + \mathcal{O}\left(\frac{1}{n}\right)\right]$$

where the constant in $\mathbb{O}(1/n)$ depends only on λ and $\bar{m}$. Therefore

$$
n^{-\Sigma m_i} M(\nu) = \det\left[\left(\frac{\nu}{n}\right)^{m_j}, \ldots, \left(\frac{\nu}{n}+q\right)^{m_j}; j = 0, 1, \ldots, q\right] + \mathbb{O}\left(\frac{1}{n}\right)
$$

$$(11.3.7)$$

where the constant in the $\mathbb{O}(1/n)$ term depends only on $\bar{m}$. The determinant $V(\alpha)$, $\alpha = \nu/n$, on the right in (11.3.7) is the generalized Vandermonde determinant for the system $1, x^{m_1}, \ldots, x^{m_q}$ at the points $\alpha, \alpha + 1, \ldots, \alpha + q$. By (4.6.6), $V(\alpha)$ is a continuous function bounded away from 0 on $[0, 1]$. Therefore, for positive C_1, C_2 we have

$$
C_1 \leqslant V(\alpha) \leqslant C_2, \qquad 0 \leqslant \alpha \leqslant 1.
$$

Moreover, $V(\alpha)$ is increasing, since $V'(\alpha)$ is a sum of positive multiples of determinants of the same type. We obtain (11.3.6) by taking $g(\nu) = 1/V(\nu/n)$.   □

The operators $H_{j,\lambda}(D)$ and $H_{\bar{m}}(D)^{-1}$ commute for polynomials. Operating on (11.2.10) by $H_{\bar{m}}(D)^{-1}$ and using (11.2.8), we find

$$
P_{\lambda,k,j}(z) = \frac{1}{n} \sum_{\nu=0}^{n-1} \frac{M_{j,\lambda}(\nu)}{M(\nu)} z_k^{m_j - \nu} z^\nu
$$

$$
= H_{j,\lambda}(D) H_{\bar{m}}(D)^{-1}(z_k^{m_j} l_k(z)). \qquad (11.3.8)
$$

This allows the estimate:

Lemma 11.5. *There is a constant* $C(\bar{m})$ *for which*

$$
\sum_{k=1}^{n} |H_{\bar{m}}(D)^{-1}(z_k^{m_j} l_k(z))| \leqslant C(\bar{m}) n^{-\Sigma m_i} \log n \qquad (11.3.9)
$$

for $|z| \leqslant 1$ *and* $j = 0, 1, \ldots, q$. *For* $|z| \leqslant r$, $0 < r < 1$, *the upper bound in* (11.3.9) *can be replaced by* $C(\bar{m}, r) n^{-\Sigma m_i}$.

Proof. From the summation by parts formula

$$
H_{\bar{m}}(D)^{-1}(l_k(z)) = \frac{1}{n} \sum_{\nu=0}^{n-1} \frac{(z/z_k)^\nu}{M(\nu)}
$$

$$
= \frac{1}{n} \sum_{r=0}^{n-1} \left[\sum_{\nu=0}^{r}\left(\frac{z}{z_k}\right)^\nu\right]\left[\frac{1}{M(r)} - \frac{1}{M(r+1)}\right]
$$

$$
+ \frac{1}{nM(n)} \sum_{\nu=0}^{n-1}\left(\frac{z}{z_k}\right)^\nu.
$$

Using Lemma 11.4 and summing the geometric series, we have

$$|H_{\bar{m}}(D)^{-1}(l_k(z))| \leqslant C(\bar{m})n^{-\Sigma m_l - 1} \max_{0 \leqslant r \leqslant n-1} \left| \sum_{\nu=0}^{r} \left(\frac{z}{z_k}\right)^\nu \right|$$

$$\leqslant C(\bar{m})n^{-\Sigma m_l - 1}\min(n, 2|z - z_k|^{-1}). \quad (11.3.10)$$

We may assume $z = e^{i\theta}$ where $0 \leqslant \theta \leqslant 2\pi/n$. For z_k on the left semicircle, that is, for $\pi/2 < 2\pi k/n \leqslant 3\pi/2$, the distance $|z - z_k|$ is at least as great as the radius: thus

$$\sum_{\frac{n}{4} < k < \frac{3n}{4}} 2|z - z_k|^{-1} \leqslant \frac{n}{2} \cdot 2 = n.$$

For $1 < k \leqslant n/4$,

$$|z - z_k| \geqslant \sin\left(2\pi\frac{k}{n} - \theta\right) \geqslant \frac{4(k-1)}{n},$$

whereas for $3n/4 < k < n$,

$$|z - z_k| \geqslant \left|\sin\left(2\pi\frac{k}{n} - \theta\right)\right| \geqslant \frac{4(n-k)}{n}.$$

Combining these estimates, we obtain

$$\sum_{k=1}^{n} \min(n, 2|z - z_k|^{-1}) \leqslant n(3 + \log n).$$

The estimate (11.3.9) follows easily from this and (11.3.10). If $|z| \leqslant r$, $0 < r < 1$, the last sum does not exceed $2n(1-r)^{-1}$.   □

It is now possible to estimate the fundamental polynomials.

Theorem 11.6. *The fundamental polynomials* $\alpha_{\bar{m}, k, j}(z)$ *of the* $E_{\bar{m}}, Z_n$ *interpolation satisfy*

$$\sum_{k=1}^{n} |\alpha_{\bar{m}, k, j}(z)| \leqslant C(\bar{m})n^{-m_j}\log n \qquad (11.3.11)$$

for $|z| \leqslant 1$ *and* $j = 0, 1, \ldots, q$, *where* $C(\bar{m})$ *is independent of* n. *For* $|z| \leqslant r$, $0 < r < 1, j = 0, \ldots, q$, *they satisfy*

$$\sum_{k=1}^{n} |\alpha_{\bar{m}, k, j}(z)| \leqslant C(\bar{m}, r)n^{-m_j}. \qquad (11.3.12)$$

Proof. For fixed z_0, $|z_0| \leqslant 1$, choose ε_k with $|\varepsilon_k| = 1$ so that $\varepsilon_k P_{\lambda, k, j}(z_0) = |P_{\lambda, k, j}(z_0)|$. Then from (11.3.8) we have

$$\sum_{k=1}^{n} |P_{\lambda, k, j}(z_0)| = H_{j, \lambda}(D)\left[\sum_{k=1}^{n} \varepsilon_k H_{\bar{m}}(D)^{-1}(z_k^{m_j}l_k(z))\right]\Bigg|_{z = z_0}.$$

But $\sum_{k=1}^{n} \varepsilon_k H_{\bar{m}}(D)^{-1}(z_k^{m_j} l_k(z))$ is a polynomial of degree $n-1$ that is bounded by (11.3.9). Therefore, by Lemma 11.3

$$\sum_{k=1}^{n} |P_{\lambda, k, j}(z_0)| \leqslant C(\bar{m}) n^{-m_j} \log n.$$

The estimate (11.3.11) now follows from (11.3.1).   □

Exercises
1. Show that

$$|H_{\bar{m}}(D)^{-1}P(z)| \leqslant C(\bar{m}) n^{-\Sigma m_j} \log n \max_{|w| \leqslant 1} |P(w)| \qquad (11.3.13)$$

for $|z| \leqslant 1$.

2. Prove that the Lebesgue constant Λ_n of the Lagrange interpolation for the points Z_n satisfies

$$C_1 \log n \leqslant \Lambda_n = \max_{|z|=1} \sum_{k=1}^{n} |l_k(z)| \leqslant C_2 \log n.$$

Remark. For $\bar{m} = (0, m)$ and for some special cases of $\bar{m} = (0, m_1, m_2)$ it is known that (11.3.13) is true without the $\log n$ on the right-hand side. We do not know whether $\log n$ can be omitted in general.

§11.4. CONVERGENCE OF INTERPOLATION AT THE ROOTS OF UNITY

In this section we prove convergence of the interpolation operators to the generating function. The first theorems of this type were due to Kis [80].

We discuss the classes of functions for which this convergence can be established. Let $\mathcal{C}$ consist of all functions $f(z)$ that are analytic in $|z| < 1$ and continuous in $|z| \leqslant 1$. The boundary values of $f \in \mathcal{C}$ for $z = e^{it}$ are $g(t) = f(e^{it})$, $t \in \mathbf{T}$. The Dini class $\mathcal{D}$ consists of all functions $f \in \mathcal{C}$ for which the modulus of continuity $\omega(f, \delta)$ of $f(e^{it})$ satisfies

$$\omega(f, \delta) \log(1/\delta) \to 0 \quad \text{as} \quad \delta \to 0+. \qquad (11.4.1)$$

The boundary value $f(e^{it})$ of a function $f \in \mathcal{C}$ is not an arbitrary continuous function. From the relations $\int_{|z|=r} f(z) z^k \, dz = 0$, $k = 0, 1, \ldots$, $0 < r < 1$, we derive that

$$\int_{-\pi}^{\pi} g(t) e^{ikt} \, dt = 0, \qquad k = 1, 2, \ldots;$$

that is, the Fourier coefficients a_k, b_k of the function $g(t) = f(e^{it})$ must satisfy

$$a_k + i b_k = 0, \qquad k = 1, 2, \ldots. \qquad (11.4.2)$$

(One can show that these conditions are also sufficient.) In particular, if $T(t) = \sum_{k=0}^{n}(a_k \cos kt + b_k \sin kt)$ is a trigonometric polynomial with the property (11.4.2), then it has the representation $T(t) = P(e^{it})$, where $P(z)$ is an algebraic polynomial of degree $\leq n$.

This allows an application of trigonometric methods to the boundary values of functions $f \in \mathcal{C}$. We shall need the Jackson integral [I, p. 55]

$$J_n(g, x) = \int_{-\pi}^{\pi} g(x+t) K_n(t)\, dt = \int_{-\pi}^{\pi} g(t) K_n(x-t)\, dt \quad (11.4.3)$$

where

$$K_n(t) = \frac{1}{\lambda_n}\left(\frac{\sin(nt/2)}{\sin(t/2)} \right)^4, \qquad \lambda_n \approx n^3.$$

The kernel $K_n(t)$ is an even trigonometric polynomial of degree $2n-2$, normalized by $\|K_n\|_1 = 1$. It follows that $J_n(g, x)$ is a trigonometric polynomial of degree $2n-2$ that satisfies (11.4.2) if g does.

Lemma 11.7. *For each continuous function $g \in C(\mathbf{T})$ and its Jackson polynomials $J_n(g, x)$ we have (with some constant C)*

$$|g(x) - J_n(g, x)| \leq C\omega(g, 1/n), \tag{11.4.4}$$

$$|J_n'(g, x)| \leq Cn\omega(g, 1/n). \tag{11.4.5}$$

Proof. We establish only the second inequality (see [I, p. 56] for the first inequality). Since

$$J_n'(g, x) = \int_{-\pi}^{\pi} g(t) K_n'(x-t)\, dt$$

$$= \int_0^{\pi} [g(x+t) - g(x-t)] K_n'(t)\, dt,$$

we see that we must estimate the derivative of $K_n(t)$,

$$K_n'(t) = \frac{4}{\lambda_n}\left[\frac{\sin(nt/2)}{\sin(t/2)} \right]^3 \left[\frac{\sin(nt/2)}{\sin(t/2)} \right]'.$$

We do this by means of the inequality $|\sin nu/\sin u| \leq n$ and its corollary $|(\sin nu/\sin u)'| \leq n^2$. This leads to the estimates

$$|K_n'(t)| \leq \text{const min}(n^2, n^{-1}t^{-3}). \tag{11.4.6}$$

For any modulus of continuity ω, we have $\omega(\lambda t) \leq (\lambda+1)\omega(t)$, $t, \lambda > 0$. Therefore

$$|g(x+t) - g(x-t)| \leq \omega(g, 2t) \leq (2tn+1)\omega(g, 1/n). \tag{11.4.7}$$

By means of (11.4.6) and (11.4.7) we obtain the estimate (11.4.5). $\qquad \square$

Making the substitution $e^{it} = z$, we obtain for each $f \in \mathcal{Q}$ the algebraic polynomials $J_n(f, z)$ of degree $2n - 2$. These polynomials satisfy

$$|f(z) - J_n(f, z)| \leqslant C\omega(f, 1/n), \tag{11.4.8}$$

$$|J_n^{(m)}(f, z)| \leqslant C_m n^m \omega(f, 1/n) \tag{11.4.9}$$

for $|z| \leqslant 1$. The inequality (11.4.8) follows from (11.4.4) by means of the maximum modulus principle; the inequality (11.4.9) for $m = 1$ and $|z| = 1$ follows from (11.4.5) because

$$J_n'(f, z) = -ie^{-it}\frac{d}{dt}J_n(f, e^{it}).$$

For higher derivatives we use (11.3.2).

For functions $f \in \mathcal{Q}$ with m_q times differentiable $f(e^{it})$, we can construct the $E_{\bar{m}}, Z_n$ interpolation operator in the form

$$Q_{\bar{m},n}(f, z) = \sum_{j=0}^{q} \sum_{k=1}^{n} f^{(m_j)}(z_k)\alpha_{\bar{m},k,j}(z). \tag{11.4.10}$$

This is the unique polynomial of degree at most $(n+1)q - 1$ that interpolates f for $E_{\bar{m}}, Z_n$. For functions that are not differentiable, the following operator is meaningful

$$\tilde{Q}_{\bar{m},n}(f, z) = \sum_{k=1}^{n} \left[f(z_k)\alpha_{\bar{m},k,0}(z) + \sum_{j=1}^{q} \beta_{k,j}(n)\alpha_{\bar{m},k,j}(z) \right]$$

$$\tag{11.4.11}$$

where $\beta_{k,j}(n)$ are 0 or some arbitrary constants that are not too large. The advantage of the operator $\tilde{Q}_{\bar{m},n}$ is that it is defined for all $f \in \mathcal{Q}$.

For certain special cases, mentioned in §11.2, the following result was obtained by Kis [80] and Sharma [158, 159]. The general case is treated in [140].

Theorem 11.8. (i) *If for fixed* $\bar{m}$ *the constants* $\beta_{k,j}(n)$ *satisfy*

$$B_j = \max_{1 \leqslant k \leqslant n} |\beta_{k,j}(n)| = o\left(\frac{n^{m_j}}{\log n}\right), \qquad n \to \infty, \quad j = 0, \dots, q,$$

$$\tag{11.4.12}$$

then for each $f \in \mathcal{D}$ *we have*

$$\lim_{n \to \infty} \tilde{Q}_{\bar{m},n}(f, z) = f(z) \tag{11.4.13}$$

uniformly in $|z| \leqslant 1$.
  (ii) *If*

$$B_j = o(n^{m_j}), \qquad n \to \infty, \quad j = 0, 1, \dots, q, \tag{11.4.14}$$

then for each $f \in \mathcal{Q}$, (11.4.13) *holds uniformly in* $|z| \leqslant r$ *for each* r, $0 < r < 1$.

Proof. We prove only (i). Let $J_n(f, z)$ be the Jackson polynomial corresponding to $f(z)$. Since $J_n(f, z)$ has degree $2n - 1$, it is preserved by all the operators $Q_{\bar{m}, n}$, that is,

$$J_n(f, z) = Q_{\bar{m}, n}(J_n f, z).$$

Therefore,

$$f(z) - \tilde{Q}_{\bar{m}, n}(f, z) = f(z) - J_n(f, z)$$

$$- \sum_{k=1}^{n} \left\{ [f(z_k) - J_n(f, z_k)] \alpha_{\bar{m}, k, 0}(z) \right.$$

$$\left. - \sum_{j=1}^{q} [\beta_{k, j}(n) - J_n^{(m_j)}(f, z_k)] \alpha_{\bar{m}, k, j}(z) \right\}.$$

$$(11.4.15)$$

The absolute value of the first term on the right does not exceed $C\omega(f, 1/n) \to 0$. Using (11.4.8) and (11.3.11), we can estimate the second term by

$$C\omega(f, 1/n)C(\bar{m})\log n \to 0,$$

since $f \in \mathfrak{D}$. Finally, the absolute value of the last sum does not exceed

$$\{B_j + Cn^{m_j}\omega(f, 1/n)\}C(\bar{m})n^{-m_j}\log n \to 0,$$

by (11.4.12), (11.4.9), (11.3.11), and (11.4.1).   □

As a special case of this theorem, we obtain a statement about the operators $Q_{\bar{m}, n}$:

Corollary 11.9.   *If $f \in \mathcal{C}$ and if $f(e^{it}) \in C^{m_q}(\mathbf{T})$, then*

$$\lim_{n \to \infty} Q_{\bar{m}, n}(f, z) = f(z)$$

uniformly for $|z| \leqslant 1$.

§11.5.   TRIGONOMETRIC INTERPOLATION

There are theorems of interpolation to 2π periodic functions at the equidistant knots $\Pi_n = \{2\pi k / n; \ k = 0, \ldots, n - 1\}$ by trigonometric polynomials that strongly parallel the theory of the last three sections. The early papers dealt with very special cases of the interpolation; for example, Kis [81] treated the $(0, 2)$ case and Sharma and Varma [165] extended this to $(0, m_1)$ interpolation. The regularity of $\bar{m} = (0, m_1, \ldots, m_q)$ interpolation was proved by Cavaretta, Sharma, and Varga [13].

In this section, our system of functions is the full trigonometric system of $2N + 1$ functions $\mathcal{G}$: $\{1, \cos x, \sin x, \ldots, \cos Nx, \sin Nx\}$ or the system

$$\mathcal{G}^* = \{e^{ikx}; \ k = -N, \ldots, N\}.   (11.5.1)$$

The knots are the set Π_n, and the matrix $E_{\bar{m}}$ is the $n \times (m_q + 1)$ matrix of lacunary interpolation that corresponds to the set $\bar{m} = \{0, m_1, \ldots, m_q\}$ with $0 = m_0 < m_1 < \cdots < m_q$. The necessary condition of Theorem 2.11 is satisfied. The trigonometric polynomials have $2N + 1$ coefficients, the matrix has $(q + 1)n$ ones, consequently we must have $(q + 1)n = 2N + 1$. In particular, *n must be odd and q even*. Instead of these conditions, several authors, if necessary, omit one of the functions $\cos Nx, \sin Nx$ from the system $\mathcal{G}$.

In the regularity theorem, we have (see Theorem 2.13) the condition

$$r(\bar{m}) = s(\bar{m}) + 1 \qquad (11.5.2)$$

where $r(\bar{m})$, or $s(\bar{m})$, stand for the number of even, or odd, integers in the sequence $\bar{m}$.

Theorem 11.10 [13].   *The pair $E_{\bar{m}}, \Pi_n$ is regular for trigonometric interpolation if and only if $\bar{m}$ satisfies (11.5.2).*

Proof.   Let $q = 2r$. We write the trigonometric polynomial annihilated by $E_{\bar{m}}, \Pi_n$ in the form

$$T(x) = \sum_{\lambda = -r}^{r} T_\lambda(x) e^{i\lambda nx} \qquad (11.5.3)$$

where $T_\lambda(x)$ are trigonometric polynomials of degree $n = 2p + 1$. As in §§11.1–11.3 the interpolation equations lead to a system of differential equations, and upon isolating T_λ we see that each T_λ must satisfy

$$\tilde{H}_{\bar{m}}(D)(T_\lambda)(x) \equiv 0 \qquad (11.5.4)$$

where $\tilde{H}_{\bar{m}}(D)$ is the determinant

$$\tilde{H}_{\bar{m}}(D) = \det\left[(D - irn)^{m_j}, \ldots, D^{m_j}, \ldots, (D + irn)^{m_j}; j = 0, \ldots, q\right]. \qquad (11.5.5)$$

If we define the determinant $\Delta(\alpha) = \Delta(\alpha, \bar{m})$ by

$$\Delta(\alpha) = \det\left[(\alpha - r)^{m_j}, \ldots, \alpha^{m_j}, (\alpha + 1)^{m_j}, \ldots, (\alpha + r)^{m_j}; j = 0, \ldots, q\right], \qquad (11.5.6)$$

then equations (11.5.4) may be written in the form

$$\tilde{H}_{\bar{m}}(D)(T_\lambda)(x) = (in)^{\Sigma m_j} \sum_{\nu = -p}^{p} c_{\nu, \lambda} \Delta(\nu/n) e^{i\nu x} \equiv 0. \qquad (11.5.7)$$

It will follow that $T_\lambda \equiv 0$ for $\lambda = -r, \ldots, r$ (and consequently, that the pair $E_{\bar{m}}, \Pi_n$ is regular) if all the determinants $\Delta(\nu/n)$ are different from 0.

The determinant $\Delta(\nu/n)$ has the form of the generalized Vandermonde determinant except that some of its entries are negative. To show that $\Delta(\nu/n)$ is nonetheless nonzero when $\bar{m}$ satisfies (11.5.2) represents the major departure from the proof in the roots of unity case; this will be a consequence of the next lemma.

Lemma 11.11 [163]. *Let* $0 = k_0 < k_1 < \cdots < k_l$ *be even integers and* $0 < k_{l+1} < \cdots < k_q$ *be odd integers. For any* $q + 1$ *numbers* $0 < t_1 < t_2 < \cdots < t_{q+1}$, *the determinant*

$$B = \det\left[t_1^{k_j}, (-t_2)^{k_j}, \ldots, \left((-1)^q t_{q+1}\right)^{k_j}; j = 0, \ldots, q\right]$$

is different from 0. *In fact*

$$\operatorname{sgn} B = (-1)^{(q-l)(q+l+1)/2}.$$

Moreover, if q *is even this is true for* $0 = t_1 < t_2 = t_3 < \cdots < t_q = t_{q+1}$ *or for* $0 < t_1 = t_2 < t_3 = t_4 < \cdots < t_{q-1} = t_q < t_{q+1}$ *if and only if* $2l = q$.

Proof. The proof uses the Laplace expansion for a determinant and the positivity of the generalized Vandermonde determinant on positive points (see (4.6.6)). This is easily done by expanding B about its first $l + 1$ rows. Then the determinants involved can always be reduced to generalized Vandermonde determinants on positive points, since powers of -1 can be factored out of the rows corresponding to the odd powers. The result will follow by showing that the sign of each term is the same and is given by the desired formula.

For the cases when $t_{2j} = t_{2j+1}$ or $t_{2j} = t_{2j-1}$, elementary row operations can be used to introduce a large number of 0's into B. In both cases, B will be a constant times the product of two generalized Vandermonde determinants on positive points. We leave the details as an exercise. $\qquad\square$

Corollary 11.12. *The determinant* $\Delta(\alpha)$ *is a polynomial in* α *and is different from* 0 *for* $0 < |\alpha| < \frac{1}{2}$. *Moreover,* $\Delta(\alpha) \neq 0$ *at* $\alpha = 0, \pm\frac{1}{2}$ *if and only if* (11.5.2) *is satisfied. Thus* $\Delta(\alpha)^{-1}$ *is of bounded variation on* $[-\frac{1}{2}, \frac{1}{2}]$ *when* (11.5.2) *is satisfied.*

Proof. We apply Lemma 11.11 to the points

$$t_1 = \alpha, \quad t_2 = 1 - \alpha, \quad t_3 = 1 + \alpha, \ldots, \quad t_q = r - \alpha, \quad t_{q+1} = r + \alpha$$

for $0 < \alpha < \frac{1}{2}$. This shows that $\Delta(\alpha)$, $0 < \alpha < \frac{1}{2}$, is never 0. Also, $\Delta(-\alpha) = (-1)^{s(\bar{m}) + q(q+1)/2} \Delta(\alpha)$. If $\alpha = 0$ or $\frac{1}{2}$, then the special case of the lemma will apply, since for even q, (11.5.2) implies that $2l = q$.

The function $\Delta(\alpha)^{-1}$ is of bounded variation since it has a bounded derivative on $[-\frac{1}{2}, \frac{1}{2}]$ if (11.5.2) holds. $\qquad\square$

We next discuss the fundamental polynomials for the $E_{\bar{m}}$, Π_n interpolation problem. The trigonometric polynomial

$$\tilde{l}(x) = \frac{1}{n} \sum_{\nu=-p}^{p} e^{i\nu x} = \frac{1}{n} \frac{\sin(nx/2)}{\sin(x/2)}, \qquad n = 2p + 1, \qquad (11.5.8)$$

has the values $\tilde{l}(2\pi k/n) = 1$ if $k = 0$, and equals 0 if $k = 1, \ldots, n - 1$. It is the fundamental polynomial for Lagrange interpolation at Π_n. The interpolation properties of the fundamental polynomials $\tilde{\alpha}_{\bar{m},j,0}, j = 0, 1, \ldots, q$, can be described in terms of $\tilde{l}(x)$; $\tilde{\alpha}_{\bar{m},j,0}(x)$ is the unique trigonometric polynomial of degree N that satisfies

$$\tilde{\alpha}_{\bar{m},j,0}^{(m_s)}(2\pi k/n) = \begin{cases} \tilde{l}(2\pi k/n), & s = j, \\ 0 & \text{otherwise.} \end{cases} \tag{11.5.9}$$

The remaining fundamental polynomials are given by translation:

$$\tilde{\alpha}_{\bar{m},j,k}(x) = \alpha_{\bar{m},j,0}(x - 2\pi k/n), \qquad k = 1, \ldots, n-1. \tag{11.5.10}$$

As in §11.3, we obtain estimates on the sums of $\tilde{\alpha}_{\bar{m},j,k}$. This is done in a completely analogous way by writing $\tilde{\alpha}_{\bar{m},j,k}$ in the form (11.5.3) and showing that the polynomials T_λ must satisfy

$$T_\lambda(x) = \tilde{H}_{j,\lambda}(D)\tilde{H}_{\bar{m}}(D)^{-1}[\tilde{l}(x - 2\pi k/n)].$$

The basic inequality is that

$$\sum_{k=0}^{n-1} \left| \tilde{H}_{\bar{m}}(D)^{-1} \tilde{l}\left(x - \frac{2\pi k}{n}\right) \right| \leqslant C(\bar{m}) n^{-\Sigma m_j} \log n. \tag{11.5.11}$$

As a consequence we obtain

$$\sum_{k=0}^{n-1} |\tilde{\alpha}_{\bar{m},j,k}(x)| \leqslant C(\bar{m}) n^{-m_j} \log n. \tag{11.5.12}$$

Once the bound (11.5.12) is established, the method of §11.4 can be used to obtain a convergence theorem. This theorem will apply to the trigonometric interpolation operators given by

$$\tilde{R}_{\bar{m},n}(f,x) = \sum_{k=0}^{n-1} \left\{ f\left(\frac{2\pi k}{n}\right) \tilde{\alpha}_{\bar{m},0,k}(x) + \sum_{j=1}^{q} \beta_{k,j}(n) \tilde{\alpha}_{\bar{m},j,k}(x) \right\}. \tag{11.5.13}$$

Theorem 11.13 [141]. *If*

$$\max_{0 \leqslant k \leqslant n-1} |\beta_{k,j}(n)| = o\left(\frac{n^{m_j}}{\log n}\right), \qquad n \to \infty, \quad j = 1, \ldots, q, \tag{11.5.14}$$

then, for any $f \in C(\mathbf{T})$ satisfying the Dini–Lipschitz condition

$$\omega(f, \delta) \log \frac{1}{\delta} \to 0, \qquad \delta \to 0+,$$

the operators $\tilde{R}_{\bar{m},n}(f,x)$ converge uniformly to $f(x)$ on $[0, 2\pi]$.

In conclusion, we give examples of the theorems that have been obtained when the restrictions n odd, q even are replaced by other restrictions. The differences are already noticeable in the $(0, m)$ case [165]. If m is odd, then the interpolation problem is regular for any $n \geq m$ provided we use the system

$$\{1, \cos x, \sin x, \ldots, \cos(N-1)x, \sin(N-1)x, \sin Nx\}. \quad (11.5.15)$$

The uniform convergence of $\tilde{R}_{\bar{m},n}(f, x)$ to $f(x)$ in this case requires (11.5.14), but is valid for any continuous 2π periodic function.

On the other hand, if m is even, then the interpolation problem is regular only for odd $n \geq m$ and for the system

$$\{1, \cos x, \sin x, \ldots, \cos(N-1)x, \sin(N-1)x, \cos Nx\}. \quad (11.5.16)$$

This time the uniform convergence of $\tilde{R}_{\bar{m},n}(f, x)$ to $f(x)$ requires

$$\max_{0 \leq k \leq n-1} |\beta_{k,1}(n)| = o(n^{m-1}), \qquad n \to \infty, \qquad (11.5.17)$$

and is only valid for functions satisfying the Zygmund condition

$$f(x+h) - 2f(x) + f(x-h) = o(h), \qquad h \to 0. \quad (11.5.18)$$

The proof can be found in [165, 141].

§11.6. NOTES

11.6.1. Cavaretta, Sharma, and Varga [13] first proved the general regularity theorem for the $E_{\bar{m}}$, Π_n trigonometric interpolation problem by a long induction proof. A simpler proof appears in [163], and the proof in §11.5 was sketched in [142]. The general theorem has two remaining cases, which are determined by the choice of the system $\mathcal{G}$. I. For the system (11.5.15), $E_{\bar{m}}$, Π_n is regular if and only if $q = 2r+1$, $r(\bar{m}) = s(\bar{m})+1$, and $N = nr + n$. II. For the system (11.5.16), $E_{\bar{m}}$, Π_n is regular if and only if either (a) $n = 2p+1$, $q = 2r+1$, $r(\bar{m}) = s(\bar{m})+2$, and $N = nr + n$, or (b) $n = 2p$, $q = 2r$, $r(\bar{m}) = s(\bar{m})+1$ and $N = nr + p$.

11.6.2. General convergence theorems for $E_{\bar{m}}$, Π_n trigonometric interpolation were proved by Riemenschneider, Smith, and Sharma [141]. It was shown that Theorem 11.15 remains valid in cases I and IIb. In case IIa, however, the Dini–Lipschitz condition is replaced by the Zygmund condition (11.5.18), and (11.5.14) is replaced by (11.5.17), but even then an additional condition on $\bar{m}$ is required.

11.6.3. Special cases of the general results on lacunary trigonometric cases appeared earlier in the literature, starting with that of O. Kis [81]. New ideas were introduced by Sharma and Varma [165, 166], Varma [181, 183], and Zeel [191]. Sometimes the estimates obtained in the special cases gave better results for convergence (e.g., $(0, m)$ with m odd) than were obtained in the general theorems of [141]. These cases have been included in a general theorem by Sharma and Varma [167].

11.6.4. For a function analytic in $|z| < \rho$ with a singularity on $|z| = \rho$, $1 < \rho < \infty$, Walsh [P, p. 153] showed that the difference between its Lagrange polynomial interpolant at the nth roots of unity and the $(n-1)$st partial sum of its power series convergences to zero uniformly in $|z| \leqslant \rho^*$, $\rho^* < \rho^2$. This result has been extended to Hermite and even $(0, m_1, \ldots, m_q)$ interpolation at the roots of unity by Cavaretta, Sharma and Varga [203] and Saxena, Sharma and Ziegler [204], but with the convergence necessarily restricted to $|z| < \rho^\alpha$, for some $\alpha = \alpha(q)$, $1 < \alpha < 2$.

Chapter 12

Turán's Problem of (0, 2) Interpolation

§12.1. REGULARITY OF SETS OF KNOTS

In this chapter, P_n will denote the nth Legendre polynomial, normalized by $P_n(1) = 1$. The polynomials P_n are orthogonal on $[-1, 1]$ with weight $w(x) \equiv 1$. They are also given by the Rodrigues formula

$$P_n(x) = \frac{1}{2^n n!} \frac{d^n}{dx^n} (x^2 - 1)^n. \tag{12.1.1}$$

For example, $P_0 = 1$, $P_1 = x$, $P_2 = \frac{3}{2} x^2 - \frac{1}{2}$. We list some of their properties (see [L, J, M]). We have $|P_n(x)| \leqslant 1$ on $[-1, +1]$; $P_n(-1) = (-1)^n$; the zeros of P_n are simple and contained in $(-1, +1)$. We have the identities

$$(1 - x^2) P_n'' - 2x P_n' + n(n+1) P_n = \left[(1 - x^2) P_n' \right]' + n(n+1) P_n = 0; \tag{12.1.2}$$

$$(n+1) P_{n+1} = (2n+1) x P_n - n P_{n-1}; \tag{12.1.3}$$

$$(1 - x^2) P_n' = n P_{n-1} - n x P_n; \tag{12.1.4}$$

$$n P_n = x P_n' - P_{n-1}'; \tag{12.1.5}$$

$$(n+1) P_n = P_{n+1}' - x P_n'. \tag{12.1.6}$$

Since our formulas will often be complicated, we will sometimes omit the argument x under the function sign [e.g., in (12.1.2)].

Let E be the $n \times 2n$ matrix with rows $(1\ 0\ 1\ 0\ 0\ \cdots\ 0)$, that is, Turán's matrix E_n of §2.3. Let X be the set of knots

$$-1 = x_1 < x_2 < \cdots < x_{n-1} < x_n = 1, \qquad (12.1.7)$$

where x_i, $i = 2,\ldots,n-1$, are the zeros of P'_{n-1}. We have [3]:

Theorem 12.1. *The pair E, X is regular if n is even and singular if n is odd.*

Proof. The points (12.1.7) are zeros of the polynomial of degree n, $\Pi_n(x) = (1 - x^2)P'_{n-1}(x)$. By (12.1.2) we have the formulas

$$\Pi_n(x) = (1 - x^2)P'_{n-1}(x) = -n(n-1)\int_{-1}^{x} P_{n-1}(t)\,dt, \quad (12.1.8)$$

$$\Pi'_n(x) = -n(n-1)P_{n-1}(x). \qquad (12.1.9)$$

Let Q be a polynomial of degree $\leqslant 2n - 1$ annihilated by E, X. We have to show that $Q \equiv 0$. Since Q vanishes at the knots (12.1.7), we have $Q = \Pi_n q$, where q has degree $\leqslant n - 1$. By differentiation,

$$Q''(x) = \Pi''_n(x)q(x) + 2\Pi'_n(x)q'(x) + \Pi_n(x)q''(x). \quad (12.1.10)$$

We put $x = x_i$, $i = 2,\ldots,n-1$. Then $Q''(x_i) = 0$, further $\Pi_n(x_i) = 0$, $\Pi'_n(x_i) \neq 0$, and $\Pi''_n(x_i) = 0$ by (12.1.9). This yields $q'(x_i) = 0$. The zeros of q' being the same as those of P'_{n-1}, we have, with a constant C, $q' = CP'_{n-1}$, and by integration

$$q = CP_{n-1} + C_1. \qquad (12.1.11)$$

We have still to use the conditions $Q''(\pm 1) = 0$. We recall that $P_n(-1) = (-1)^n P_n(1) = (-1)^n$. Here and later on we need the following values:

$$\Pi'_n(1) = -n(n-1), \qquad \Pi'_n(-1) = n(n-1)(-1)^n,$$

$$P'_{n-1}(1) = \tfrac{1}{2}n(n-1), \qquad P'_{n-1}(-1) = \tfrac{1}{2}n(n-1)(-1)^n, \qquad (12.1.12)$$

$$\Pi''_n(1) = -\tfrac{1}{2}n^2(n-1)^2, \qquad \Pi''_n(-1) = -\tfrac{1}{2}n^2(n-1)^2(-1)^n.$$

The first line follows from (12.1.9); the second from (12.1.2) for $x = \pm 1$; the third from the relation $\Pi''_n = -n(n-1)P'_{n-1}$.

The conditions $Q''(\pm 1) = 0$ give, with the help of (12.1.11),

$$\Pi''_n(1)(C + C_1) + 2\Pi'_n(1)P'_{n-1}(1)C = 0,$$

$$\Pi''_n(1)\big(C(-1)^{n-1} + C_1\big) + 2\Pi'_n(1)P'_{n-1}(1)C(-1)^{n-1} = 0.$$

If n is odd, the equations are identical and have nonzero solutions C, C_1. If n is even, however, we get $C = C_1 = 0$. $\qquad\square$

Next we consider the singular $(n+2) \times 4(n+1)$ matrix

$$
\begin{bmatrix}
1 & 1 & 0 & 0 & 0 & 0 & \cdots \\
1 & 1 & 1 & 0 & 1 & 0 & \cdots \\
\cdot & \cdot & \cdot & \cdot & \cdot & \cdot & \cdots \\
1 & 1 & 1 & 0 & 1 & 0 & \cdots \\
1 & 1 & 0 & 0 & 0 & 0 & \cdots
\end{bmatrix}
$$

of the $(0,1,2,4)$ interpolation, modified to $(0,1)$ interpolation at the extreme knots. Let X be the set of zeros of $(1-x^2)P_n(x)$.

Theorem 12.2 (Varma [182]). *The pair E, X is regular if n is even, singular if n is odd.*

Proof. Let

$$X: -1 = x_1 < x_2 < \cdots < x_{n+1} < x_{n+2} = 1$$

be the zeros of $(1-x^2)P_n(x)$. An essential relation for the feasibility of all calculations is

$$\left[(1-x^2)P_n(x)^2\right]'''\big|_{x=x_i} = 0, \qquad i = 2,\ldots,n+1. \qquad (12.1.13)$$

Indeed, by Leibnitz' formula we have the congruence modulo the polynomial P_n,

$$
\begin{aligned}
\left[(1-x^2)P_n(x)^2\right]''' &\equiv 3(-2x)2P_n'(x)^2 + (1-x^2)6P_n'(x)P_n''(x) \\
&\equiv 6P_n'(x)\left[-2xP_n'(x) + (1-x^2)P_n''(x)\right] \equiv 0.
\end{aligned}
$$
$$(12.1.14)$$

Let Q be a polynomial of degree $\leqslant 4n+3$ annihilated by E, X. Since $Q(x_i) = Q'(x_i) = 0$ for all i, $Q''(x_i) = 0$, $i = 2,\ldots,n+1$, we put

$$Q(x) = (1-x^2)^2 P_n(x)^3 q(x) \qquad (12.1.15)$$

where q is a polynomial of degree $\leqslant n-1$. We compute $Q^{(4)}(x)$ for $x = x_i$, $i = 2,\ldots,n+1$. Again modulo P_n we have, in view of (12.1.14) and (12.1.2),

$$
\begin{aligned}
Q^{(4)} &\equiv 4\left[(1-x^2)^2 P_n^3\right]''' q' + \left\{\left[(1-x^2)P_n^2\right]\left[(1-x^2)P_n\right]\right\}^{(4)} q \\
&\equiv 4(1-x^2)^2 6P_n'^3 q' + 6\left[(1-x^2)P_n^2\right]''\left[(1-x^2)P_n\right]'' q \\
&\equiv 24(1-x^2)P_n'^3\left[(1-x^2)q' - xq\right].
\end{aligned}
$$

We see that $(1-x^2)q' - xq$ vanishes at the knots x_i, $i = 2,\ldots,n+1$. Hence

$$(1-x^2)q'(x) - xq(x) = CP_n(x) \qquad (12.1.16)$$

where C is some constant. Conversely, if a polynomial q of degree $\leqslant n-1$ satisfies this relation, then Q given by (12.1.15) is annihilated by E, X. The question is whether (12.1.16) implies that q is identically 0. Since the P_k are linearly independent, we have, for some constants a_k,

$$q(x) = \sum_{k=0}^{n-1} a_k P_k(x). \qquad (12.1.17)$$

We substitute (12.1.17) into (12.1.16). From relations (12.1.3), (12.1.4),

$$(1 - x^2) P_k' - x P_k = \frac{1}{2k+1} \left[k^2 P_{k-1} - (k+1)^2 P_{k+1} \right]$$

and we get

$$\sum_{k=0}^{n-2} \frac{(k+1)^2}{2k+3} a_{k+1} P_k - \sum_{k=1}^{n} \frac{k^2}{2k-1} a_{k-1} P_k = C P_n.$$

Comparing the coefficients, we obtain $0 = a_1 = a_3 = \cdots$ and $0 = a_{n-2} = a_{n-4} = \cdots$. If n is even, this gives $a_k = 0$ for all k. But for odd n, the even-numbered coefficients $a_2, \ldots, a_{n-1}$ are not necessarily 0, and E, X is singular. $\square$

§12.2. FURTHER PROPERTIES OF LEGENDRE POLYNOMIALS

We collect here, partly without proof, some properties of the Legendre polynomials P_{n-1} which will be needed in §12.5.

It is known that the zeros $\xi_1 < \cdots < \xi_{n-1}$ of $P_{n-1}(x)$ are real, simple, and contained in the open interval $(-1, +1)$. Some of their properties are best formulated by the substitution $x = \cos \theta$, $0 \leqslant \theta \leqslant \pi$.

Let $\xi_i = \cos \theta_i$, $0 < \theta_i < \pi$; then $\pi > \theta_1 > \cdots > \theta_{n-1} > 0$. We have (see [L, p. 122])

Lemma 12.3. *The numbers θ_i satisfy the inequalities*

$$\frac{2(n-i)-1}{2n-1} \pi < \theta_i < \frac{2(n-i)}{2n-1} \pi, \qquad i = 1, \ldots, n-1. \qquad (12.2.1)$$

We denote the zeros of P_{n-1}' by x_i, $i = 2, \ldots, n-1$. They satisfy $\xi_{i-1} < x_i < \xi_i$, $i = 2, \ldots, n-1$. If we write $x_i = \cos \alpha_i$, then $\theta_i < \alpha_i < \theta_{i-1}$ and because of (12.2.1),

$$\pi - \frac{2i}{2n-1} \pi = \frac{2(n-i)-1}{2n-1} \pi < \alpha_i < \frac{2(n-i+1)}{2n-1} \pi = \pi - \frac{2i-3}{2n-1} \pi. \qquad (12.2.2)$$

Returning to x_i, we have $\sqrt{1 - x_i^2} = \sin \alpha_i$; hence as a corollary of (12.2.2), with absolute constants $c_1, c_2 > 0$,

$$c_1 \frac{i}{n} \leqslant \sqrt{1 - x_i^2} \leqslant c_2 \frac{i}{n}, \qquad 2 \leqslant i \leqslant \frac{n}{2},$$

$$c_1 \frac{n-i}{n} \leqslant \sqrt{1 - x_i^2} \leqslant c_2 \frac{n-i}{n}, \qquad \frac{n}{2} < i \leqslant n - 1. \tag{12.2.3}$$

Lemma 12.4. *For the polynomial Π_n of (12.1.8) we have*

$$|\Pi_n(x)| \leqslant 3\sqrt{n} \tag{12.2.4}$$

Proof. From the known inequality [J, Vol. II, p. 94]

$$|P_n(x)| \leqslant \sqrt{\frac{\pi}{2n}} \frac{1}{\sqrt{1 - x^2}}$$

we obtain $|\sin\theta P_n(\cos\theta)| \leqslant \sqrt{\pi/2n}$, and from here, by Bernstein's inequality for the derivative of a trigonometric polynomial,

$$|P_{n-1}'(\cos\theta)\sin^2\theta - \cos(\theta)P_{n-1}(\cos\theta)| \leqslant \sqrt{\frac{\pi n}{2}} .$$

Hence $|P_{n-1}'(\cos\theta)\sin^2\theta| \leqslant \sqrt{\pi n/2} + 1 \leqslant 3\sqrt{n}$, which is identical with (12.2.4). (Bernstein [A, p. 67] proves even that $|\Pi_n(x)| \leqslant \text{const} \sqrt{n}$ $\times (1 - x^2)^{1/4}$.) $\qquad\qquad \square$

Lemma 12.5 [3]. *For some constant $c > 0$ and all even n,*

$$|P_{n-1}(x_i)| \geqslant \frac{c}{\sqrt{i}}, \quad 1 < i \leqslant \frac{n}{2}, \qquad |P_{n-1}(x_i)| \geqslant \frac{c}{\sqrt{n-i}}, \quad \frac{n}{2} < i < n. \tag{12.2.5}$$

Proof. It is enough to prove the first inequality. We start with the functions

$$u(\theta) = \sqrt{\sin\theta} \, P_{n-1}(\cos\theta), \qquad 0 \leqslant \theta \leqslant \pi,$$

$$\Phi(\theta) = \left(n - \tfrac{1}{2}\right)^2 + \frac{1}{4\sin^2\theta} .$$

We differentiate the function u two times and take into account the differential equation (12.1.2) of the Legendre polynomials, which in the present situation takes the form

$$\sin^2\theta P_{n-1}''(\cos\theta) - 2\cos\theta P_{n-1}'(\cos\theta) + n(n-1)P_{n-1}(\cos\theta) = 0.$$

In this way we obtain the differential equation of the function u,

$$u''(\theta) + \Phi(\theta)u(\theta) = 0.$$

We then see that the positive function

$$g(\theta) = u(\theta)^2 + \frac{1}{\Phi(\theta)}u'(\theta)^2 \tag{12.2.6}$$

has the derivative $g' = - \Phi'\Phi^{-2}u'^2$, which is negative for $\frac{1}{2}\pi < \theta < \pi$, since the function Φ increases on this interval. Thus, *the function $g(\theta)$ is decreasing on $[\pi/2, \pi]$.*

We want to estimate the values $M_i = |P_{n-1}(x_i)|$ for $2 \leqslant i \leqslant n/2$. We have $x_i = \cos \alpha_i$, where the α_i belong to the interval $[\pi/2, \pi]$. Since $P'_{n-1}(\cos \alpha_i) = 0$, we obtain from (12.2.6) and the definition of the function Φ,

$$g(\alpha_i) \leqslant M_i^2 \sin \alpha_i + \frac{1}{\Phi(\alpha_i)} \left(\frac{\cos \alpha_i}{2\sqrt{\sin \alpha_i}} \right)^2 M_i^2$$

$$\leqslant M_i^2 \sin \alpha_i + 4\sin^2\alpha_i \frac{\cos^2\alpha_i}{4\sin\alpha_i} M_i^2$$

$$\leqslant 2 M_i^2 \sin \alpha_i.$$

Hence

$$M_i \geqslant \left(\frac{g(\alpha_i)}{2\sin\alpha_i} \right)^{1/2} \geqslant \frac{g(\alpha)^{1/2}}{\sqrt{2\sin\alpha_i}}$$

where α, $\alpha_2 \leqslant \alpha < \pi$, is arbitrary. This leads to

$$M_i \geqslant \frac{|u(\alpha)|}{\sqrt{2\sin\alpha_i}} = \sqrt{\frac{\sin\alpha}{2\sin\alpha_i}} |P_{n-1}(\cos\alpha)|, \qquad 2 \leqslant i \leqslant \frac{n}{2}. \quad (12.2.7)$$

We select $\alpha = \pi - n^{-1}$. This is possible, because by (12.2.2), $\alpha_2 < \pi - \pi/(2n-1) < \alpha$.

Next, we have $P_{n-1}(-1) = -1$. The inequality

$$|1 + P_{n-1}(\cos\alpha)| \leqslant \|P'_{n-1}\| |-1-\cos\alpha| < n^2 2\sin^2\frac{1}{2n} < \frac{1}{2}$$

shows that $|P_{n-1}(\cos\alpha)| \geqslant \frac{1}{2}$. Moreover,

$$\sin\alpha \geqslant \text{const}\frac{1}{n} \quad \text{and} \quad \alpha_i \leqslant \pi - (2i-3)\frac{\pi}{(2n-1)} + \mathcal{O}\left(\frac{1}{n}\right),$$

so that $\sin\alpha_i \leqslant \text{const}(i/n)$. Collecting these estimates, we obtain (12.2.5). $\square$

Let $x_1 < \cdots < x_n$ be arbitrary real interpolation knots. The fundamental functions $l_i(x)$ of the Lagrange interpolation are then

$$l_i(x) = \frac{\omega(x)}{(x - x_i)\omega'(x_i)}, \qquad 1 \leqslant i \leqslant n,$$

where ω is the polynomial $\omega(x) = \prod_{i=1}^{n}(x - x_i)$. We shall need properties of these functions when the x_i are the roots of the polynomial

$$\Pi_n(x) = (1 - x^2) P'_{n-1}(x).$$

In this case,

$$l_i(x) = \frac{\Pi_n(x)}{(x - x_i)\Pi'_n(x_i)}, \qquad 1 \leqslant i \leqslant n. \qquad (12.2.8)$$

Lemma 12.6. *The functions* (12.2.8) *satisfy*

$$\sum_{i=1}^{n} l_i(x)^2 \leqslant 1, \qquad -1 \leqslant x \leqslant 1. \qquad (12.2.9)$$

Proof. This is best derived from the formula of the $(0,1)$ Hermite interpolation. First let the knots x_i be arbitrary. For a differentiable function f, this formula gives the interpolating polynomial

$$Q_{2n-1}(x) = \sum_{i=1}^{n} f(x_i)s_i(x) + \sum_{i=1}^{n} f'(x_i)\sigma_i(x), \qquad (12.2.10)$$

and the fundamental polynomials s_i, σ_i are given by [L, p. 330]:

$$s_i(x) = v_i(x)l_i(x)^2, \qquad \sigma_i(x) = (x - x_i)l_i(x)^2,$$

$$v_i(x) = 1 - \frac{\omega''(x_i)}{\omega'(x_i)}(x - x_i), \qquad i = 1, \ldots, n.$$

If f is a polynomial of degree $\leqslant 2n - 1$, formula (12.2.10) reproduces f. In particular, for $f \equiv 1$,

$$\sum_{i=1}^{n} v_i(x)l_i(x)^2 = 1. \qquad (12.2.11)$$

We compute the linear functions v_i in the particular case $\omega = \Pi_n$. Since $\Pi''_n(x_i) = 0$, $1 < i < n$, by (12.1.9) we have for these i, $v_i(x) \equiv 1$. For $i = 1$ and $i = n$, we use the values (12.1.12) for $\Pi'_n(\pm 1)$, $\Pi''_n(\pm 1)$ and obtain

$$v_1(x) = 1 + \frac{n(n-1)}{2}(1 + x) \geqslant 1,$$

$$v_n(x) = 1 + \frac{n(n-1)}{2}(1 - x) \geqslant 1,$$

for $-1 \leqslant x \leqslant 1$. We see that (12.2.11) implies (12.2.9).   □

Corollary 12.7. *If the knots x_i are the zeros of Π_n, we have*

$$l'_i(x_i) = 0, \qquad 1 < i < n. \qquad (12.2.12)$$

Indeed, from (12.2.9), $l_i(x)^2 \leqslant 1$, and since $l_i(x_i) = 1$, the polynomial $l_i(x)$ attains an extremum (equal to ± 1) at the interior point x_i of $[-1, +1]$.   □

§12.3. FUNDAMENTAL POLYNOMIALS ρ_i

The problem of $(0,2)$ interpolation in the roots of polynomials $(1 - x^2)$ $\times P'_{n-1}(x)$ is solvable (see §12.1) for even n. If $-1 = x_1 < \cdots < x_n = +1$

are the zeros of this polynomial, and $\alpha_i, \beta_i, i = 1,\ldots,n$, are given, the polynomial Q of degree $\leqslant 2n - 1$ that assumes the values $Q(x_i) = \alpha_i, Q''(x_i) = \beta_i, i = 1,\ldots,n$, is of the form

$$Q(x) = \sum_{i=1}^{n} \alpha_i r_i(x) + \sum_{i=1}^{n} \beta_i \rho_i(x) \qquad (12.3.1)$$

where the fundamental polynomials r_i, ρ_i are given by the conditions

$$r_i(x_j) = \begin{cases} 0, & j \neq i, \\ 1, & j = i, \end{cases} \qquad r_i''(x_j) = 0, \qquad j = 1,\ldots,n, \quad (12.3.2)$$

$$\rho_i(x_j) = 0, \qquad j = 1,\ldots,n, \qquad \rho_i''(x_j) = \begin{cases} 0, & j \neq i, \\ 1, & j = i. \end{cases} \quad (12.3.3)$$

In this section we shall find explicit formulas for the ρ_i. In what follows, n will be even.

Theorem 12.8.   *We have*

$$\rho_1(x) = \frac{\Pi_n(x)}{n^2(n-1)^2}\left[-1 + \frac{1}{3}P_{n-1}(x)\right], \qquad (12.3.4)$$

$$\rho_n(x) = -\frac{\Pi_n(x)}{n^2(n-1)^2}\left[1 + \frac{1}{3}P_{n-1}(x)\right], \qquad (12.3.5)$$

and for $1 < i < n$,

$$\rho_i(x) = \frac{\Pi_n(x)(1 - x_i^2)}{2\Pi_n'(x_i)^2}\left[\int_{-1}^{x} \frac{P_{n-1}'(t)\,dt}{t - x_i}\right.$$

$$\left. -\left(\frac{1}{6}I_i + \frac{2}{3}\frac{x_i}{1 - x_i^2}\right)P_{n-1}(x) - \left(\frac{1}{2}I_i + \frac{2}{1 - x_i^2}\right)\right], \quad (12.3.6)$$

where I_i *stands for the integral*

$$I_i = \int_{-1}^{+1} \frac{P_{n-1}'(t)\,dt}{t - x_i}, \qquad 1 < i < n. \qquad (12.3.7)$$

   Proof.   To find a formula for ρ_i, we start with the fact that $\rho_i(x_j) = 0$, $j = 1,\ldots,n$. Hence ρ_i must be of the form $\rho_i = \Pi_n q$, where q is some polynomial (which depends on i) of degree $\leqslant n - 1$. Differentiating this relation, putting $x = x_j$, and using the fact that $\Pi_n(x_j) = 0$ for all j, we get

$$\Pi_n''(x_j)q(x_j) + 2\Pi_n'(x_j)q'(x_j) = \begin{cases} 0 & \text{if} \quad j \neq i, \\ 1 & \text{if} \quad j = i. \end{cases} \qquad (12.3.8)$$

Since $\Pi_n''(x_j) = 0, j = 2,\ldots,n - 1$, and $\Pi_n'(x_j) \neq 0$ for all j, we get

$$q'(x_j) = 0, \qquad j \neq i, \quad j = 2,\ldots,n - 1. \qquad (12.3.9)$$

It is particularly simple to compute ρ_1 and ρ_n; we do this for ρ_1 (leaving the case of ρ_n to the reader). If $i = 1$, q' vanishes at the zeros of $\Pi_n''(x)$. Hence $q' = C\Pi''$, and

$$q(x) = C\Pi_n'(x) + C_1. \tag{12.3.10}$$

As in the proof of Theorem 12.1, the constants C, C_1 can be determined from the conditions $\rho_1''(x_1) = 1$, $\rho_1''(x_n) = 0$, because the values $\Pi_n'(\pm 1)$, $\Pi_n''(\pm 1)$ are known [see (12.1.12), where now n is even]. Substituting these values into (12.3.10), we obtain (12.3.4).

Determination of ρ_i, $i = 2, \ldots, n - 1$, in the same way first leads to a formula of different structure. From (12.3.9) we see that q' vanishes at the zeros of P_{n-1}', except for the zero x_i. Hence we must have

$$q'(x) = \frac{P_{n-1}'(x)}{x - x_i}\left[C(x - x_i) + C_1\right] = CP_{n-1}'(x) + C_1\frac{P_{n-1}'(x)}{x - x_i}. \tag{12.3.11}$$

From (12.3.8) for $x = x_i$, $q'(x_i) = (2\Pi_n'(x_i))^{-1}$. On the other hand, from (12.3.11), $q'(x_i) = P_{n-1}''(x_i)C_1$. Remembering that

$$P_{n-1}''(x_i) = -\frac{n(n-1)}{1 - x_i^2}P_{n-1}(x_i) = \frac{\Pi_n'(x_i)}{1 - x_i^2},$$

we obtain for C_1,

$$C_1 = \frac{1 - x_i^2}{2\Pi_n'(x_i)^2}. \tag{12.3.12}$$

Integrating (12.3.11) yields

$$q(x) = CP_{n-1}(x) + C_1\int_{-1}^{x}\frac{P_{n-1}'(t)\,dt}{t - x_i} + C_2. \tag{12.3.13}$$

To determine C and C_2, we use the equations $\rho_i''(\pm 1) = 0$. Equation (12.3.8) and values of $\Pi_n'(\pm 1)$, $\Pi_n''(\pm 1)$ from (12.1.12), with even n, give

$$\begin{aligned}n(n-1)q(-1) - 4q'(-1) &= 0,\\ n(n-1)q(1) + 4q'(1) &= 0.\end{aligned} \tag{12.3.14}$$

The formulas for q, q' and the values of $P_n'(\pm 1)$ from (12.1.12) yield

$$q(-1) = -C + C_2, \qquad q(1) = C + C_1 I_i + C_2,$$

$$q'(-1) = \frac{1}{2}n(n-1)\left(C + \frac{C_1}{-1 - x_i}\right), \qquad q'(1) = \frac{1}{2}n(n-1)\left(C + \frac{C_1}{1 - x_i}\right),$$

where I_i is the integral (12.3.7). Substituting this into (12.3.14), we obtain the values

$$C = -C_1\left(\frac{1}{6}I_i + \frac{2}{3}\frac{x_i}{1 - x_i^2}\right), \qquad C_2 = -C_1\left(\frac{1}{2}I_i + \frac{2}{1 - x_i^2}\right).$$

Substituting this and the value of C_1 into (12.3.13), and remembering that $\rho_i = \Pi_n q$, we get the required formula (12.3.6).   □

Lemma 12.9. *We have, for even n and $1 < i < n$,*

$$I_i = \int_{-1}^{+1} \frac{P'_{n-1}(t)\,dt}{t - x_i} = \frac{2x_i}{1 - x_i^2} - \frac{2}{1 - x_i^2}\frac{1}{P_{n-1}(x_i)}. \qquad (12.3.15)$$

Proof. It is easy to see that the derivatives P'_n of Legendre polynomials form an orthogonal (but not normalized) system on $[-1, +1]$ with respect to the weight $1 - x^2$. To each orthogonal system of polynomials corresponds a Christoffel–Darboux formula [J, p. 67]. We prefer to derive this formula for P'_n directly as follows. From (12.1.5) and (12.1.6) we eliminate P_n and obtain (after replacing x by t and n by $k - 1$) the recursion formula for the P'_n:

$$P'_k(t) = \frac{2k - 1}{k - 1} t P'_{k-1}(t) - \frac{k}{k - 1} P'_{k-2}(t), \qquad k = 2, \ldots .$$

We multiply this with $P'_{k-1}(x)$, then interchange x and t in the obtained formula and subtract the results:

$$\frac{1}{k}\left[P'_k(t)P'_{k-1}(x) - P'_k(x)P'_{k-1}(t)\right]$$

$$= \frac{2k - 1}{k(k - 1)}(t - x)P'_{k-1}(t)P'_{k-1}(x)$$

$$+ \frac{1}{k - 1}\left[P'_{k-1}(t)P'_{k-2}(x) - P'_{k-1}(x)P'_{k-2}(t)\right].$$

Summation with respect to k gives the required formula

$$\frac{P'_{n-1}(t)P'_{n-2}(x) - P'_{n-1}(x)P'_{n-2}(t)}{(n - 1)(t - x)} = \sum_{k=2}^{n-1} \frac{2k - 1}{k(k - 1)} P'_{k-1}(t)P'_{k-1}(x).$$

$$(12.3.16)$$

To compute the integral I_i, we note that

$$(1 - x_i^2)I_i = \int_{-1}^{+1}(1 - t^2)\frac{P'_{n-1}(t)}{t - x_i}\,dt + \int_{-1}^{+1}(t^2 - x_i^2)\frac{P'_{n-1}(t)}{t - x_i}\,dt.$$

Because n is even, the second integral is equal to

$$\int_{-1}^{+1}(t + x_i)P'_{n-1}(t)\,dt = x_i\int_{-1}^{1}P'_{n-1}(t)\,dt = 2x_i.$$

Therefore

$$I_i = \frac{2x_i}{1 - x_i^2} + \frac{1}{1 - x_i^2}\int_{-1}^{+1}(1 - t^2)\frac{P'_{n-1}(t)\,dt}{t - x_i}. \qquad (12.3.17)$$

To compute this, we put in (12.3.16) $x = x_i$, multiply by $1 - t^2$, and integrate. It is to be noted that by integration by parts,

$$\int_{-1}^{+1}(1-t^2)P'_{k-1}(t)\,dt = \int_{-1}^{+1}2tP_{k-1}(t)\,dt = 0$$

for $k = 3, 4, \ldots$, since P_{k-1} is orthogonal to P_1. If $k = 2$, this integral is equal to $\frac{4}{3}$. Hence

$$\frac{1}{n-1}P'_{n-2}(x_i)\int_{-1}^{+1}(1-t^2)\frac{P'_{n-1}(t)\,dt}{t-x_i} = \frac{3}{2}\frac{4}{3}P'_1(x_i) = 2. \quad (12.3.18)$$

Combining (12.3.17), (12.3.18) and observing that because of (12.1.5), $P'_{n-2}(x_i) = -(n-1)P_{n-1}(x_i)$, we obtain the lemma.   □

§12.4. FUNDAMENTAL POLYNOMIALS r_i

Theorem 12.10.   *For even n, we have the formulas*

$$r_i(x) = l_i(x)^2 + \frac{\Pi_n(x)}{\Pi'_n(x_i)}\left(\int_x^1 \frac{l'_i(t)\,dt}{t-x_i} + \frac{P_{n-1}(x)-3}{6}I'_i\right),$$

$$1 < i < n, \quad (12.4.1)$$

$$r_1(x) = \frac{3-x}{4}l_1^2(x) + \frac{1-x^2}{4}l_1(x)l'_1(x)$$

$$+ \left[\frac{5}{16} + \frac{1}{8n(n-1)}\right]\Pi_n(x)\left(1 - \frac{1}{3}P_{n-1}(x)\right), \quad (12.4.2)$$

$$r_n(x) = \frac{3+x}{4}l_n(x)^2 - \frac{1-x^2}{4}l_n(x)l'_n(x)$$

$$+ \left[\frac{5}{16} + \frac{1}{8n(n-1)}\right]\Pi_n(x)\left(1 + \frac{1}{3}P_{n-1}(x)\right). \quad (12.4.3)$$

Here I'_i is the integral

$$I'_i = \int_{-1}^{+1}\frac{l'_i(t)}{t-x_i}\,dt. \quad (12.4.4)$$

Proof.   The integral I'_i exists because $l'_i(x_i) = 0$ by Corollary 12.7. The polynomials r_i of degree $2n-1$ are uniquely determined by the equations (12.3.2). We first assume $1 < i < n$. Since the fundamental polynomials of Lagrange interpolation, given by

$$l_i(x) = \frac{\Pi_n(x)}{(x-x_i)\Pi'_n(x_i)}, \quad i = 1, \ldots, n, \quad (12.4.5)$$

as well as their squares satisfy the first group of these equations, it is natural to seek r_i in the form

$$r_i(x) = l_i(x)^2 + \frac{\Pi_n(x)}{\Pi'_n(x_i)}q(x) \quad (12.4.6)$$

where $q(x)$ is a polynomial (that depends on i) of degree $n-1$. We must select $q(x)$ so that the conditions $r_i''(x_j) = 0$, $j = 1, \ldots, n$, are satisfied. Differentiating (12.4.6) twice and putting $x = x_j$, we must remember that $\Pi_n(x_j) = 0$ for all j, $\Pi_n''(x_j) = 0$ for $1 < j < n$ [see (12.1.8), (12.1.9)] and that $l_i'(x_i) = 0$. Separating the cases when $j = i$, $x_j = \pm 1$, and all other values of j, we obtain

$$l_i'(x_j)^2 + \frac{\Pi_n'(x_j)}{\Pi_n'(x_i)} q'(x_j) = 0, \qquad 1 < j < n, \quad j \neq i, \qquad (12.4.7)$$

$$l_i''(x_i) + q'(x_i) = 0, \qquad 1 < i < n, \qquad (12.4.8)$$

$$2l_i'(1)^2 + 2\frac{\Pi_n'(1)}{\Pi_n'(x_i)} q'(1) + \frac{\Pi_n''(1)}{\Pi_n'(x_i)} q(1) = 0,$$
$$(12.4.9)$$
$$2l_i'(-1)^2 + 2\frac{\Pi_n'(-1)}{\Pi_n'(x_i)} q'(-1) + \frac{\Pi_n''(-1)}{\Pi_n'(x_i)} q(-1) = 0, \qquad 1 < i < n.$$

Because

$$\frac{\Pi_n'(x)}{\Pi_n'(x_i)} = l_i'(x)(x - x_i) + l_i(x),$$

the equations (12.4.7) will certainly be satisfied if we have $l_i'(x_j) + (x_j - x_i)q'(x_j) = 0$ for $1 < j < n$, $j \neq i$; in other words, if $q'(x) + l_i'(x)/(x - x_i)$ vanishes at all these x_j. It follows that conditions (12.4.7) are satisfied if

$$q'(x) = -\frac{l_i'(x)}{x - x_i} + CP_{n-1}'(x) \qquad (12.4.10)$$

where C is an arbitrary constant. For $x \to x_i$, using $l_i'(x_i) = 0$, we see that (12.4.8) is also satisfied. From (12.4.10) we get

$$q(x) = \int_x^1 \frac{l_i'(t)\, dt}{t - x_i} + CP_{n-1}(x) + C_1. \qquad (12.4.11)$$

It remains to find C, C_1 from conditions (12.4.9). If I_i' is the integral (12.4.4), we get from the representation of q and q',

$$q(1) = C + C_1, \qquad q(-1) = I_i' - C + C_1,$$

$$q'(1) = -\frac{l_i'(1)}{1 - x_i} + CP_{n-1}'(1), \qquad q'(-1) = \frac{l_i'(-1)}{1 + x_i} + CP_{n-1}'(-1).$$

We use the fact that $l_i'(1) = \Pi_n'(1)[(1 - x_i)\Pi_n'(x_i)]^{-1}$, and a similar relation that prevails at the point -1; then the substitution of these values into the equations (12.4.9) yields

$$2\Pi_n'(1)CP_{n-1}'(1) + \Pi_n''(1)(C + C_1) = 0,$$
$$2\Pi_n'(-1)CP_{n-1}'(-1) + \Pi_n''(-1)(I_i' - C + C_1) = 0.$$

Substituting the values of $\Pi'_n(\pm 1), P'_{n-1}(\pm 1), \Pi''_n(\pm 1)$ from (12.1.12), we get the equations $-C - \frac{1}{2}(C + C_1) = 0$, $C - \frac{1}{2}(I'_i - C + C_1) = 0$, which have the unique solution $C = \frac{1}{6}I'_i$, $C_1 = -\frac{1}{2}I'_i$, proving (12.4.1).

The computation of r_1 and r_n can be done similarly, or we can check directly that the expressions (12.4.2) and (12.4.3) are polynomials of required degree satisfying (12.3.2). We leave the details to the reader (see also [3]).   □

We shall use the following formulas for the integrals of Theorem 12.10.

Lemma 12.11.   *We have, for even n and $1 < i < n$,*

$$I'_i = - \frac{1}{(1 - x_i^2) P_{n-1}(x_i)^2}, \qquad (12.4.12)$$

and for $x_i < x \leqslant 1$,

$$l_i(x)^2 + \frac{\Pi_n(x)}{\Pi'_n(x_i)} \int_x^1 \frac{l'_i(t)\,dt}{t - x_i} = \frac{\Pi_n(x)}{\Pi'_n(x_i)} \int_x^1 \frac{l_i(t)\,dt}{(t - x_i)^2}, \qquad (12.4.13)$$

while for $-1 \leqslant x < x_i$ the left-hand side of (12.4.13) is equal to

$$\frac{\Pi_n(x)}{\Pi'_n(x_i)} \left[I'_i - \int_{-1}^x \frac{l_i(t)\,dt}{(t - x_i)^2} \right]. \qquad (12.4.14)$$

Proof.   To establish the first formula, we find a relation between integrals I_i, I'_i. From

$$l_i(t)(t - x_i) = \frac{\Pi_n(t)}{\Pi'_n(x_i)}, \qquad 1 < i < n, \qquad (12.4.15)$$

and the relation $(1 - t^2)\Pi''_n(t) + n(n-1)\Pi_n(t) = 0$, we obtain the differential equation of the fundamental function l_i:

$$(1 - t^2)(t - x_i)l''_i(t) + 2(1 - t^2)l'_i(t) + n(n-1)(t - x_i)l_i(t) = 0. \qquad (12.4.16)$$

Dividing by $2(1 - t^2)(t - x_i)$ and integrating between -1 and $+1$, we obtain

$$\frac{1}{2}[l'_i(1) - l'_i(-1)] + I'_i + \frac{1}{2}n(n-1)\int_{-1}^{+1} \frac{l_i(t)\,dt}{1 - t^2} = 0.$$

Since

$$\frac{l_i(t)}{1 - t^2} = \frac{P'_{n-1}(t)}{(t - x_i)\Pi'_n(x_i)},$$

using (12.1.9) we obtain

$$I'_i = - \frac{1}{2}[l'_i(1) - l'_i(-1)] + \frac{1}{2 P_{n-1}(x_i)} I_i. \qquad (12.4.17)$$

By differentiation of (12.4.15) we have

$$l_i'(1) = \frac{\Pi_n'(1)}{(1-x_i)\Pi_n'(x_i)} = \frac{P_{n-1}(1)}{(1-x_i)P_{n-1}(x_i)} = \frac{1}{(1-x_i)P_{n-1}(x_i)}$$

and similarly

$$l_i'(-1) = \frac{1}{(1+x_i)P_{n-1}(x_i)}.$$

Substituting this and the value of I_i from Lemma 12.9, we obtain the required expression for I_i'. Relation (12.4.13) and the corresponding formula for $-1 \leqslant x < x_i$ follow by partial integration; in case $-1 \leqslant x < x_i$, we first use

$$\int_x^1 \frac{l_i'(t)\, dt}{t-x_i} = I_i' - \int_{-1}^x \frac{l_i'(t)\, dt}{t-x_i}.$$ □

§12.5. PROPERTIES OF THE INTERPOLATION OPERATOR; CONVERGENCE

We first derive some estimates for the fundamental polynomials r_i, ρ_i.

Theorem 12.12. *For the functions ρ_i we have the estimates, with an absolute constant C and with $l_i(x)$ given by (12.2.8), for even n,*

$$|\rho_i(x)| \leqslant \begin{cases} C\{n^{-7/2}i^{3/2} + n^{-4}i^{5/2}|l_i(x)|\}, & 1 \leqslant i \leqslant \dfrac{n}{2}, \\ C\{n^{-7/2}(n+1-i)^{3/2} + n^{-4}(n+1-i)^{5/2}|l_i(x)|\}, & \dfrac{n}{2} < i \leqslant n, \end{cases}$$

(12.5.1)

$$\sum_{i=1}^n |\rho_i(x)| \leqslant Cn^{-1}.$$ (12.5.2)

Proof. The second inequality can be derived from the first, since, for example,

$$\sum_{i=1}^{n/2} |\rho_i(x)| \leqslant \text{const}\left[n^{-7/2}n^{5/2} + n^{-4}\sum_{i=1}^{n/2} i^{5/2}|l_i(x)|\right],$$

and by Cauchy's inequality and Lemma 12.6,

$$\sum_1^{n/2} i^{5/2}|l_i(x)| \leqslant \left(\sum_1^{n/2} i^5\right)^{1/2}\left(\sum_1^{n/2} l_i(x)^2\right)^{1/2} \leqslant \text{const } n^3.$$

We prove the first of the estimates (12.5.1). For $i=1$ this follows from (12.3.4) and Lemma 12.4. For the functions ρ_i, $1 < i \leqslant n/2$, we have the representation (12.3.6), with the integrals I_i given by (12.3.15). From Lemma 12.5 we have for these integrals

$$|I_i| \leqslant \frac{2}{1-x_i^2} + \frac{2}{1-x_i^2} C\sqrt{i} \leqslant \text{const} \frac{\sqrt{i}}{1-x_i^2}. \qquad (12.5.3)$$

Therefore from (12.3.6)

$$|\rho_i(x)| \leqslant \frac{1}{2}\Lambda(x)\left[\mathcal{O}\left(\frac{\sqrt{i}}{1-x_i^2}\right) + \left|\int_{-1}^{x} \frac{P_{n-1}'(t)\,dt}{t-x_i}\right|\right],$$

$$\Lambda(x) = \frac{|\Pi_n(x)|(1-x_i^2)}{\Pi_n'(x_i)^2}.$$

We estimate the last integral. Let first $-1 \leqslant x < x_i$. By partial integration

$$\int_{-1}^{x} \frac{P_{n-1}'(t)\,dt}{t-x_i} = \frac{P_{n-1}(x)}{x-x_i} - \frac{1}{1+x_i} + \int_{-1}^{x} \frac{P_{n-1}(t)\,dt}{(t-x_i)^2}.$$

The absolute value of the last term does not exceed

$$\int_{-1}^{x} (t-x_i)^{-2}\,dt = (x_i-x)^{-1} - (1+x_i)^{-1} \leqslant (x_i-x)^{-1}.$$

Therefore

$$\left|\int_{-1}^{x} \frac{P_{n-1}'(t)\,dt}{t-x_i}\right| \leqslant \frac{1}{x_i-x} + \frac{1}{x_i-x} \leqslant \frac{2}{x_i-x}.$$

If $x_i < x < 1$, then

$$\int_{-1}^{x} \frac{P_{n-1}'(t)\,dt}{t-x_i} = I_i - \int_{x}^{1} \frac{P_{n-1}'(t)\,dt}{t-x_i},$$

and the last integral does not exceed $2(x-x_i)^{-1}$. Hence for all $x \neq x_i$,

$$\left|\int_{-1}^{x} \frac{P_{n-1}'(t)\,dt}{t-x_i}\right| \leqslant |I_i| + \frac{2}{|x-x_i|}. \qquad (12.5.4)$$

We see that

$$|\rho_i(x)| \leqslant \text{const}\left(\Lambda(x)\frac{\sqrt{i}}{1-x_i^2} + \Lambda(x)\frac{1}{|x-x_i|}\right). \qquad (12.5.5)$$

For $\Lambda(x)$ we have two alternative estimates. On the one hand, since $\Pi_n' = -n(n-1)P_{n-1}$, we have, by Lemmas 12.4 and 12.5,

$$\Lambda(x) \leqslant \text{const}\frac{\sqrt{n}(1-x_i^2)}{n^4|P_{n-1}(x_i)|^2} \leqslant \text{const}\frac{i(1-x_i^2)}{n^{7/2}}.$$

On the other hand, using the $l_i(x)$, again Lemma 12.5, and (12.2.3),

$$\Lambda(x) = \left| \frac{\Pi_n(x)}{(x - x_i)\Pi'_n(x_i)} \right| \frac{|x - x_i|}{|\Pi'_n(x_i)|} (1 - x_i^2)$$

$$\leqslant \text{const}|l_i(x)| \frac{i^{5/2}}{n^4}|x - x_i|.$$

Substituting these estimates into (12.5.5), we obtain (12.5.1).   □

Theorem 12.13. *For the polynomials r_i we have*

$$|r_i(x)| \leqslant \begin{cases} C\sqrt{n/i} & \text{if} \quad 1 \leqslant i \leqslant n/2, \\ C\sqrt{n/(n+1-i)} & \text{if} \quad n/2 \leqslant i \leqslant n. \end{cases} \qquad (12.5.6)$$

$$\sum_{i=1}^{n} |r_i(x)| \leqslant Cn. \qquad (12.5.7)$$

Proof. Obviously, the second inequality follows from the first.

For the r_i, $2 \leqslant i \leqslant n/2$, we have the representation (12.4.1) with the formula (12.4.12) for the integral I'_i. Let first $x > x_i$. Using (12.4.13), we obtain $r_i(x) = A + B$, where for $\Lambda^* = \Lambda^*(x) = \Pi_n(x)/\Pi'_n(x)$,

$$A = \Lambda^* \int_x^1 \frac{l_i(t)\,dt}{(t - x_i)^2}, \qquad B = \Lambda^* \frac{P_{n-1}(x) - 3}{6} I'_i.$$

By (12.4.12), Lemma 12.5, and (12.2.3),

$$|I'_i| \leqslant \frac{1}{(1 - x_i^2)|P_{n-1}(x_i)|^2} \leqslant \text{const} \frac{n^2}{i};$$

moreover, by Lemma 12.6,

$$\left| \int_x^1 \frac{l_i(t)\,dt}{(t - x_i)^2} \right| \leqslant \int_x^\infty \frac{dt}{(t - x_i)^2} = \frac{1}{x - x_i}.$$

For Λ^* we have the two estimates (the second by Lemmas 12.4 and 12.5)

$$|\Lambda^*(x)| = |(x - x_i)l_i(x)| \leqslant x - x_i,$$

$$|\Lambda^*| \leqslant \text{const} \frac{\sqrt{n}\sqrt{i}}{n^2} = \text{const}\, n^{-3/2}i^{1/2}.$$

This gives $|A| \leqslant 1$, $|B| \leqslant \text{const}\sqrt{n/i}$. If $x < x_i$, we argue as before, but replace (12.4.13), by (12.4.14).

The case in which $i = 1$ is treated separately. From (12.4.2), Lemma 12.4, and Lemma 12.6,

$$|r_1(x)| \leqslant \text{const}\sqrt{n} + \tfrac{1}{4}(1 - x^2)|l'_1(x)|.$$

Differentiating the relation $l_1(x)(x+1) = \Pi_n(x)/\Pi_n'(-1)$ we obtain [using 12.1.9) and (12.1.12)]

$$|l_1'(x)|(x+1) \le \left| \frac{\Pi_n'(x)}{\Pi_n'(-1)} \right| + |l_1(x)| \le 2,$$

so that $|r_1(x)| \le \text{const } \sqrt{n}$. □

After these preparations, we can study the *interpolation operator*

$$Q_{2n-1}(f,x) = \sum_{i=1}^{n} r_i(x)f(x_i) + \sum_{i=1}^{n} \rho_i(x)f''(x_i) \qquad (12.5.8)$$

which is defined for $f \in C^2[-1,1]$. It reproduces polynomials of degree $\le 2n-1$. Our estimates of r_i, ρ_i yield the interesting relation

$$\|Q_{2n-1}(f)\| \le C\left(n\|f\| + \frac{1}{n}\|f''\| \right). \qquad (12.5.9)$$

Functions $f \in C^2$ have good approximation properties by polynomials R_n of degree $\le n$. From this it is easy to derive that $Q_n(f) \to f$ if $f \in C^2$ (see Corollary 12.15).

To obtain convergence for wider classes of functions, Balázs and Turán replace the operator Q_{2n-1} by

$$\tilde{Q}_{2n-1}(f,x) = \sum_{1}^{n} r_i(x)f(x_i) + \sum_{1}^{n} \beta_{i,n}\rho_i(x) \qquad (12.5.10)$$

where $\beta_{i,n} = 0$ or at least are not too large. To obtain better results, we consider *local approximation* of f when the deviation $|f(x) - R_n(x)|$ is estimated not only in terms of n but also taking account of the position of x in the interval $[-1, +1]$ [see (12.5.12)]. Balázs and Turán proved the convergence $\tilde{Q}_{2n-1}f \to f$ for functions f with derivative f' that is Dini continuous. It has been pointed out by Freud [44] that their conclusion holds for the wider class of functions $\mathfrak{Z}$ defined below.

Let $\omega_2(h)$ be the *modulus of smoothness* of a function $f \in C[-1, +1]$:

$$\omega_2(h) = \max_{x, |t| \le h} |f(x+t) - 2f(x) + f(x-t)|. \qquad (12.5.11)$$

(For properties of ω_2 see [I].) A function belongs to the Zygmund class $\mathfrak{Z}$ if $\omega_2(h) = h\varepsilon(h)$, $\varepsilon(h) \to 0$ for $h \to 0$. For functions $f \in \mathfrak{Z}$, Dzyadyk and Freud (see [N, p. 266]) have proved the existence of a sequence of polynomials R_n of degree $\le n$ with the property

$$|f(x) - R_n(x)| \le K\omega_2(\Delta_n(x)) \le K_1\varepsilon_n\Delta_n(x), \qquad (12.5.12)$$

$$\Delta_n(x) = n^{-1}\sqrt{1-x^2} + n^{-2},$$

where $\varepsilon_n > 0$ is a sequence decreasing to 0. Since $\Delta_n(x) \le 4\Delta_{2n}(x)$, from (12.5.12) and the properties of ω_2 we obtain

$$|f(x) - R_{2^k}(x)| \le K_2\varepsilon_n\Delta_n(x) \qquad \text{if} \quad 2^k \le n < 2^{k+1}.$$

We need an estimate of $R_n''(x)$ which follows from this. Let k_0 be defined by $2^{k_0} \leqslant n < 2^{k_0+1}$; let $S_k = R_{2^k} - R_{2^{k-1}}$, $1 \leqslant k \leqslant k_0$, and $S_{k_0+1} = R_n - R_{2^{k_0}}$. Then $|S_k(x)| \leqslant \text{const } \varepsilon_{2^k} \Delta_{2^k}(x)$, and by a known inequality [I, p. 71], this implies $|S_k'(x)| \leqslant \text{const } \varepsilon_{2^k}$. The inequalities of Markov and Bernstein yield two upper bounds for $|S_k''(x)|$,

$$\text{const } \varepsilon_{2^k} 2^{2k}, \qquad \text{const } \varepsilon_{2^k} \frac{2^k}{\sqrt{1-x^2}}, \qquad 1 \leqslant k \leqslant k_0. \qquad (12.5.13)$$

For $|S_{k_0+1}''(x)|$ these bounds hold with 2^k replaced by n.

Since $R_n''(x) = \sum_{k=1}^{k_0+1} S_k''(x)$, by summing the inequalities we obtain

$$|R_n''(x)| \leqslant K_3 \bar{\varepsilon}_n \max\left(n^2, \frac{n}{\sqrt{1-x^2}} \right)$$

where $\bar{\varepsilon}_n \to 0$ as $n \to \infty$. For $x = x_i$ and $2 \leqslant i \leqslant n/2$, by (12.2.3) we can replace $\sqrt{1-x_i^2}$ by i/n. Therefore, we have

$$|R_n''(x_i)| \leqslant K_3 \bar{\varepsilon}_n n^2/i, \qquad 1 \leqslant i \leqslant n/2. \qquad (12.5.14)$$

There is a similar inequality for $n/2 < i \leqslant n$.

We can now state our main theorem.

Theorem 12.14 (Convergence theorem of Balázs and Turán, improved by Freud). *If $|\beta_{i,n}| \leqslant \delta_n$, $n^{-1}\delta_n \to 0$, then for each function $f \in \mathfrak{X}$,*

$$\tilde{Q}_{2n-1}(f, x) \to f(x) \quad \text{uniformly for} \quad -1 \leqslant x \leqslant 1. \qquad (12.5.15)$$

Proof. With the polynomials R_n we have

$$\tilde{Q}_{2n-1}(f, x) - f(x) = \tilde{Q}_{2n-1}(f, x) - Q_{2n-1}(R_n, x) + R_n(x) - f(x)$$

$$= \sum_{i=1}^{n} r_i(x)(f(x_i) - R_n(x_i))$$

$$+ \sum_{i=1}^{n} (\beta_{i,n} - R_n''(x_i))\rho_i(x) + o(1).$$

By (12.5.12) and (12.5.7), the absolute value of the first sum does not exceed

$$K_1 \frac{1}{n} \varepsilon_n \sum_{1}^{n} |r_i(x)| \leqslant CK\varepsilon_n \to 0.$$

Also, by (12.5.2),

$$\left| \sum_{1}^{n} \beta_{i,n} \rho_i(x) \right| \leqslant \delta_n \frac{1}{n} C \to 0.$$

The remaining sum we break into two parts, for i satisfying $1 \leqslant i \leqslant n/2$ and

$n/2 \leqslant i \leqslant n$:

$$\left| \sum_{i=1}^{n/2} R_n''(x_i)\rho_i(x) \right| \leqslant \sum_{1}^{n/2} K'\bar\varepsilon_n \frac{n^2}{i} \left(n^{-7/2}i^{3/2} + n^{-4}i^{5/2}|l_i(x)| \right)$$

$$\leqslant K'\bar\varepsilon_n \left[n^{-3/2} \sum_{1}^{n/2} i^{1/2} + n^{-2} \sum_{1}^{n/2} i^{3/2}|l_i(x)| \right]$$

$$\leqslant \text{const } \bar\varepsilon_n \to 0$$

where we have used (12.5.14), (12.5.1) and Cauchy's inequality.   □

Corollary 12.15.   *For a function* $f \in C^2[-1,+1]$,

$$Q_{2n-1}(f,x) \to f(x) \tag{12.5.16}$$

uniformly in $[-1,+1]$ *as* $n \to \infty$.

§12.6. NOTES

12.6.1. Theorem 12.14 of Balázs and Turán, in the form improved by Freud, can be shown to be essentially the best possible. This follows from results of Vértesi [186]. In another paper, Vértesi [187] shows that this applies also to the L_2 convergence, and even to convergence in measure, instead of pointwise convergence. This is important because for the Lagrange interpolation polynomial $L_n(f,x)$ the situation is different. For some functions we have divergence everywhere; on the other hand, by a theorem of Erdös and Turán we have $L_n(f) \to f$ in L_2 (see [J, Vol. III, pp. 43,55]). For L_p, $p > 2$, this is again different, see for example Nevai [202, pp. 180–183].

12.6.2. Inequalities (12.5.2) and (12.5.7) show that for each poly-nomial Q of degree $\leqslant 2n-1$ that at the knots x_i of (12.1.7) satisfies the inequalities

$$|Q(x_i)| \leqslant A, \qquad |Q''(x_i)| \leqslant B \tag{12.6.1}$$

we have

$$|Q(x)| \leqslant C(An + Bn^{-1}), \qquad -1 \leqslant x \leqslant 1, \tag{12.6.2}$$

where C is some absolute constant. In [3], Balázs and Turán derive estimates of the derivatives ρ_i', r_i' of the fundamental polynomials and show in this way that (12.6.1) implies also

$$|Q'(x)| \leqslant C'(An^{5/2} + Bn^{1/2}), \qquad -1 \leqslant x \leqslant 1. \tag{12.6.3}$$

It should be noted that Markov's inequality, if applied to (12.6.2), does not yield (12.6.3).

12.6.3. Several authors have studied interpolation of types $(0,1,3)$ and $(0,1,2,4)$ at knots (12.1.7). We mention only the papers of Saxena and Sharma [146, 147] (somewhat improved by Vértesi [186]) for the first type of interpolation and of Varma [182] for the second type. Additional references can be found in the review of Sharma [160].

12.6.4. Zeros of some other classical polynomials can be treated in a similar way. For example, let X_α: $x_1 < \cdots < x_n$ be the zeros in $[-1, +1]$ of the nth ultraspherical polynomial P_n^α of order $\alpha > -1$ (see [L, p. 29]). (For $\alpha = 0$ we obtain Legendre polynomials.) For even n, Surányi and Turán [173] prove that X_α is regular for the $(0,2)$ interpolation if $\alpha + \frac{1}{2}$ is not an even integer and $n \geqslant 4$, and singular in the opposite case for $n \geqslant \alpha + \frac{5}{2}$.

Interpolation of types $(0,2)$ and $(0,1,3)$ has also been studied at the roots of the Chebyshev polynomials [180, 184]. This appears to be simpler than interpolation at the knots (12.1.7).

12.6.5. In his paper "On some open problems of approximation theory" [179], Turán formulates several open questions for the $(0,2)$ interpolation. For example, he suggests that knots (12.1.7) must be in some sense extremal for this interpolation.

12.6.6. In contrast to Chapter 11, we have no theorems that apply to interpolation matrices with a modest degree of generality. The reason is probably that it is very difficult to obtain useful formulas for the fundamental interpolation polynomials. This applies also to interpolation matrices that are regular. Sharma mentions that convergence facts for the $(0,2,3)$ interpolation are not known even for the knots (12.1.7). In [161], however, Sharma and Leeming give explicit expressions for the fundamental polynomials of degree $\leqslant 3n - 1$ for the $(0, n-1, n)$ case and for arbitrary real knots.

12.6.7. If the interpolating functions are not polynomials, lacunary interpolation may assume a very different character. We shall mention only the paper of Meir and Sharma [117] about quintic splines (see §14.7), and the papers of Linden, Pittnauer, and Wyrwich [88] (where also further references can be found) and Shull [168] about lacunary interpolation of entire functions.

Chapter 13

Birkhoff Interpolation by Splines

§13.1. INTRODUCTION

Interpolation by spline functions is a more complex subject than polynomial interpolation. This will be evident from the preliminary material of this chapter (definitions, duality, Pólya conditions, and decomposition theorems) when compared with the material of Chapter 1. The problem of regularity of spline interpolation and its applications to self-dual matrices, to problems of best quadrature, and to extremal problems will be considered in Chapter 14.

Among the first to consider spline interpolation were Schoenberg and Whitney [155], Karlin and Ziegler [74], Karlin [64], and Schoenberg [152]. Using the second of these papers as an example, we shall explain how the interpolation problem and regularity theorems have been formulated in them. Let

$$E = [e_{i,k}]_{i=1, k=0}^{m, n}, \qquad E^* = [e_{p,q}^*]_{p=1, q=0}^{\mu, n} \qquad (13.1.1)$$

be matrices of 0's and 1's, and let

$$X: -1 = x_1 < x_2 < \cdots < x_m = +1, \qquad X^*: -1 < x_1^* < \cdots < x_\mu^* < 1 \qquad (13.1.2)$$

be fixed sets of knots.

The pair E, X characterizes the set of interpolation functionals for the problem $L_{i,k}(g) = g^{(k)}(x_i)$, $e_{i,k} = 1$. The pair E^*, X^* will determine the space of interpolating splines $\mathbb{S}_n$: It is the linear span of powers $1, x/1!, \ldots,$ $x^n/n!$ and of the truncated power functions $(x_p^* - x)_+^{n-q}/(n-q)!$ for $e_{p,q}^* = 1$ in E^*. Since these functions are linearly independent on $[-1,1]$, the dimension of $\mathbb{S}_n$ is $n+1+|E^*|$. The *splines* $S \in \mathbb{S}_n$ have the unique representations

$$S(x) = \sum_{q=0}^{n} \alpha_q \frac{x^q}{q!} + \sum_{e_{p,q}^* = 1} \alpha_{p,q} \frac{\left(x_p^* - x\right)_+^{n-q}}{(n-q)!}.$$

If the matrix E^* is empty or has only zero entries, then $\mathbb{S}_n = \mathcal{P}_n$ and we return to the problem of Birkhoff interpolation by polynomials.

Two additional assumptions are needed. First, the kth derivative of $(x_p^* - x)_+^k/k!$ at $x = x_p^*$ is not defined. Therefore we require

$$e_{i,k} = e_{p,n-k}^* = 1 \quad \text{is impossible if} \quad x_i = x_p^*, \quad k = 0, \ldots, n. \quad (13.1.3)$$

Second, we require that the number of interpolation functionals be equal to the dimension of $\mathbb{S}_n$:

$$|E| = n + 1 + |E^*|. \quad (13.1.4)$$

Then the tuple E, X, E^*, X^* is called *normal*. The problem is called *regular* if there is a unique $S \in \mathbb{S}_n$ for each choice of values $c_{i,k}$ for the functionals $L_{i,k}(S) = c_{i,k}, e_{i,k} = 1$.

Example. Let $n = 0$,

$$E = \begin{bmatrix} 1 \\ 1 \\ 1 \end{bmatrix}, \qquad E^* = \begin{bmatrix} 1 \\ 1 \end{bmatrix},$$

$$X: \ -1 = x_1 < x_2 < x_3 = 1 \qquad \text{and} \qquad X^*: \ -1 < x_1^* < x_2^* < 1.$$

The space $\mathbb{S}_0$ consists of step functions that can have jumps only at x_1^*, x_2^*. For $x_2 \neq x_1^*, x_2 \neq x_2^*$ the tuple E, X, E^*, X^* defines a normal spline interpolation problem. It is regular if and only if $x_1^* < x_2 < x_2^*$.

This simple regularity condition is typical of what is needed for the regularity when both E and E^* are Hermite matrices. In this case we rewrite the knot sequence X as $t_1 \leqslant t_2 \leqslant \cdots \leqslant t_{|E|}$ where each knot x_i appears according to its multiplicity as indicated by the length of the Hermite sequence of row i in E. Similarly, we extend the sequence X^* to $t_1^* \leqslant t_2^* \leqslant \cdots \leqslant t_{|E^*|}^*$ in the same way. The *interlacing conditions* [155, 74] for X, X^* are defined by the inequalities

$$t_i < t_i^* < t_{i+n+1}, \qquad i = 1, \ldots, |E^*|. \quad (13.1.5)$$

Theorem 13.1 (Karlin and Ziegler [74]). *A normal interpolation problem E, X, E^*, X^* with Hermite matrices E, E^* is regular if and only if the interlacing conditions (13.1.5) are satisfied.*

We shall discuss the early history of the spline interpolation problem in §14.1.

§13.2. BASIC DEFINITIONS

We shall adopt a different method for the description of spline interpolation. The interlacing condition of §13.1 shows that relative positions of
knots x_i, x_p^* should be taken into account, for instance, by ordering them.
Instead of this, we can order the rows of the matrices E, E^* of §13.1 into a
single sequence. However, we prefer to construct a "master matrix" F
(introduced in [61]) which combines information provided by both E and
E^*.

By definition, a *spline interpolation matrix* is any matrix

$$F = [f_{i,k}]_{i=1,k=0}^{m,\ n}, \qquad f_{i,k} = 0, 1, \text{ or } -1. \qquad (13.2.1)$$

A matrix F generates two $m \times (n+1)$ matrices $E = [e_{i,k}]$, $E^* = [e_{i,k}^*]$ of 0's
and 1's given by $e_{i,k} = 1$ if $f_{i,k} = 1$, $e_{i,k} = 0$ otherwise, and $e_{i,k}^* = 1$ if
$f_{i,n-k} = -1$, $e_{i,k}^* = 0$ otherwise. Then

$$F = E - \tilde{E}^* \qquad (13.2.2)$$

where $\tilde{E}^*$ has entries $\tilde{e}_{i,k}^* = e_{i,n-k}^*$. Many properties of interpolation are
better formulated in terms of F (see the Pólya conditions in §13.4; condition
(13.1.3) becomes redundant). Sometimes the matrices E, E^* will be more
appropriate.

For technical reasons (duality in §13.3, horizontal decomposition in
§13.5) we shall always assume that the exterior rows of F contain no 0's:

$$f_{i,k} = 1 \text{ or } -1 \qquad \text{if} \quad i = 1 \text{ or } i = m. \qquad (13.2.3)$$

The sum of all entries of F is denoted by $|F|$:

$$|F| = \sum_{i,k} f_{i,k}. \qquad (13.2.4)$$

For an arbitrary matrix F the matrix F^+ is obtained by replacing in
the rows $i = 1$, $i = m$ of F, all -1's by 0's. A matrix F is *normal* if
$|F^+| = n+1$; this is equivalent to $|E| = n+1+|\mathring{E}^*|$, where $\mathring{E}^*$ is the matrix
consisting of the interior rows of E^*. [Because of the assumption (13.2.3), it
is $\mathring{E}^*$ that corresponds to the E^* of §13.1.]

For an arbitrary set of knots X: $-1 = x_1 < x_2 < \cdots < x_m = 1$, we
consider the interpolation functionals $L_{i,k} : g \to g^{(k)}(x_i)$ for all $f_{i,k} = 1$, and
the spline space $\mathbb{S}_n(F, X)$ [also denoted by $\mathbb{S}_n(E^*, X)$] which is the linear
span of the functions

$$\mathcal{G} = \left\{ \frac{x^q}{q!}, \frac{(x_p - x)_+^q}{q!}; f_{p,q} = -1, 1 < p < m, q = 0, \ldots, n \right\}. \qquad (13.2.5)$$

The functions $g \in \mathcal{G}$ are linearly independent on $[-1, +1]$. Hence the splines $S \in \mathcal{S}_n(F, X)$ have the unique representation

$$S = \sum_{q=0}^{n} \alpha_q \frac{x^q}{q!} + \sum_{\substack{f_{p,q} = -1 \\ 1 < p < m}} \alpha_{p,q} \frac{(x_p - x)_+^q}{q!}. \tag{13.2.6}$$

If in the expression for S or $S^{(k)}$ there appears a term $\alpha(x_p - x)_+^l / l!$, with $\alpha \neq 0$, $l = 0$, then this function is not defined at $x = x_p$. For $l < 0$, this term is interpreted as 0.

A different, intrinsic definition of the space $\mathcal{S}_n(F, X) = \mathcal{S}_n(\mathring{E}^*, X)$ is given by Proposition 7.8. It follows from this that all derivatives of $S \in \mathcal{S}_n(F, X)$ must be continuous at $x = -1$ or 1.

There is an important relation to Birkhoff splines $S \in \mathcal{S}_n^0(E^*, X)$, defined in §7.3. Let $F = E - \tilde{E}^*$. Then we have: *Birkhoff splines $S \in \mathcal{S}_n^0(E^*, X)$ for the pair E^*, X which are annihilated by E, X are identical with the extensions by 0 to $(-\infty, +\infty)$ of splines $S \in \mathcal{S}_n(F, X)$ which are annihilated by E, X.* The assumption (13.2.3) is essential for this equivalence.

A pair F, X is *regular for spline interpolation* if for each choice of values $\{c_{i,k}: f_{i,k} = 1\}$ there is a unique spline $S \in \mathcal{S}_n(F, X)$ for which

$$S^{(k)}(x_i) = c_{i,k}, \qquad f_{i,k} = 1. \tag{13.2.7}$$

Equivalently, F, X is regular if $S \in \mathcal{S}_n(F, X)$ and $S^{(k)}(x_i) = 0$ for all $f_{i,k} = 1$ in F implies $S \equiv 0$. This can only happen if the functionals $L_{i,k}, f_{i,k} = 1$, are linearly independent on the space $\mathcal{S}_n(F, X)$, and if their number is equal to the dimension of this space, that is, only if F *is normal*.

Let $\mathcal{G}$ be any basis for $\mathcal{S}_n(F, X)$, written in some order. The system (13.2.7) has the coefficient matrix

$$A = A(F, X, \mathcal{G}) = \left[L_{i,k}(g); f_{i,k} = 1, g \in \mathcal{G} \right] \tag{13.2.8}$$

with the rows corresponding to (i, k), $f_{i,k} = 1$, ordered lexicographically. The regularity of F, X is then equivalent to the condition

$$D(F, X, \mathcal{G}) = \det A(F, X, \mathcal{G}) \neq 0. \tag{13.2.9}$$

The matrix F is regular if the pair F, X is regular for each X.

There are *singular*, or nonregular, pairs F, X even for quite simple matrices F. For example, it is easy to check directly that for the first matrix (13.3.8), the pair F, X with $X: -1 < x < 1$, is singular if and only if $x = 0$.

§13.3. DUALITY

Each spline interpolation matrix defines a *dual matrix* F^* and a corresponding *dual interpolation problem* (Jetter [56]). The matrix F^* is obtained from the representation $F = E - \tilde{E}^*$ by interchanging E and E^*, $F^* = E^* - \tilde{E}$.

Equivalently, F^* is given by

$$F^* = [f_{i,k}^*], \qquad f_{i,k}^* = -f_{i,n-k}. \qquad (13.3.1)$$

Clearly, $(F^*)^* = F$ and we easily check that F^* is normal if F is normal.

The pair F^*, X defines an interpolation problem connected with the space of splines $S_n(F^*, X)$ which is spanned by

$$\mathcal{G} = \left\{ 1, \ldots, \frac{x^n}{n!}, \frac{(x-x_i)_+^{n-k}}{(n-k)!}; \qquad f_{i,k} = 1, 1 < i < m, 0 \leqslant k \leqslant n \right\}$$

$$(13.3.2)$$

and with interpolation functionals $L_{p,q}: g \to g^{(q)}(x_p)$ for $f_{p,n-q} = -1$.

It is an interesting and useful fact that regularity is invariant under duality. The following *duality theorem* is due to Jetter [56]; the proof here is from [61].

Theorem 13.2. *A pair F, X is regular if and only if the dual pair F^*, X is regular. Moreover, we have (13.3.7). In particular, F is regular if and only if F^* is regular.*

Proof. For the Taylor kernel $K(x,t) = (x-t)_+^n/n!$ we have

$$\frac{\partial^{n-q}}{\partial t^{n-q}} K(x,t) = (-1)^{n-q} \frac{(x-t)_+^q}{q!}.$$

Thus the functions

$$\frac{(x-x_1)^q}{q!}, \quad 0 \leqslant q \leqslant n, \qquad \frac{\partial^{n-q}}{\partial t^{n-q}} K \bigg|_{t=x_p}, \qquad \text{for} \quad f_{p,q} = -1, \quad 1 < p < m,$$

form a basis for $S_n(F, X)$ for a given X. Equivalently, on $[x_1, x_m]$ we have the basis

$$\mathcal{G} = \left\{ \frac{(x-x_1)^q}{q!} \ (f_{1,q} = 1), \quad \frac{\partial^{n-q}}{\partial t^{n-q}} K \bigg|_{t=x_p} \ (f_{p,q} = -1, 1 \leqslant p < m) \right\}.$$

$$(13.3.3)$$

To the functions $g \in \mathcal{G}$ we apply the functionals $L_{i,k}(g) = g^{(k)}(x_i), f_{i,k} = 1$; for $i = 1$ the result will be 0 unless g is the function $(x-x_1)^k/k!$, in which case it is 1. Hence

$$A(F, X, \mathcal{G}) = \begin{bmatrix} 1 & & 0 & \\ & \ddots & & 0 \\ 0 & & 1 & \\ \hline & * & & A_2 \end{bmatrix}$$

where

$$A_2 = \left[\frac{\partial^k}{\partial x^k} \frac{\partial^{n-q}}{\partial t^{n-q}} K \bigg|_{x=x_i, t=x_p} \right]. \qquad (13.3.4)$$

The rows of A_2 are enumerated by pairs (i, k) with $f_{i,k} = 1$, $1 < k \leqslant m$, the columns by pairs (p, q) for which $f_{p,q} = -1$, $1 \leqslant p < m$.

In the same way, since

$$\frac{\partial^k K}{\partial x^k} = \frac{(x - t)_+^{n-k}}{(n-k)!},$$

a basis for the dual space $\mathcal{S}_n(F^*, X)$ is given by the functions of t,

$$\mathcal{G}^* = \left\{ \left. \frac{\partial^k K}{\partial x^k} \right|_{x = x_i} (f_{i,k} = 1, 1 < i \leqslant m), \frac{(x_m - t)^{n-k}}{(n-k)!} (f_{m,k} = -1) \right\}.$$

$$(13.3.5)$$

To functions $g \in \mathcal{G}^*$ we apply functionals $L_{p,q}^*(g) = g^{(q)}(x_p)$, $f_{p, n-q} = -1$. For $p = m$ the result is 0, unless g is the function

$$\frac{(x_m - t)^q}{q!}, \qquad f_{m, n-q} = -1,$$

when it is $\varepsilon_q = (-1)^q$. Consequently,

$$A(F^*, X, \mathcal{G}^*) = \begin{bmatrix} A_2^* & & * & \\ \hline & \ddots & & 0 \\ 0 & & \varepsilon_q & \\ & 0 & & \ddots \end{bmatrix}.$$

Here

$$A_2^* = \left[\left. \frac{\partial^{n-q}}{\partial t^{n-q}} \frac{\partial^k}{\partial x^k} K \right|_{x = x_i, t = x_p} \right] \qquad (13.3.6)$$

with rows enumerated by (p, q) with $f_{p,q} = -1$, $1 \leqslant p < m$, and columns by pairs (i, k), $f_{i,k} = 1$, $1 < i \leqslant m$. Since pairs $(i, k), (p, q)$ in A_2 and A_2^* cannot be equal, we can interchange the order of differentiation. Hence the matrices A_2, A_2^* are the transposes of each other. Consequently,

$$D(F, X, \mathcal{G}) = \pm D(F^*, X, \mathcal{G}^*). \qquad \square \quad (13.3.7)$$

As an application, we can derive many regular or singular spline interpolation matrices from normal polynomial matrices $E = [e_{i,k}]$, $e_{i,k} = 0, 1$. Assume that F is obtained from E by completing its exterior rows by -1's. Then F and also F^* are regular if and only if E is regular for polynomials. For example, the first two of the following matrices are singular [see (1.3.7) and §2.3], and the last two are regular [Theorem 1.6 and

(1.3.7)]:

$$
\begin{bmatrix} 1 & 1 & -1 \\ 0 & -1 & 0 \\ 1 & 1 & -1 \end{bmatrix},
\quad
\begin{bmatrix} 1 & 1 & 1 & 1 & 1 & -1 & 1 & -1 \\ 0 & 0 & 0 & 0 & 0 & -1 & 0 & -1 \\ 0 & 0 & 0 & 0 & 0 & -1 & 0 & -1 \\ 1 & 1 & 1 & 1 & 1 & -1 & 1 & -1 \end{bmatrix},
$$

(13.3.8)

$$
\begin{bmatrix} 1 & 1 & 1 & -1 \\ 0 & -1 & -1 & 0 \\ 1 & 1 & 1 & -1 \end{bmatrix},
\quad
\begin{bmatrix} 1 & 1 & 1 & 1 & -1 & -1 \\ 0 & -1 & 0 & 0 & -1 & 0 \\ 1 & 1 & 1 & 1 & -1 & -1 \end{bmatrix}.
$$

(13.3.9)

All these matrices define two-row interpolations on the spline spaces given by the interior rows.

§13.4. PÓLYA CONDITIONS FOR SPLINE INTERPOLATION

Many different necessary conditions for the regularity of F have been used by several authors in this field. In Schoenberg and Whitney [155] and Karlin and Ziegler [74] they are formulated as interlacing conditions, bearing on the knots x_i and x_p^*. In this form they should be used only for Hermite matrices E, E^*. More general conditions of Melkman [120] and Jetter [57], the so-called local Pólya conditions, involve both the knots and the entries of the matrices E, E^*. So do the very general Pólya conditions of Goodman [47]. We find it convenient [61] to reformulate Goodman's conditions in terms of the matrix F alone. The notion of the upper and lower sets, of the boundary, and of the operation F^1 on a matrix F are necessary for this reformulation. In this way we obtain what we call the *Goodman–Pólya conditions*.

Let Φ be a lower set (see §7.3) and F a spline interpolation matrix; let $F_\Phi = [f_{i,k}]_{(i,k) \in \Phi}$ be the restriction of F to Φ. We need also the following operations on F_Φ, which change only elements $f_{i,k}$ on the horizontal boundary of Φ: For F_Φ^+, the -1's on the horizontal boundary are replaced by 0's; for F_Φ^-, the 1's are replaced by 0's; for F_Φ^1, 0's are replaced by 1's, and -1's are replaced by 0's. Finally, $\mathring{F}_\Phi$ is the submatrix of F_Φ restricted to pairs $(i, k) \in \Phi$ that are not on the horizontal boundary of Φ.

First, we prove a theorem about the space of splines that "live" on a lower set Φ. Let X be a set of knots in $[-1, +1]$. This produces a set of intervals X_Φ; *intervals of the kth level* of X_Φ are $[x_j, x_{j'}]$ for each interval $[j, j']$ of the kth level of Φ. The space $\mathbb{S} = \mathbb{S}_{\Phi, X}$, associated with Φ, X, consists of all S on $[-1, +1]$ that satisfy the following conditions: S is a polynomial of degree $\leqslant n$ on each $[x_{i-1}, x_i]$; the derivative $S^{(k)}$ vanishes outside of the intervals of level k of X_Φ; and $S^{(k)}$ can be discontinuous at an *interior* point x_i of an interval of level k only if $f_{i,k} = -1$. No restriction is made on the continuity of $S^{(k)}$ at the end points of intervals of level k.

Theorem 13.3. *For each lower set Φ and each set of knots X the dimension of $\mathbb{S}_{\Phi, X}$ is*

$$\dim \mathbb{S}_{\Phi, X} = L(\Phi) + \left(\text{number of } -1\text{'s in } \mathring{F}_\Phi\right). \qquad (13.4.1)$$

Proof. We prove this by induction in h—the number of columns that contain points of Φ. Let $I = [x_{i_1}, x_{i_2}]$ be one of the p intervals of X_Φ of level 0. The derivatives S' of splines $S \in \mathbb{S}$ on I form a space $\mathbb{S}' = \mathbb{S}_{\Phi', X}$ where the lower set Φ' is the part of Φ contained in $i_1 \leqslant i \leqslant i_2$, $1 \leqslant k \leqslant n$. Let $\tilde{S}$ be integrals of $S' \in \mathbb{S}'$, normalized by $\tilde{S}(x_{i_1}) = 0$. If $d' = \dim \mathbb{S}'$, and $S'_1, \dots, S'_{d'}$ is a basis for $\mathbb{S}'$, then their integrals $\tilde{S}_1, \dots, \tilde{S}_{d'}$ form a basis for all $\tilde{S}$.

On the other hand, splines of degree 0 on I are spanned by 1 and by the step functions $(x - x_j)^0_+$ corresponding to all $f_{j,0} = -1$, $i_1 < j < i_2$. Since each S is uniquely representable on I by the sum of a spline of degree 0 and an integral $\tilde{S}$, the dimension of the restriction of $\mathbb{S}$ to I is $d' + 1 + ($number of $f_{i,0} = -1$, $i_1 < i < i_2)$. The total dimension of $\mathbb{S}$ is

$$\sum d' + p + \left(\text{number of } f_{i,0} = -1 \text{ in } \mathring{F}_\Phi\right). \qquad (13.4.2)$$

Here p is the portion of the number $L = L(\Phi)$ for the column 0. Substituting the expressions (13.4.1) for the d' into (13.4.2), we obtain our statement. $\qquad \square$

The most general Pólya conditions for a spline interpolation matrix F are

$$|F^1_\Phi| \geqslant L(\Phi) \quad \text{for each lower set } \Phi, \qquad (13.4.3)$$

$$|F^+_\Psi| \leqslant L(\Psi) \quad \text{for each upper set } \Psi. \qquad (13.4.4)$$

We call them *the lower and the upper Goodman–Pólya conditions* (they appear in Goodman [47] in a very different form). They are not quite independent: If we add the condition that F is normal, $|F^+| = n + 1$, then any two of the conditions imply the third. This follows from the identity

$$|F^1_\Phi| + |F^+_\Psi| + (n + 1) = |F^+| + L(\Phi) + L(\Psi). \qquad (13.4.5)$$

Here Ψ is the complement of Φ. This means that the intervals of kth level of Ψ are complements of intervals of the same level of Φ. In other words, if $[j_1, j'_1], \dots, [j_p, j'_p]$ are all intervals of level k of Φ, then those of level k of Ψ are $[j'_1, j_2], \dots, [j'_{p-1}, j_p]$ and also $[1, j_1]$, if $1 < j_1$ and $[j'_p, m]$, if $j'_p < m$.

Theorem 13.4. *If the pair F, X is regular for some set of knots X, then F satisfies the Goodman–Pólya conditions.*

Proof. We subject the splines $S \in \mathbb{S}_{\Phi, X}$ to the conditions $L_{i,k}(S) = 0$ for all $f_{i,k} = 1$ in F^1_Φ, where $L_{i,k}$ are the differentiation functionals $g \rightarrow$

$g^{(k)}(x_i)$. If x_i is one of the end points of the intervals X_Φ of level k, this should be interpreted as the limit of the derivative $g^{(k)}(x)$ from the inside of the interval. Splines S satisfying these equations will be elements of $\mathfrak{S}_n(F, X)$ annihilated by all $L_{i,k}$, $f_{i,k} = 1$ in F. If (13.4.3) is violated, the dimension of $\mathfrak{S}_{\Phi, X}$ given by (13.4.1) is greater than the number of 1's in F_Φ^1, that is, greater than the number of the equations. Then there is a nontrivial $S \in \mathfrak{S}_n(F, X)$ annihilated by F, X, and the pair is singular.

The necessity of (13.4.4) can be derived from (13.4.5) and the normality of F.   □

The lower and the upper Goodman–Pólya conditions of different kinds, (13.4.3) and (13.4.4), are dual to each other. More precisely, let Ψ and $\Phi = \Psi^*$ be dual to each other, in the sense that $[j, j']$ is an interval of level k of Φ if and only if it is an interval of level $n - k$ of Ψ. Then *condition* (13.4.4) *for* $F = [f_{i,k}]$ *and* Ψ *is equivalent to* (13.4.3) *for* $F^* = [-f_{i,n-k}]$ *and* Φ.

Indeed, inequality $|F_\Psi^+| \leqslant L(\Psi)$ can be written $-|F_\Phi^{*-}| \leqslant L(\Phi)$, and since for any matrix F, $|F_\Phi^1| - |F_\Phi^-| = 2L(\Phi)$, the last inequality is $|F_\Phi^{*1}| \geqslant L(\Phi)$.

A disadvantage of the Goodman–Pólya conditions is that there are so many of them. Indeed, for an $m \times (n + 1)$ matrix, the number of different lower sets Φ is equal to $(n + 2)^{m-1}$. For we can select the heights arbitrarily on each of the $m - 1$ intervals. Sometimes it is possible to reduce the number of relevant conditions.

Let R, R' be the rectangles $R: 0 \leqslant k \leqslant r, i_1 \leqslant i \leqslant i_2$, and $R': r \leqslant k \leqslant n$, $i_1 \leqslant i \leqslant i_2$. The *lower* and *upper rectangular Pólya conditions*,

$$|F_R^1| \geqslant r + 1, \qquad\qquad 0 \leqslant r \leqslant n, \quad 1 \leqslant i_1 \leqslant i_2 \leqslant m, \qquad (13.4.6)$$
$$|F_{R'}^+| \leqslant n + 1 - r, \qquad\qquad\qquad\qquad\qquad\qquad\qquad\qquad (13.4.7)$$

appear for the first time in Melkman [120] and Jetter [57]. They are also called the *local Pólya conditions*.

Theorem 13.5. *If $F = E - \tilde{E}^*$, and either E or E^* is quasi-Hermitian, then the rectangular Pólya conditions together imply the Goodman-Pólya conditions.*

Proof. We prove a little more, namely, that *conditions* (13.4.6) *imply* (13.4.3) *if E is quasi-Hermitian*. By duality it would follow then that (13.4.7) imply (13.4.4) for quasi-Hermitian E^*.

A given lower set Φ decomposes in a natural way into rectangles. The proof of the theorem is by induction on the number N of these rectangles. Let $N > 1$. Then Φ has an inside corner (i_0, k_0), for which $(i_0, k_0 + 1)$ but not (i_0, k_0) is on the horizontal boundary of Φ.

First assume that $f_{i_0, k_0} \neq 1$. Then all entries $f_{i_0, k}$, $k \geqslant k_0$, are 0's or -1's. Without loss of generality, let i_0, k_0 be an upper inside corner. From

Φ we derive another lower set $\Phi^* \supset \Phi$ which satisfies

$$|F^1_{\Phi^*}| - L(\Phi^*) \leq |F^1_\Phi| - L(\Phi). \qquad (13.4.8)$$

Let (i_0, k_1), $k_1 > k_0$ be the last point of row i_0 on the boundary of Φ. If the first row following i_0 with a piece of boundary of Φ is $i_0 + 1$, and if this boundary is given by points $(i_0 + 1, k)$, $k_0 < k \leq k_2$, then we add to Φ intervals $[i_0, i_0 + 1]$ of levels k, $k_0 < k \leq k^*$, $k^* = \min(k_1, k_2)$. Then $L(\Phi^*)$ $= L(\Phi) - (k^* - k_0)$, $|F^1_{\Phi^*}| \leq |F^1_\Phi| - (k^* - k_0)$, so that we get (13.4.8). In all other cases, we add to Φ intervals $[i_0, i_0 + 1]$ of levels $k_0 < k \leq k_1$. This will not change $L(\Phi)$, and will not increase $|F^1_\Phi|$.

After a finite number of operations, we either decrease N, or obtain an (upper) inside corner with $f_{i_0, k_0} = 1$.

In the latter case, we have $f_{i_0, k} = 1$, $0 \leq k \leq k_0$. Let Φ_1 and Φ_2 be parts of Φ corresponding to the intervals $1 \leq i \leq i_0$ and $i_0 \leq i \leq m$, respectively. They consist of fewer than N rectangles. Adding the relations $|F^1_{\Phi_j}| \geq L(\Phi_j)$, $j = 1, 2$, we obtain (13.4.3). $\square$

Exercises

1. The relation between the interlacing conditions (13.1.5) and the rectangular conditions is given by the following theorem. A normal matrix $F = E - \tilde{E}^*$, for which E and E^* are Hermitian, satisfies the interlacing conditions if and only if it satisfies (13.4.6) and (13.4.7) with $r = n$ and $r = 0$ respectively and with either $i_1 = 1$ or with $i_2 = m$.

2. Prove that conditions (13.4.6) do not imply (13.4.3) if E^*, but not E, is quasi-Hermitian.

3. There is a dual proposition to Theorem 13.3. Let $\mathcal{T}_{\Psi, X}$ be the space of the restrictions of the derivatives $S^{(k)}$ of splines $S \in \mathbb{S}_n(F, X)$ to X_Ψ, where Ψ is an upper set. Then $\dim \mathcal{T}_{\Psi, X} = L(\Psi) + $ (number of -1's in $\mathring{F}_\Psi$). Directly from this derive the necessity of (13.4.4) for the regularity of F.

§13.5. DECOMPOSITION OF MATRICES

There are two decomposition theorems for spline interpolation; see Melkman [120] when E^* is Hermitian, and Jetter [57] for the general case.

For a matrix $F = [f_{i, k}]$ we denote by $F(r_1, r_2; i_1, i_2)$ its restriction to the rectangle $r_1 \leq k \leq r_2$, $i_1 \leq i \leq i_2$.

We have the identities

$$|F^+(0, n; i_1, i_2)| = |F^+(0, r-1; i_1, i_2)| + |F^+(r, n; i_1, i_2)|, \qquad (13.5.1)$$

where $0 < r \leq n$, $1 \leq i_1 < i_2 \leq m$, and for $i_1 < i < i_2$.

$$|F^+(0, n; i_1, i_2)| = |F^+(0, n; i_1, i)| + |F^+(0, n; i, i_2)| - \sum_{k=0}^{n} |f_{i, k}|.$$

$$(13.5.2)$$

If for a normal matrix F and $0 < r \leqslant n$ we have

$$|F^+(r, n; 1, m)| = n + 1 - r, \qquad (13.5.3)$$

then we say that F is *vertically decomposable between columns* $r - 1$ *and* r (or simply, *decomposable at* r). We put then $F_1 = F(0, r - 1; 1, m)$; $F_2 = F(r, n; 1, m)$ and write $F = F_1 \oplus F_2$. It follows from this definition and (13.5.1) that F_1 and F_2 are also normal.

Theorem 13.6. *A vertically decomposable matrix* $F = F_1 \oplus F_2$ *is regular if and only if* F_1 *and* F_2 *are regular.*

The proof is the same as that of Theorem 1.4.

Horizontal decomposition may happen *along* a row that, with some alterations, is *assigned to both* F_1 *and* F_2. If for a normal matrix F

$$|F^+(0, n; 1, i_1)| = n + 1, \qquad 1 < i_1 < m, \qquad (13.5.4)$$

we say that F is *horizontally decomposable along row* i_1; more precisely, it is $[1, i_1]$ *decomposable*. In this case, F_1 is defined to be $F(0, n; 1, i_1)$, with all 0's in row i_1 replaced by -1's, and F_2 to be $F(0, n; i_1, m)$, with 0's of row i_1 replaced by 1's. We call F_1 the *primary matrix* and F_2 the *secondary matrix of the decomposition*. (No precise definition of horizontal decomposition appears in the literature.)

There is the symmetric case of $[i_1, m]$ *decomposability*, when

$$|F^+(0, n; i_1, m)| = n + 1, \qquad 1 < i_1 < m, \qquad (13.5.5)$$

with the upper and lower matrices, F_1, F_2, defined accordingly. In both cases we have the identity

$$|F_1^+| + |F_2^+| = |F^+| + (n + 1), \qquad (13.5.6)$$

which shows that F_1 and F_2 are normal.

In particular, let F satisfy the rectangular Pólya conditions (13.4.7). We can then describe the case when both (13.5.4) and (13.5.5) are satisfied.

Proposition 13.7. *Let the normal matrix* F *satisfy* (13.4.7).

(i) *The matrix* F *is both* $[1, i_2]$ *and* $[i_1, m]$ *decomposable with* $i_1 < i_2$, *if and only if*

$$|F^+(0, n; i_1, i_2)| = n + 1.$$

(ii) *The matrix* F *is both* $[1, i]$ *and* $[i, m]$, $1 < i < m$, *decomposable if and only if it has no entries* 0 *in the* ith *row.*

Proof. To prove (i) we use the identity

$$|F^+| + |F^+(0, n; i_1, i_2)| = |F^+(0, n; 1, i_2)| + |F^+(0, n; i_1, m)| \qquad (13.5.7)$$

for $i_1 < i_2$. We have here $|F^+| = n + 1$, while all other terms are $\leqslant n + 1$. To prove (ii) we similarly use identity (13.5.2) for $i_1 = 1$, $i_2 = m$. □

Theorem 13.8 (Melkman [120]; Jetter [57]). *If a normal matrix F decomposes horizontally into F_1 and F_2, then F is regular if and only if both F_1 and F_2 are regular.*

Proof. Let, for example, F be $[1, i_0]$ decomposable. For the basis

$$\mathcal{G} = \left\{ \frac{x^q}{q!}, \frac{(x_p - x)_+^q}{q!}, 1 < p < i_0, \right.$$

$$\left. \frac{(x - x_p)_+^q}{q!}, i_0 \leq p < m, f_{p,q} = -1, 0 \leq q \leq n \right\}$$

we arrange the rows of $A = A(F, X, \mathcal{G})$ so that the first rows correspond to $f_{i,k} = 1$ in $F^+(0, n; 1, i_0)$. Then

$$A = \begin{bmatrix} A_1 & 0 \\ \hline * & A_2 \end{bmatrix} \tag{13.5.8}$$

where $A_1 = A(F_1, X, \mathcal{G}_1)$ is by (13.5.4) a square matrix if $\mathcal{G}_1$ is the first part of $\mathcal{G}$ relevant to F_1 (functions corresponding to $f_{p,q} = -1$, $i_0 \leq p < m$, are omitted).

Now consider the matrix $A(F_2, X, \mathcal{G}_2)$ corresponding to the basis for $\mathcal{S}_n(F_2, X)$ given by

$$\mathcal{G}_2 = \left\{ \frac{(x - x_{i_0})^q}{q!}, 0 \leq q \leq n; \frac{(x - x_p)_+^q}{q!}, i_0 < p < m, f_{p,q} = -1 \right\}.$$

With proper arrangement of rows and columns in $A(F_2, X, \mathcal{G}_2)$ we have

$$A(F_2, X, \mathcal{G}_2) = \begin{bmatrix} 1 & & 0 & & 0 \\ & \ddots & & & \\ 0 & & 1 & & \\ \hline & * & & & A_2 \end{bmatrix}. \tag{13.5.9}$$

Therefore

$$D(F, X, \mathcal{G}) = \pm D(F_1, X, \mathcal{G}_1) D(F_2, X, \mathcal{G}_2). \quad \square \tag{13.5.10}$$

Concerning the Pólya conditions under decomposition we have the easily checked

Proposition 13.9. *If a normal matrix F is vertically or horizontally decomposable into matrices F_1, F_2, then F satisfies the Goodman–Pólya conditions if and only if F_1 and F_2 satisfy them.*

Not so simple is the situation with the rectangular Pólya conditions. It is easy to see that if F_1 and F_2 satisfy (13.4.7), then F does also. The inverse is not true. An interesting example is

$$F = \begin{bmatrix} -1 & -1 & -1 & 1 & 1 & -1 \\ 1 & 1 & 0 & 0 & 0 & -1 \\ 0 & -1 & -1 & 1 & 1 & 0 \\ \hline 1 & 1 & 1 & -1 & -1 & 0 \\ 0 & 0 & -1 & -1 & 1 & 1 \\ \hline 0 & 0 & 0 & 0 & 0 & -1 \\ 1 & 1 & 1 & -1 & -1 & -1 \end{bmatrix} \qquad (13.5.11)$$

This matrix satisfies both (13.4.6) and (13.4.7). For the lower set Φ indicated here, F_Φ does not satisfy the Goodman–Pólya condition (13.4.3). The matrix F decomposes horizontally along row $i = 4$, but F_2 does not satisfy (13.4.7) (because in F_2, 0 at the end of row 4 of F has to be replaced by 1). Also, for vertical decomposition it is easy to construct matrices for which condition (13.4.7) is not preserved.

For some special matrices (13.4.7) is preserved under decomposition:

Exercise (*see* [120, 57, 134]). If $F = E - \tilde{E}^*$ is normal, E^* is quasi-Hermitian, and if F decomposes (vertically or horizontally) into matrices F_1, F_2, then F_1 and F_2 satisfy (13.4.7) if F does.

§13.6. NOTES

13.6.1. There is a relation between kernels of dual interpolation problems (Jetter [56]). For a regular pair F, X, let $U: C^{n+1}[-1,1] \to \tilde{S}_n(F, X)$ be the operator that assigns to each function g the unique spline in $\tilde{S}_n(F, X)$ that satisfies $S^{(k)}(x_i) = g^{(k)}(x_i)$, $f_{i,k} = 1$ in F. Assume that there exists an interpolation kernel $K(x, t)$ of F, X such that for all $g \in C^{n+1}$

$$g(x) - U(g, x) = \int_{-1}^{+1} K(x, t) g^{(n+1)}(t)\, dt. \qquad (13.6.1)$$

If $K^*(x, t)$ is the kernel of F^*, X, then

$$K(x, t) = (-1)^{n+1} K^*(t, x),\ x, t \in (-1, +1). \qquad (13.6.2)$$

In particular, the kernel of a self-dual problem is symmetric.

13.6.2. Melkman [120], Pence [134], and Goodman [47] also discuss interpolation with one-sided derivatives $S^{(k)}(x_i -)$ and $S^{(k)}(x_i +)$ prescribed at some points x_i with $f_{i,k} = -1$. Melkman [119] also admits mixed conditions, which contain linear combinations of point evaluation functionals.

Chapter 14

Regularity Theorems and Self-Dual Problems

§14.1. HISTORICAL REMARKS

The first papers containing regularity results for spline interpolation were [155, 74, 64]. Their characteristic features are similar: The interlacing conditions (13.1.5) are necessary and sufficient for regularity; the proof is based on the total positivity of certain matrices. This approach gives much more than regularity, and this is perhaps the reason why no stronger regularity results have been achieved this way. The strongest theorem (regularity when E, E^* are quasi-Hermitian) is stated in Karlin [64, 66].

The next important step was achieved by Melkman [120]; he established the regularity when E is conservative, E^* Lagrangian, and announced that he also had a proof when E^* is Hermitian. This was soon established by Jetter [57], who sketched a proof relying on Theorem 7.13 of Birkhoff and Lorentz. Pence [134] somewhat improved Melkman's conditions. Like Melkman, Goodman [47] also relied on Budan–Fourier theorems to derive his remarkable theorem.

The use of Rolle's theorem for splines (Lemma 7.10) was inaugurated by Lorentz [93]; de Boor [B, 10] used the *ordinary* Rolle theorem to derive theorems of [155] and [74]. In §14.2 we give a general theorem [61] that uses diagrams of splines, from which the theorem of Goodman follows easily (by means of a relatively simple Lemma 14.5).

Several methods of proof of regularity theorems are known. In this chapter we illustrate: an application of the zero count theorem in this section, of Rolle's lemmas in §§14.2–14.3, and a completely different, more elementary approach that works for self-dual interpolation problems in §14.4. The reader can consult in [47] the original proof of theorem 14.4, based on the classical Budan–Fourier Theorem 2.2. The last two sections contain examples where self-dual matrices arise naturally in the solution of problems concerning polynomial interpolation.

The theorem of Melkman–Jetter, which follows, can be derived easily from Theorem 7.13 (if the duality theorem is known) or from Goodman's Theorem 14.4 (if Theorem 13.5 is known).

Theorem 14.1. *A normal matrix $F = E - \tilde{E}^*$ is regular if it satisfies the upper rectangular Pólya conditions (13.4.7), if E is conservative, and if E^* is quasi-Hermitian.*

Proof. Instead of the regularity of F, we prove that of $F^* = E^* - \tilde{E}$. Then F is regular by the duality theorem.

Assume that F^* is not regular. Then for some set of knots X there is a nontrivial spline $S \in \mathbb{S}_n(F^*, X)$ annihilated by F^*, X. We can assume that S has a one-interval support. Let $[a, b] = [x_{i_1}, x_{i_2}]$ be the support of S, and q its degree.

For a matrix $E = [e_{i,k}]_{i=1, k=0}^{m, \quad n}$, we denote by $E(r_1, r_2; i_1, i_2)$ its submatrix restricted to $i_1 \leqslant i \leqslant i_2$, $r_1 \leqslant k \leqslant r_2$, and by $\mathring{E}$ the matrix of interior rows of E, $1 < i < m$.

The spline S is annihilated by $E^*(0, q; i_1, i_2)$, X and belongs to the space $\mathbb{S}_q(F', X)$ where $F' = F^*(0, q; i_1, i_2) = E^*(0, q; i_1, i_2) - E(n - q, n; i_1, i_2)$. Then $S \in \mathbb{S}_q^0(E', X)$, $E' = E(n - q, n; i_1, i_2)$, and by Theorem 7.13 with $\gamma(E') = 0$,

$$Z = Z(S; (a, b)) \leqslant |E(n - q, n; i_1, i_2)| - q - 2.$$

It is obvious that $|\mathring{E}^*(0, q; i_1, i_2)| \leqslant Z$. This gives

$$|F^+(n - q, n; i_1, i_2)| = |E(n - q, n; i_1, i_2)| - |\mathring{E}^*(0, q; i_1, i_2)| \geqslant q + 2,$$

in contradiction to (13.4.7). $\qquad\qquad\square$

The theorem also follows from Goodman's Theorem 14.4. Indeed, in the assumptions of Theorem 14.1, all interior odd blocks of F are not supported from the left; and F satisfies the Goodman–Pólya conditions by Theorem 13.5.

Pence [134] gave an improved version of the Melkman–Jetter theorem. A row i_0 forms an *obstruction to the support from the left above* (the row i_0) if there are no 1's in the rectangle $F(0, k_0 - 1; 1, i_0)$ and if $f_{i_0, k} = -1$ for $k_0 \leqslant k \leqslant n$. Obstruction to the support from the left below by row i_0 is similarly defined. A sequence has unobstructed support from the left (above or below) if there is no obstructing row between it and some supporting 1.

Exercise [134]. Theorem 14.1 remains true if the requirement "E is conservative" is replaced by "E has no odd sequences that have unobstructed support in F from both left above and left below." [*Hint*: If i_0 is the first interior row that forms an obstruction from the left above, then the matrix F is $[i_0, m]$ decomposable into matrices, each of which has no odd sequences with unobstructed support. Then use induction.]

§14.2. ESTIMATION OF $|F_\Phi^1|$

The next theorem, due to the present authors [61], is of the same character as Theorem 7.11, the key theorem of Chapter 7. Both have very similar proofs. (See an example at the end of the next section which shows that Theorem 7.11 cannot be strengthened to include the present theorem.)

For a spline interpolation matrix $F = [f_{i,k}]_{i=1, k=0}^{m, \ n}$ and a lower set Φ, we study the matrices F_Φ, F_Φ^1. This time we consider *blocks*. A maximal sequence of $+1$'s and -1's in a row of F_Φ^1

$$B: f_{i,s}, \ldots, f_{i,s+t} \tag{14.2.1}$$

is a block in F_Φ^1. The block is *interior* to F_Φ^1 if it contains none of the horizontal or vertical boundary of F_Φ^1. A block B in F_Φ^1 is *supported on the left* (or simply supported) if there are $f_{i_1, k_1} = f_{i_2, k_2} = 1$ in F_Φ^1 for which $i_1 < i < i_2$, $k_1, k_2 < s$. For an interior supported block, elements $f_{i,s-1}$ and $f_{i,s+t+1}$ belong to F_Φ^1 and are 0. A block is *odd* (or *even*) if the number of its entries is odd (or even). The block B is *supported on the right* in F if there are $f_{i_3, k_3} = f_{i_4, k_4} = -1$ in F for which $i_3 < i < i_4$, $k_3, k_4 > s + t$. By Lemma 7.9, an interior block of F_Φ^1, where $\Phi = \Phi(S)$ is the diagram of a spline, is supported on the right in F. As in Theorem 7.11, we consider splines S from $\mathbb{S}_n^0(F, X)$, or equivalently from $\mathbb{S}_n(F, X)$, that are annihilated by F, X.

Theorem 14.2. *Let F be an $m \times (n+1)$ spline interpolation matrix and let X: $-1 = x_1 < \cdots < x_m = 1$ be a set of knots. If a spline $S \in \mathbb{S}_n(F, X)$ with interval support is annihilated by F, X, and if $\Phi = \Phi(S)$ is the diagram of S, then*

$$|F_\Phi^1| \leqslant L(\Phi) - 1 + \gamma(F_\Phi^1) \tag{14.2.2}$$

where $\gamma(F_\Phi^1)$ is the number of interior odd blocks in F_Φ^1 that are supported on the left.

Proof. As before, we apply Lemma 7.10 to a spline S of degree q, and derive numbers μ_k, $k = 0, \ldots, q$, which are lower bounds for the number of continuous Rolle zeros ξ of $S^{(k)}$, not stipulated by F_Φ^1. We start with $\mu_0 = 0$.

We need slightly different notations from those in the proof of Theorem 7.11. Let $l_k :=$ number of intervals of $\Phi(S)$ (or of X_S) of level k (in particular, $l_0 = 1$); $u_k :=$ number of 1's in F_Φ^1 in column k; $v_k :=$ number of

-1's in the interior of F_Φ^1 in column k; $\varepsilon_k :=$ number of blocks (14.2.1) supported on the left in F_Φ^1 and that begin in column k with a 1; $\bar\varepsilon_k :=$ number of sequences of -1's in F_Φ^1 that end at a point (i,k), with $(i, k+1)$ in the interior of one of the intervals of Φ of level $k+1$ (in particular, points on the vertical boundary of Φ do not contribute to $\bar\varepsilon_k$); $\eta_k :=$ number of -1's in column k that are not supported on the left or that are preceded by a 1; $w_{k+1} :=$ number of interior blocks in F_Φ^1 that end with a 1 in column k, for which $S^{(k+1)}(x_i) = 0$.

At step k, there will be $\mu_k + u_k$ continuous zeros of $S^{(k)}$, with $\mu_k + u_k - 1$ intervals between them. If we omit at most $l_{k+1} - 1$ intervals that intersect complementary intervals to those of X_S of level $k+1$, we obtain at least $\mu_k + u_k - l_{k+1}$ intervals I of level k to which we can apply Lemma 7.10. On each of them, $S^{(k)}$ and $S^{(k+1)}$ can vanish only in isolated points. For an interval $I = (\xi_1, \xi_2)$, Lemma 7.10 may produce a continuous zero ξ of $S^{(k+1)}$. This would be a special case of (A), Lemma 7.10. This $\xi = x_i$ can be specified by F_Φ^1 only if (α) $f_{i, k+1} = 1$, x_i is not a zero of $S^{(k)}$, and with (i,k) counted either by ε_{k+1} or $\bar\varepsilon_k$. Unspecified continuous zeros will be counted toward μ_{k+1}.

Otherwise, the lemma will produce a ξ at which $S^{(k)}$ and $S^{(k+1)}$ alternate and that is either (β) a discontinuous zero x_i of $S^{(k+1)}$, corresponding to $f_{i, k+1} = -1$, or (γ) a discontinuous zero x_i of $S^{(k)}$ with $f_{i,k} = -1$. If $f_{i, k+1} = -1$, the disjoint cases (β) and (γ) are counted in $v_{k+1} - \eta_{k+1}$; if $f_{i, k+1} = 1$ or 0, the disjoint cases (α) and (γ) are counted in $\bar\varepsilon_k + \varepsilon_{k+1}$. Thus, if we omit $\varepsilon_{k+1} + \bar\varepsilon_k + (v_{k+1} - \eta_{k+1})$ intervals I of level k, the remaining intervals will each contribute to μ_{k+1}.

It is important that we can improve this estimate. We call a pair (i,k) in Φ confirming if $\xi = x_i$ can be obtained from Lemma 7.10 applied to $S^{(k)}$ under (α), (β), or (γ). Otherwise, it is nonconfirming. In particular, (i,k) is nonconfirming if (i,k) is on the horizontal or vertical boundary of Φ, if $S^{(k)}$, $S^{(k+1)}$ are not alternating at x_i or if $(i, k+1)$ is counted by ε_{k+1} but either $(i, k+1)$ is not supported or $S^{(k)}(x_i) = 0$.

Let N_k be the number of nonconfirming pairs (i,k) among those counted by $(v_{k+1} - \eta_{k+1})$ or $\bar\varepsilon_k + \varepsilon_{k+1}$. For $k = 0, \dots, q-1$ we will have examined, without duplication, all positions (i,k) that correspond to a -1, that precede a -1, or that precede a block beginning with a 1.

Let $N = \Sigma_{k=0}^{q-1} N_k$ and $W = \Sigma_{k=1}^q w_k$. Instead of $\varepsilon_{k+1} + \bar\varepsilon_k + (v_{k+1} - \eta_{k+1})$ intervals I under (α), (β), or (γ), we need to remove at most $\varepsilon_{k+1} + \bar\varepsilon_k + (v_{k+1} - \eta_{k+1}) - N_k$ of them which contain confirming (i,k). This gives a lower estimate for μ_{k+1}. However, zeros of $S^{(k+1)}$ counted by w_{k+1} cannot be obtained this way. Therefore

$$\mu_{k+1} - w_{k+1} \geqslant \mu_k + u_k - l_{k+1} - \bar\varepsilon_k - (v_{k+1} - \eta_{k+1}) - \varepsilon_{k+1} + N_k.$$

$$(14.2.3)$$

Summing (14.2.3) for $k = 0, \dots, q-1$, we obtain, since $\varepsilon_0 = 0$, $\eta_0 = v_0$,

$\bar\varepsilon_q = 0,\ u_q = 0,$

$$0 \geqslant \mu_q = \sum_0^q u_k - \sum_0^q v_k - \sum_1^q l_k - \sum_0^q (\bar\varepsilon_k - \eta_k + \varepsilon_k) + N + W; \quad (14.2.4)$$

in other words,

$$|F_\Phi^1| \leqslant L(\Phi) - 1 + \Delta - N - W \qquad (14.2.5)$$

where

$$\Delta = \sum_0^{q-1} \bar\varepsilon_k - \sum_0^q \eta_k + \sum_0^q \varepsilon_k$$

is the last sum in (14.2.4).

We estimate the contributions Δ_B, N_B, W_B to Δ, N, W for each block (14.2.1). The block can end at the vertical boundary of F_Φ^1 only with $f_{i,s+t} = -1$.

A maximal subsequence of 1's in B contributes nothing to Δ_B unless it begins at $k = s$, and unless $f_{i,s} = 1$ is supported. Then the contribution is 1. Thus, the 1's contribute 0 or 1 to Δ_B, the last being the case exactly when the block is supported and begins with a 1. If the block begins with $s = 0$, this part of the contribution is 0.

A maximal sequence of -1's in B can contribute a -1 to Δ_B at the beginning of the sequence, by the way of η_k, and a $+1$ at its end through $\bar\varepsilon_k$. The latter will not happen if the sequence ends at the vertical boundary; the former will not happen exactly if the block B begins with a supported -1. In conclusion, -1's in B will contribute $\leqslant 1$ to Δ_B; they will contribute 1 exactly when the block begins with a supported -1 and does not end at the boundary. Thus, we have $\Delta_B \leqslant 1,\ N_B \geqslant 0$ for each B. We have $\Delta_B \leqslant 0$ *except for interior blocks B which are supported on the left in F_Φ^1.*

Let B be an interior block (14.2.1) with $s > 0$, $f_{i,s-1} = f_{i,s+t+1} = 0$, that is supported. We can assume that all pairs $(i, s-1),\ldots,(i, s+t)$ counted by $\varepsilon_{k+1} + \bar\varepsilon_k + (v_{k+1} - \eta_{k+1})$ are confirming, for otherwise we would have $N_B \geqslant 1$. Then Lemma 7.10 and properties of zero multiplicities show that $S^{(j)}$ are alternating at x_i for $j = s-1,\ldots,s+t$. Since $(i, s-1)$ is confirming, x_i is not a zero of the continuous derivative $S^{(s-1)}$ under (α), (β), or (γ); hence x_i is an even point of $S^{(s-1)}$. Now assume that the block B is *even*. Then x_i is an even point of $S^{(s+t)}$. This implies $f_{i,s+t} \neq -1$ for $(i, s+t)$ can be confirming only in the case (γ) when x_i is an odd point of $S^{(s+t)}$; but if $f_{i,s+t} = 1$, $W_B = 1$. Thus, we need to consider only the *odd* supported blocks and obtain $\Delta - N - W \leqslant \gamma(F_\Phi^1)$. $\square$

Example (14.3.2) shows that Theorem 14.2 cannot be improved. However, there is a variant of this theorem in which $|F_\Phi^1|$ is increased by the number of changes of sign of S, but also $\gamma(F_\Phi^1)$ is replaced by a larger number. This theorem will be used in §14.6.

Let $\tilde{S}_n^0(F, X)$ consist of splines $S \in \tilde{S}_n(F, X)$ annihilated by F, X. Let $\mathfrak{S}^-(S)$ be the number of changes of sign of S on its support.

Theorem 14.3. *If F, X, S, and Φ are as in Theorem 14.2, then*

$$\mathfrak{S}^-(S) \leqslant L(\Phi) - |F_\Phi^1| - 1 + \tilde{\gamma}(F_\Phi^1) \qquad (14.2.6)$$

where $\tilde{\gamma}(F_\Phi^1)$ is the number of odd interior blocks in F_Φ^1. Moreover, if F satisfies the upper Goodman–Pólya conditions, then

$$\mathfrak{S}^-(S) \leqslant n - |F^+| + \tilde{\gamma}(F) \qquad (14.2.7)$$

where $\tilde{\gamma}(F)$ is the number of odd blocks supported on the right in F.

Proof. The proof parallels that of Theorem 14.2. This time we begin with $\mu_0 :=$ number of points ξ where S has a continuous change of sign that is not specified as a zero by a 1 in F_Φ. Then in place of (14.2.5), we have

$$\mu_0 \leqslant L(\Phi) - |F_\Phi^1| - 1 + \Delta - N - W. \qquad (14.2.8)$$

Let $\mathfrak{S}_B^-$ be the contribution of the block B of (14.2.1) to $\mathfrak{S}^-(S)$; then $\mathfrak{S}_B^- = 1$ if $s = 0$ and S changes sign at x_i, and $\mathfrak{S}_B^- = 0$ otherwise. Clearly, $\mathfrak{S}^-(S) = \mu_0 + \Sigma \mathfrak{S}_B^-$. To obtain (14.2.6), we must show that

$$\Delta_B - N_B - W_B \leqslant \tilde{\gamma}_B - \mathfrak{S}_B^-. \qquad (14.2.9)$$

We know from the proof of Theorem 14.2 that (14.2.9) holds if $s \neq 0$ or if $\mathfrak{S}_B^- = 0$, or if $s = 0$ and B is odd.

Accordingly, we assume that $s = 0$, $\mathfrak{S}_B^- = 1$, and B is even. We have $\Delta_B \leqslant 0$. If B extends to the vertical boundary, then $(i, s + t)$ is nonconfirming, $N_B \geqslant 1$, and (14.2.9) follows. Thus, we may assume that the block B is interior. Since x_i is an odd point of S, it is an even point of $S^{(s+t)}$. This implies that $W_B = 1$. Again we have (14.2.9).

By means of the identity (13.4.5), we can write the upper Goodman–Pólya conditions in the form $L(\Phi) - |F_\Phi^1| \leqslant n + 1 - |F^+|$. This and the remark that odd interior blocks of F_Φ are supported on the right in F gives (14.2.7). $\qquad \square$

§14.3. GOODMAN'S THEOREM

We can now state and prove Goodman's theorem.

Theorem 14.4 (Goodman [47]). *Let the matrix F have no odd blocks that are supported from both left and right. Then F is regular for spline interpolation if and only if it satisfies the Goodman–Pólya conditions.*

We need only prove the *sufficiency* of the conditions. We assume that S is a nontrivial spline in $\tilde{S}_n(F, X)$ annihilated by F, X and obtain a contradiction. If we could replace $\gamma(F_\Phi^1)$ in (14.2.2) by $\gamma(F_\Phi)$, this would be easy. This is not always possible. But we can save this argument by slightly changing the spline S and the matrix F.

Lemma 14.5. *If, in addition to the assumptions of Theorem* 14.2, *the matrix* F *satisfies the lower Goodman–Pólya conditions, then there is a restriction* S^* *of* S *to an interval* $[x_{i_3}, x_{i_4}]$ *and a matrix* F^* *so that with* $\Phi^* = \Phi(S^*)$, (i) S^* *is not trivial in* $\mathbb{S}_n(F^*, X^*)$ *and is annihilated by* F^*, X^*; (ii) F^* *satisfies the lower Goodman–Pólya conditions;* (iii) $\gamma(F_{\Phi^*}^1) \leqslant \rho(F)$, *the number of odd blocks in* F *supported from the left and from the right.*

Proof. A block B of F_Φ^1 given by (14.2.1) may be supported from the left above by a 1 from F_Φ^1, $f_{i_0, k_0}^1 = 1$, without being supported from the left above by a 1 in F. If this happens for some B, we say that $(i_0, k_0) \in U$. Similarly we define $(i_0, k_0) \in V$ for the support of blocks B from the left below.

If $(i_0, k_0) \in U$, then none of the $f_{i,k}$ in the rectangle R: $1 \leqslant i \leqslant i_0$, $0 \leqslant k \leqslant k_0$, is 1; consequently $f_{1,k} = -1$, $k = 0, \ldots, k_0$, by the definition of F. From the lower Goodman–Pólya conditions for R, $|F_R^1| \geqslant k_0 + 1$, it follows that all $f_{i,k} = 0$, $1 < i \leqslant i_0$, $0 \leqslant k \leqslant k_0$. There is a similar conclusion if $(i_0, k_0) \in V$. We define i_3 to be the largest i_0 with $(i_0, k_0) \in U$ (if U is empty, this is to be interpreted as $i_3 = 1$), and i_4 to be the smallest i_0 with $(i_0, k_0) \in V$ (and $i_4 = m$ if V is empty). Then $i_3 < i_4$, for otherwise the lower Goodman–Pólya conditions would be violated for some rectangle $1 \leqslant i \leqslant m$, $0 \leqslant k \leqslant r$. Let k_3 (or k_4) be the largest k_0 with $(i_3, k_0) \in U$ [or with $(i_4, k_0) \in V$]. If q^* is the degree of S on $[x_{i_3}, x_{i_3+1}]$, we put $k_3^* = \min(k_3, q^*)$, and define k_4^* similarly using the degree of S on $[x_{i_4-1}, x_{i_4}]$.

Let F^* be the matrix with rows i, $i_3 \leqslant i \leqslant i_4$, obtained from the corresponding rows of F by changing the 0's in $i = i_3$, $0 \leqslant k \leqslant k_3^*$, and $i = i_4$, $0 \leqslant k \leqslant k_4^*$, to -1's.

Since $i_3 < i_4$, S^* is not trivial; it is annihilated by F^*, X, because the 1's in rows $i_3 \leqslant i \leqslant i_4$ of F^* and F are the same. This proves (i).

If Ψ^* is any lower set in $i_3 \leqslant i \leqslant i_4$, $0 \leqslant k \leqslant n$, we extend it to the left by intervals $1 \leqslant i < i_3$ for any k, $k \leqslant k_3^*$, for which (i_3, k) is on the boundary of Ψ^*, and similarly on the right. This gives a lower set Ψ for F with $L(\Psi) = L(\Psi^*)$, $|F_\Psi^1| = |F_{\Psi^*}^1|$. Hence (ii) follows from the lower Goodman–Pólya conditions for F. An interior odd block of $F_{\Phi^*}^1$ is also a block of this type in F_Φ^1. If it is supported from the left above in $F_{\Phi^*}^1$ by some 1 in position (i_0, k_0), then this is also a 1 in F_Φ^1, hence $(i_0, k_0) \notin U$ and the block is supported from the left above in F. This proves (iii). $\square$

Proof of the theorem. Let $S \in \mathbb{S}_n(F, X)$ be a nonzero spline annihilated by F, X, and let S^*, F^* be given by Lemma 14.5. Applying Theorem 14.2 to S^*, we get

$$|F_{\Phi^*}^1| - L(\Phi^*) \leqslant -1 + \gamma(F_{\Phi^*}^1). \tag{14.3.1}$$

From Lemma 14.5 (iii), $\gamma(F_{\Phi}^{*!}) \leqslant \gamma(F) = 0$. Relation (14.3.1) contradicts (ii) of the lemma.   □

The following example shows that Theorems 7.11, 14.2, and 14.3 cannot be improved.

Example. For the matrix

$$
F = \begin{bmatrix}
1 & 1 & -1 & -1 & -1 \\
1 & 0 & 0 & 0 & 0 \\
1 & -1 & -1 & -1 & 0 \\
1 & 0 & 0 & 0 & 0 \\
1 & 1 & -1 & -1 & -1
\end{bmatrix} \tag{14.3.2}
$$

let $X = \{-1, -\frac{1}{2}, 0, \frac{1}{2}, 1\}$, let $\Phi = \Gamma$ (so that $F_{\Phi} = F$) and let S be the spline

$$
S(x) = (x+1)^4 - \tfrac{3}{2}(x+1)^3 + \tfrac{1}{2}(x+1)^2 - 5x_+^3 - x_+. \tag{14.3.3}
$$

The spline S is annihilated by F, X, while $S^{(4)}(0) = 4!$. If we replace $f_{3,4} = 0$ by 1, we obtain (by Theorem 14.4) a regular matrix. Hence there is only one S satisfying these conditions. Therefore, $S(x) = S(-x)$ by the symmetry in the interpolation, and this gives multiplicity 2 for 0 in $Z(S)$. Thus, $Z(S) = 4$, $\mathfrak{S}^-(S) = 2 = \tilde{\gamma}(F)$, $|F_{\Phi}^1| = 4$, $L(\Phi) - 1 = 4 = n$, and we have equality in Theorems 7.11, 14.2, and 14.3.

§14.4. REGULARITY OF SELF-DUAL MATRICES

Because of formula (14.4.2), properties of a self-dual interpolation matrix F reduce to those of its first half. We explore this in Theorem 14.7 to reduce the regularity of F to the completeness of F_1. Formula (14.4.2) also shows how a normal self-dual interpolation problem is constructed for applications. Crucial for this is the elementary Lemma 14.6, which establishes a duality relation for certain linear functionals defined on two different spaces of derivatives.

The results of §14.4 are of independent interest, but they also provide a correct setting and tools for §§14.5–14.6 where some extremal problems are solved. In §§14.4–14.5 we use the proofs of Schoenberg to obtain more general theorems; he considered matrices F_1 that consist only of 0's and 1's. Accordingly (see §14.5), his theorem was about quadrature formulas (14.5.1) exact for polynomials while we have quadrature formulas exact for splines.

In §14.6 we study the problem of Birkhoff interpolation by a function g with minimal derivative $g^{(n+1)}$ (this is a generalization of the polynomial interpolation problem where $g^{(n+1)} \equiv 0$). Here the method of proof of de Boor, employed by him for the Hermite case, carries us in a natural way to a theorem of Goodman.

An $m \times (n+1)$ spline interpolation matrix $F = [f_{i,k}]$ is *self-dual* if

$$f_{i,k} = -f_{i,n-k}, \qquad i = 1,\ldots,m, \quad k = 0,\ldots,n. \qquad (14.4.1)$$

Then the number n must be odd, for otherwise $f_{i,n/2} = 0$ is a contradiction to (13.2.3) for $i = 1$ or m. For this reason, we replace n by $2n+1$ and split the matrix F vertically into two $m \times (n+1)$ matrices, $F = F_1 \oplus F_2$ (we use this notation for self-dual matrices even if F_1 and F_2 are not normal). We see that relation (14.4.1) is equivalent to $F_2 = F_1^*$. Thus, the formula

$$F = F_1 \oplus F_1^* \qquad (14.4.2)$$

establishes a one-to-one correspondence between $m \times (n+1)$ spline interpolation matrices F_1 and $m \times (2n+2)$ self-dual matrices. It follows also that *each self-dual matrix is normal*.

Let $W_p^{n+1}[-1,1]$, $1 \le p \le \infty$, be the space of all functions on $[-1,1]$ with n absolutely continuous derivatives and with the $(n+1)$st derivative in $L_p[-1,1]$. Let $F = [f_{i,k}]$ be an $m \times (n+1)$ spline interpolation matrix, and let X: $-1 = x_1 < \cdots < x_m = +1$ be a set of knots. The space $W_p^{n+1}[-1,+1] + \mathbb{S}_n(F, X)$ consists of functions g with derivatives as above, except that $g^{(k)}(x)$ is allowed to have a finite jump discontinuity at x_i if $f_{i,k} = -1$. These spaces appear in the following lemma, which establishes a relation between integrals on subspaces of these spaces:

Lemma 14.6. *If* $g \in W_1^{n+1}[-1,1] + \mathbb{S}_n(F_1, X)$ *is annihilated by* F_1, X, *and* $h \in W_1^{n+1}[-1,1] + \mathbb{S}_n(F_1^*, X)$ *is annihilated by* F_1^*, X, *then*

$$\int_{-1}^{1} g^{(n+1)}(x) h(x)\, dx = (-1)^{n+1} \int_{-1}^{1} g(x) h^{(n+1)}(x)\, dx. \qquad (14.4.3)$$

Proof. By partial integration we obtain

$$\int_{-1}^{1} g^{(n+1)}(x) h(x)\, dx = \sum_{i=1}^{m-1} \int_{x_i}^{x_{i+1}} g^{(n+1)}(x) h(x)\, dx$$

$$= \sum_{i=1}^{m-1} \sum_{j=1}^{n+1} (-1)^{j-1} g^{(n+1-j)}(x) h^{(j-1)}(x) \Big|_{x_i^+}^{x_{i+1}^-}$$

$$+ (-1)^{n+1} \int_{-1}^{1} g(x) h^{(n+1)}(x)\, dx.$$

The fully integrated terms are

$$\sum_{j=1}^{n+1} (-1)^j \big[g^{(n+1-j)}(-1+) h^{(j-1)}(-1+) - g^{(n+1-j)}(1-) h^{(j-1)}(1-) \big]$$

$$+ \sum_{i=2}^{m-1} \sum_{j=1}^{n+1} (-1)^j \big[g^{(n+1-j)}(x_i+) h^{(j-1)}(x_i+)$$

$$- g^{(n+1-j)}(x_i-) h^{(j-1)}(x_i-) \big]. \qquad (14.4.4)$$

Since F_1^* is the dual of F_1, $f_{i,j-1}^* = -f_{i,n-j+1}$, $j = 1,\ldots,n+1$. For $i = 1$, exactly one of these entries is 1; this means that at least one of $g^{(n+1-j)}(-1+)$, $h^{(j-1)}(-1+)$ is 0, and that always

$$g^{(n+1-j)}(-1+)h^{(j-1)}(-1+) = 0.$$

Similarly, $g^{(n+1-j)}(1-)h^{(j-1)}(1-) = 0$. This argument applies also to the term in square brackets in the last sum of (14.4.4). When both $h^{(j-1)}$ and $g^{(n+1-j)}$ are continuous at x_i, the term is 0; otherwise, say $g^{(n+1-j)}$ is discontinuous, then $f^*_{i,\,j-1} = -f_{i,\,n-j+1} = 1$ so that $h^{(j-1)}$ is continuous and equal to 0 at x_i. In this way, we see that the second sum in (14.4.4) is 0.□

The pair F_1, X will be called *complete for* $\mathbb{S}_n(F_1, X)$ if the only function $S \in \mathbb{S}_n(F_1, X)$ that is annihilated by F_1, X is $S \equiv 0$. A simple consequence of Lemma 14.6 is a regularity theorem for self-dual matrices.

Theorem 14.7.   *The pair F, X is regular if and only if the pair F_1, X is complete for* $\mathbb{S}_n(F_1, X)$.

Proof.   A spline $S \in \mathbb{S}_{2n+1}(F, X)$ is a linear combination of powers $1, x, \ldots, x^{2n+1}$ and functions $(x - x_p)^q_+$, $f_{p,\,q} = -1$ in $\mathring{F}$. The -1's in $\mathring{F}_1$ produce terms of degree $\leqslant n$ in S, while those in $\mathring{F}^*_1$ correspond to terms of degree $> n$.

First assume that F, X is regular and that $S \in \mathbb{S}_n(F_1, X)$ is annihilated by F_1, X. Since functionals corresponding to 1's in F^*_1 are derivatives of order $> n$, S is annihilated by F, X and is 0.

To prove the sufficiency we observe that if $S \in \mathbb{S}_{2n+1}(F, X)$ is annihilated by F, X, then S is annihilated by F_1, X and $S^{(n+1)}$ is annihilated by F^*_1, X. Since $S^{(2n+2)} \equiv 0$, Lemma 14.6 gives $\int_{-1}^{1} |S^{(n+1)}(x)|^2\,dx = 0$. This implies $S \in \mathbb{S}_n(F_1, X)$ and hence $S \equiv 0$.   □

In the case when $F^+_1 = E$ is a matrix of 0's and 1's, the ordinary Pólya conditions (1.4.2) are necessary for F_1, X to be complete for $\mathscr{P}_n = \mathbb{S}_n(F_1, X)$. Since these conditions are also sufficient for conservative matrices, we arrive at a case treated by Schoenberg [150, 152] and Stieglitz [171]:

Corollary 14.8.   *The self-dual matrix $F = F_1 \oplus F^*_1$ is regular if $F^+_1 = E$ and E contains a conservative Pólya matrix.*

For example, the self-dual matrix

$$F_1 \oplus F^*_1 = \begin{bmatrix} 1 & -1 & -1 & -1 & 1 & 1 & 1 & -1 \\ 0 & 1 & 0 & 0 & 0 & 0 & -1 & 0 \\ 1 & 1 & 0 & 0 & 0 & 0 & -1 & -1 \\ -1 & 1 & -1 & -1 & 1 & 1 & -1 & 1 \end{bmatrix} \quad (14.4.5)$$

is regular.

§14.5.   BEST QUADRATURE FORMULAS

Let $E = [f_{i,\,k}]$ be a given $m \times (n+1)$ matrix of 0's and 1's; then for each set of knots $X := -1 = x_1 < \cdots < x_m = 1$, and a positive Borel measure dg on

$[-1,1]$, we can set up a quadrature formula Q:

$$\sum_{f_{i,k}=1} c_{i,k} f^{(k)}(x_i) = \int_{-1}^{1} f \, dg. \tag{14.5.1}$$

We are interested in formulas Q that are exact for $f \in \mathscr{P}_n = \mathbb{S}_n(F_1, X)$ and are not of Gaussian type. Then $|E| \geq n+1$. For given E, X, dg, there may be several such formulas (see Chapter 10); for example, interpolatory formulas arising from conditionally regular submatrices of E.

The *kernel* of Q is the function

$$K_Q(t) = \int_{-1}^{1} \frac{(x-t)_+^n}{n!} dg(x) - \sum_{f_{i,k}=1} c_{i,k} \frac{(x_i-t)_+^{n-k}}{(n-k)!}. \tag{14.5.2}$$

From Theorem 10.15, for any $f \in W_1^{(n+1)}[-1,1]$ we have the remainder formula

$$R(f) = \int_{-1}^{1} f \, dg - Q(f) = \int_{-1}^{1} K_Q(t) f^{(n+1)}(t) \, dt \tag{14.5.3}$$

[where $Q(f)$ is the left-hand side of (14.5.1)].

For a given triple E, X, dg we wish to minimize the remainder. For individual f this is too difficult. It is easier to find a quadrature formula that minimizes the largest remainder for a class of functions f. By Cauchy's inequality, $|R(f)| \leq \|f^{(n+1)}\|_2 \|K_Q\|_2$. Thus, at least for $f \in W_2^{(n+1)}[-1,1]$, it is natural to minimize $\|K_Q\|_2$. A quadrature formula Q^* associated with the triple E, X, dg and exact for $\mathscr{P}_n$ is called *best in the sense of Sard* if its kernel $K^* = K_{Q^*}$ is smallest in the L_2-norm, among all other formulas Q associated with E, X, dg and exact for $\mathscr{P}_n$:

$$\int_{-1}^{1} K^{*2}(t) \, dt \leq \int_{-1}^{1} K^2(t) \, dt \quad \text{for other} \quad K = K_Q. \tag{14.5.4}$$

In a series of papers of increasing generality, Greville [50], Schoenberg [152], Karlin [65], and Stieglitz [170, 171] proved the existence of best formulas when E is complete for polynomials.

Instead of this, we study quadrature formulas (14.5.1) constructed for 1's of an $m \times (n+1)$ spline interpolation matrix $F_1, |F_1^+| \geq n+1$, which are exact for the spline space $\mathbb{S}_n(F_1, X)$. This will contain the results just mentioned.

We fix the triple F_1, X, dg with absolutely continuous measure $dg = w \, dx$, $w \in L_1[-1, +1]$, and put $F = F_1 \oplus F_1^*$. Since $\mathscr{P}_n \subseteq \mathbb{S}_n(F_1, X)$, the remainder formula (14.5.3) holds for $f \in W_1^{(n+1)}[-1, +1]$, and it makes sense to ask whether there exists, among all formulas exact for $\mathbb{S}_n(F_1, X)$, a best one in the sense of Sard.

Exactness for $\mathbb{S}_n(F_1, X)$ entails properties of the kernel K_Q:

Lemma 14.9. *If the formula (14.5.1) is exact for all $S \in \mathbb{S}_n(F_1, X)$, then the kernel K belongs to the space $W_1^{n+1}[-1, +1] + \mathbb{S}_n(F_1^*, X)$ and is annihilated by F_1^*, X.*

Proof. That the kernel K belongs to the indicated space is seen from (14.5.2) and the fact that $f_{i,k} = 1 = -f^*_{i,n-k}$, $0 \leq k \leq n$. Differentiating (14.5.2), we have

$$K^{(n-q)}(t) = (-1)^{n-q}\left[\int_{-1}^{1} \frac{(x-t)^q_+}{q!} dg(x) - \sum_{f_{i,k}=1} c_{i,k}\frac{(x_i-t)^{q-k}_+}{(q-k)!}\right].$$

Since (14.5.1) is exact for all functions $(x - x_p)^q_+/q!$, $f_{p,q} = -1$, the right-hand side is 0 for $0 \leq q \leq n$ and $t = x_p$ with $f^*_{p,n-q} = 1$.   □

If the pair F_1, X is complete for $\mathbb{S}_n(F_1, X)$, then there is at least one quadrature formula Q^* that is exact for $\mathbb{S}_n(F_1, X)$. It is given by the coefficients

$$c^*_{i,k} = \int_{-1}^{1} l_{i,k} \, dg, \qquad f_{i,k} = 1 \text{ in } F_1 \qquad (14.5.5)$$

where $l_{i,k}$ are the fundamental splines in $\mathbb{S}_{2n+1}(F, X)$ for the self-dual problem F, X or $F_1 \oplus F^*_1$, X, which is regular by Theorem 14.7. Indeed, for any $S \in \mathbb{S}_n(F_1, X)$ we have (with summation restricted to $0 \leq k \leq n$)

$$S(x) = \sum_{f_{i,k}=1} S^{(k)}(x_i) l_{i,k}(x). \qquad (14.5.6)$$

Integrating this formula, we find that Q^* is exact for $\mathbb{S}_n(F_1, X)$.
Even more is true.

Theorem 14.10. *Let an $m \times (n+1)$ spline interpolation matrix F_1, a set of knots X: $-1 = x_1 < \cdots < x_m = 1$, and an absolutely continuous measure dg be given. If F_1, X is complete for $\mathbb{S}_n(F_1, X)$, then the quadrature formula Q^* with coefficients (14.5.5) is best in the sense of Sard.*

Proof. Let K^* be the kernel of Q^* and let K be the kernel of any other formula Q corresponding to the triple F_1, X, dg that is exact for $\mathbb{S}_n(F_1, X)$. The spline

$$\bar{S}(x) = \sum_{\substack{f_{i,k}=1 \\ 0 \leq k \leq n}} (c^*_{i,k} - c_{i,k})\frac{(x_i - x)^{2n+1-k}_+}{(2n+1-k)!}$$

belongs to $\mathbb{S}_{2n+1}(F, X)$ and satisfies $\bar{S}^{(n+1)} = K^* - K$. By Lemma 14.9, $\bar{S}^{(n+1)}$ is annihilated by F^*_1, X. Hence, (14.5.6) (with $0 \leq k \leq n$) holds for $\bar{S}$ and Q^* is exact for $\bar{S}$. By (14.5.3) we have

$$\int_{-1}^{1} \bar{S}^{(n+1)}(x) K^*(x) \, dx = \int_{-1}^{1} (K^* - K)(x) K^*(x) \, dx = 0.$$

This implies

$$\int_{-1}^{1} [K(x)]^2 \, dx = \int_{-1}^{1} [K^*(x)]^2 \, dx + \int_{-1}^{1} [K^*(x) - K(x)]^2 \, dx.$$

Therefore (14.5.4) holds with strict inequality if $K \neq K^*$.   □

In particular, if $F_1^+ = E$ and E, X is complete for $\mathcal{P}_n = \mathcal{S}_n(F_1, X)$ (for instance, if E contains a conservative Pólya submatrix), then Theorem 14.10 reduces to results given by Schoenberg [152] and Stieglitz [170, 171].

Exercise. If F_1, X is complete for $\mathcal{S}_n(F_1, X)$ and the quadrature formula (14.5.1) with kernel (14.5.2) is exact for $\mathcal{S}_n(F_1, X)$, then

$$c_{i,k} = (-1)^{n-k} \left[K^{(n-k)}(x_i +) - K^{(n-k)}(x_i -) \right].$$

§14.6. INTERPOLATION WITH DERIVATIVE OF MINIMAL NORM

Assume that an $m \times (n+1)$ matrix $E = [e_{i,k}]$, of 0's and 1's, a set of knots $X: -1 = x_1 < \cdots < x_m = +1$ and real numbers $c = \{c_{i,k}: e_{i,k} = 1\}$ are given. For $1 \leq p \leq +\infty$, let $\Lambda_p(c)$ be the set of functions

$$\Lambda_p(c) = \left\{ g \in W_p^{n+1}[-1, +1]: g^{(k)}(x_i) = c_{i,k}, e_{i,k} = 1 \right\}. \quad (14.6.1)$$

In this section, we study the following extremal problem:

Among all $g \in \Lambda_p(c)$, find a function for which $\|g^{(n+1)}\|_p$ is minimal.

$$(14.6.2)$$

The problem is of importance for least squares smoothing ($p = 2$, see de Boor [B, Chapter 14]) and for optimal control ($p = +\infty$, Glaeser [45], where the problem originated, and Louboutin in [154]). For $p = 2$, Schoenberg [152] showed the existence and uniqueness of the solution (Theorem 14.11). For $p = \infty$, the solution is not necessarily unique, but there is one, in the subclass of $\Lambda_\infty(c)$ consisting of perfect splines. This is our main theorem (Theorem 14.14), proved by Karlin [67] and de Boor [9] for Hermite interpolation, and by Goodman [46] in its present form. For other values of p and related problems, see Karlin [68,69], Micchelli and Rivlin [201], de Boor [11, §8], Melkman [121], and Goodman and Lee [48].

We shall always assume that E satisfies the Pólya conditions (1.4.2) (equivalently, that E contains a Pólya matrix) and has no odd supported sequences. Thus $|E| \geq n+1$, and E, X is complete for $\mathcal{P}_n$ for each set X. If $|E| = n+1$, there is a polynomial of degree $\leq n$ in $\Lambda_p(c)$ which trivially solves (14.6.2). Hence, we shall assume that $|E| > n+1$.

We construct a spline interpolation matrix F_1 so that $F_1^+ = E$ (this is achieved by replacing 0's in the exterior rows of E by -1's), and define the self-dual matrix $F = F_1 \oplus F_1^*$. Several spline spaces are connected with these matrices. We need the spaces $\mathcal{S}_{2n+1}(F, X), \mathcal{S}_n(F_1^*, X)$. There are no -1's in the interior rows of F_1, hence $\mathcal{S}_{2n+1}(F, X) \subset W_\infty^{n+1}[-1, +1]$.

If $S \in \bar{S}_{2n+1}(F, X)$, then $S^{(n+1)} \in \bar{S}_n(F_1^*, X)$. Finally, we also need the space $\bar{S}_n^0(F_1^*, X)$ of splines $S \in \bar{S}_n(F_1^*, X)$ that are annihilated by F_1^*, X. This space is identical with the space of Birkhoff splines $\bar{S}_n^0(E, X)$, since $f_{i,k}^* = -\tilde{e}_{i,k} = -f_{i,n-k}$ by (14.4.1); it may contain only the zero function.

A distinguished element of $\Lambda_p(c)$ is the spline S_0 defined as follows. By Theorem 14.7, the pair F, X is regular. Therefore, there exists a unique spline $S_0 \in \bar{S}_{2n+1}(F, X)$ for which $S_0^{(k)}(x_i) = c_{i,k}$ for $f_{i,k} = 1$ in F_1 and $S_0^{(n+1)}$ is annihilated by F_1^*, X. This leads to

Theorem 14.11. *For $p = 2$, the spline S_0 just constructed is the unique solution of* (14.6.2).

Proof. If g and S_0 both belong to $\Lambda_2(c)$, then $g - S_0$ is annihilated by F_1, X. Also, $S_0^{(n+1)}$ is annihilated by F_1^*, X, and $S_0^{(2n+2)} = 0$. By Lemma 14.6, the functions $(g - S_0)^{(n+1)}$ and $S_0^{(n+1)}$ are orthogonal in $L_2[-1, +1]$. Therefore, if $g \neq S_0$, $g \in \Lambda_2(c)$,

$$\|g^{(n+1)}\|_2^2 = \|S_0^{(n+1)}\|_2^2 + \|g^{(n+1)} - S_0^{(n+1)}\|_2^2 > \|S_0^{(n+1)}\|_2^2. \qquad \square$$

Let g_0 be any fixed element of $\Lambda_\infty(c)$, for instance, let $g_0 = S_0$. To deal with the problem (14.6.1) for $p = \infty$, we need the following lemma. We consider bounded, measurable functions h on $[-1, +1]$ that satisfy

$$\int_{-1}^{+1} Sh \, dx = \int_{-1}^{+1} S g_0^{(n+1)} \, dx \quad \text{for all} \quad S \in \bar{S}_n^0(F_1^*, X). \quad (14.6.3)$$

Lemma 14.12. *If $g \in \Lambda_\infty(c)$, then $h = g^{(n+1)}$ satisfies* (14.6.3). *Conversely, if h has this property, there is a unique $g \in \Lambda_\infty(c)$ with $g^{(n+1)} = h$.*

Proof. (a) If $g \in \Lambda_\infty(c)$, then $g - g_0$ is annihilated by F_1, X. Let $S \in \bar{S}_n^0(F_1^*, X)$ be arbitrary. Applying Lemma 14.6 to the functions $g - g_0$ and S we obtain that the difference of the integrals (14.6.3) is

$$\int_{-1}^{+1} S(g - g_0)^{(n+1)} \, dx = \pm \int S^{(n+1)}(g - g_0) \, dx = 0.$$

(b) Assume that h satisfies (14.6.3) and let H be one of its $(n+1)$st integrals. Since F, X is regular, there exists a spline $S^* \in \bar{S}_{2n+1}(F, X)$ for which $g = H - S^* \in \Lambda_\infty(c)$ and $S^{*(n+1)}$ is annihilated by F_1^*, X. Then $H - g_0 - S^*$ is annihilated by F_1, X. By Lemma 14.6,

$$\int_{-1}^{+1} (H - g_0 - S^*)^{(n+1)} S^{*(n+1)} \, dx = \pm \int_{-1}^{+1} (H - g_0 - S^*) S^{*(2n+2)} \, dx = 0.$$

By (14.6.3), this reduces to $-\int_{-1}^{+1} (S^{*(n+1)})^2 \, dx = 0$.

It follows that $S^* \in \bar{S}_n(F_1, X) = \mathcal{P}_n$ is a polynomial of degree $\leq n$. Hence $g^{(n+1)} = (H - S^*)^{(n+1)} = h$.

If $H_1, H_2 \in \Lambda_\infty(c)$ are two functions with $H_1^{(n+1)} = H_2^{(n+1)} = h$, then $H_1 - H_2$ would be a polynomial P of degree $\leq n$, annihilated by F_1, X; hence $P \equiv 0$. $\qquad \square$

There is a one-to-one correspondence between linear functionals $\lambda(f)$ defined on $L_1[-1,1]$, and bounded measurable functions h on $[-1,+1]$, given by $\lambda(f) = \int_{-1}^{+1} hf\,dx$ and $\|\lambda\| = \|h\|_\infty$. Let $\lambda_0(f)$ be the linear functional $\int_{-1}^{+1} g_0^{(n+1)} f\,dx$, restricted to the subspace $\mathbb{S}_n^0(F_1^*, X)$. Equation (14.6.3) means then that the functional λ generated by h is an extension of λ_0. The problem (14.6.2) is equivalent to finding such an extension with the minimal norm. Since this extension exists by the Hahn–Banach theorem, we get

Proposition 14.13. *For $p = \infty$, the problem (14.6.2) has at least one solution.*

Now we will try to find a solution of a special kind. A *perfect spline* on $[-1,+1]$ is any function

$$P(x) + b\left[x^{n+1} + 2\sum_{j=1}^{r}(-1)^j (x-\xi_j)_+^{n+1}\right] \tag{14.6.4}$$

where P is a polynomial of degree $\leqslant n$, b is a real number, and $-1 < \xi_1 < \cdots < \xi_r < +1$.

Theorem 14.14 (Karlin, de Boor, Goodman). *If E contains a Pólya matrix and has no odd supported sequences, then among all solutions of the problem (14.6.2) with $p = \infty$, there is a perfect spline (14.6.4) with $r \leqslant |E| - n - 2$.*

Proof. Let λ be an extension of λ_0 with minimal norm. If the functional λ is 0 [in particular, if $\mathbb{S}_n^0(F_1, X)$ consists only of the zero function], we can take $h \equiv 0$; then g is of the form (14.6.4) with $b = 0$.

Let $\|\lambda\| > 0$. By Theorem 7.14, each nonzero spline in $\mathbb{S}_n^0(F_1, X)$ has at most $|E| - n - 2$ changes of sign on $(-1, +1)$. First assume that none of the nonzero splines of $\mathbb{S}^0 := \mathbb{S}_n^0(F_1^*, X)$ vanishes on intervals. The norm of the extremal functional is achieved on $\mathbb{S}^0$, hence there is an $S_0 \in \mathbb{S}^0$ with $\|S_0\|_1 = 1$ and

$$\|\lambda\| = \|h\|_\infty = \int_{-1}^{+1} hS_0\,dx.$$

This means that $h(x) = \|\lambda\|\,\text{sign}\,S_0(x)$ a.e.

Let $r \leqslant |E| - n - 2$ be the number of sign changes of S_0. Then h is a step function with values $\pm\|\lambda\|$ and r sign changes at points ξ_j, $j = 1,\ldots,r$. Integration and Lemma 14.12 lead to expression (14.6.4) for the extremal g, with $b = \pm\|\lambda\|/(n+1)!$ and a properly assigned $\pm$.

If $\mathbb{S}^0$ does not have the postulated property, we apply a *variation diminishing* operation, for example, the Bernstein polynomials of S on $[-1, +1]$. They have the form

$$B_p(S)(x) = \sum_{k=0}^{p} S\left(-1 + \frac{2k}{p}\right) a_k (1+x)^k (1-x)^{p-k} \tag{14.6.5}$$

where $a_k > 0$ are some constants. [The values of S in (14.6.5) at points of discontinuity can be taken as limits from the right.] The polynomials (14.6.5) are uniformly bounded by $\|S\|_\infty$ and converge for $p \to \infty$ a.e. to $S(x)$; therefore, $\|B_p(S) - S\|_1 \to 0$ for $p \to \infty$. Most important for us is the property of Bernstein polynomials discovered by Pólya; the number of zeros (and hence the number of changes of sign) of $B_p(S)$ does not exceed the number of changes of sign of S. Indeed, putting $u = (1 + x)/(1 - x)$, we see that the number of zeros of $B_p(S)$ on $(-1, +1)$ is the same as the number of zeros of $Q(u) = \Sigma_0^p S(-1 + 2k/p)a_k u^k$ on $(0, +\infty)$ and this, by the Descartes rule of signs, is not more than the number of changes of sign in the coefficients of Q, that is, than $\mathfrak{S}^-\{S(-1 + 2k/p)\}_{k=0}^p$.

On the linear space $\mathfrak{S}_p = \{B_p(S): S \in \mathfrak{S}^0\}$ the extremal functional λ has a norm $n_p \leqslant \|\lambda\|$ and has a representation

$$\lambda(f) = \int_{-1}^{+1} h_p f \, dx,$$

where h_p is a step function with values $\pm n_p$ and at most $|E| - n - 2$ changes of sign. Passing to a subsequence, we can assume that $h_p \to h$ a.e., $\|h\|_\infty = \lim n_p \leqslant \|\lambda\|$. Moreover, for any $S \in \mathfrak{S}^0$ we have

$$\int_{-1}^{+1} hS \, dx = \lim_{p \to \infty} \int_{-1}^{+1} h_p B_p(S) \, dx = \lim_{p \to \infty} \lambda(B_p(S)) = \lambda(S).$$

As before, h is a step function with values $\pm\|\lambda\|$ and $r \leqslant |E| - n - 2$ changes of sign from which we can obtain the extremal perfect spline.   □

The given proof works in a more general situation, when F_1 is an arbitrary spline interpolation matrix. This time we have to assume that F_1, X is complete for $\mathfrak{S}_n(F_1, X)$, hence that F_1 satisfies the lower Goodman–Pólya conditions (proof of Theorem 13.4). The class $\Lambda_p(c)$ is now

$$\Lambda_p(c) = \{g \in W_p^{n+1}[-1, +1] + \mathfrak{S}_n(F_1, X): g^{(k)}(x) = c_{i,k}, f_{i,k} = 1 \text{ in } F_1\}.$$

In place of perfect splines, we take the functions

$$S(x) + b\left[x^{n+1} + 2\sum_{j=1}^r (-1)^j (x - \xi_j)_+^{n+1}\right] \tag{14.6.6}$$

with $S \in \mathfrak{S}_n(F_1, X)$.

The analogue of Theorem 14.11 follows as before, but for $p = \infty$ we need Theorem 14.3 to estimate the number of changes of sign of $S \in \mathfrak{S}_n^0(F_1^*, X)$. Since F_1^* satisfies the upper Goodman–Pólya condition, this number does not exceed

$$n + 1 - |F_1^{*+}| + \tilde{\gamma}(F_1^*) - 1 = |F_1^+| - n - 2 + \rho(F_1)$$

where $\rho(F_1)$ is the number of odd blocks in F_1 supported from the left. In this way we get

Theorem 14.15. *If F_1, X is complete for $\mathfrak{S}_n(F_1, X)$, then there is a spline $S^* \in \Lambda_\infty(c)$ of the form (14.6.6), with $r \leqslant |F_1^+| + \rho(F_1) - (n + 2)$, that*

is extremal:

$$\|S^{*(n+1)}\|_\infty \leqslant \|g^{(n+1)}\|_\infty \quad \textit{for all} \quad g \in \Lambda_\infty(c). \qquad (14.6.7)$$

§14.7. NOTES

14.7.1. Pence [134, 135] gives applications of Theorem 14.1 to approximation by monotone splines, obtaining results similar to those of Lorentz and Zeller [106] and R. A. Lorentz [109].

14.7.2. Spline interpolation at special sets of knots have been discussed by Meir and Sharma [117], Swartz and Varga [176], and Demko [23].

Meir and Sharma consider, for example, the interpolation problem for quintic splines given by the $m \times 6$ matrix

$$F = \begin{bmatrix} 1 & -1 & 1 & 1 & -1 & -1 \\ 1 & 0 & 1 & 0 & -1 & -1 \\ & & \vdots & & & \\ 1 & 0 & 1 & 0 & -1 & -1 \\ 1 & -1 & 1 & 1 & -1 & -1 \end{bmatrix}. \qquad (14.7.1)$$

The pair F, X is regular if m is even and the knots X are equally spaced. If U is the interpolation operator for F, X, then they obtain estimates of the form $\|f - Uf\|_\infty \leqslant Cm^{-3}\omega_4(f, 1/m)$ for $f \in C^4$. Swartz and Varga sharpen these estimates.

Demko proves the regularity of pairs F, X for equally spaced knots and for large classes of matrices that contain all F in (14.7.1). For example, let d and m be even integers, and let $F = [f_{i,k}]_{i=1,\,k=0}^{m,\,3d-1}$ have last d columns consisting of -1's with the remaining columns described by (a) for odd k, $0 \leqslant k \leqslant 2d - 1$, $f_{i,k} = 1$; (b) for $k = 4l \leqslant 2d - 1$, $f_{i,k} = 1$ if $i = 1$, m and $f_{i,k} = 0$, otherwise; for $k = 4l + 2 \leqslant 2d - 1, f_{i,k} = -1$ if $i = 1$, m and $f_{i,k} = 0$ otherwise. In this situation the pair F, X is regular and the interpolation operator satisfies $\|f - Uf\|_\infty = \mathcal{O}(m^{-3d+1})$ if $f \in C^{3d}[-1, 1]$. Many other results can be found in Demko's paper [23].

Bibliography and References

A. BOOKS ON INTERPOLATION AND RELATED TOPICS

A. Bernstein, S., *Collected Works*, Vols. I, II (in Russian), Akad. Nauk SSSR, Moscow, 1952, 1954. §12.2

B. de Boor, C., *A Practical Guide to Splines* (Applied Mathematical Sciences, Vol. 27), Springer-Verlag, 1978. §§7.2, 14.1, 14.6

C. Cheney, E. W., *Introduction to Approximation Theory*, McGraw-Hill, New York, 1966. §9.2

D. Davis, P. J., *Interpolation and Approximation*, Blaisdell, New York, 1963.

E. Gel'fand, A. O., *Calculus of Finite Differences* (in Russian), Akad. Nauk SSSR, Moscow, 1967.

F. Goncharov, V. L., *Theory of Interpolation and Approximation of Functions* (in Russian), Tehizdat, Moscow, 1954. §4.2

G. Karlin, S., *Total Positivity*, Vol. I, Stanford University Press, Stanford, CA, 1968.

H. Karlin, S., and Studden, W. J., *Tchebycheff Systems: With Applications in Analysis and Statistics*, Wiley-Interscience, New York, 1966. §§4.2, 4.6, 5.5, 10.1, 10.3

I. Lorentz, G. G., *Approximation of Functions*, Holt, Rinehart and Winston, New York, 1966. §§1.2, 9.2, 11.3, 11.4, 12.5

J. Natanson, I. P., *Constructive Function Theory*, Vols. I, II, III, Ungar, New York, 1964, 1965. §§11.1, 12.1, 12.2, 12.3, 12.6

K. Schumaker, L. L., *Spline Functions: Basic Theory*, Wiley, New York, 1981. §7.5

L. Szegö, G., *Orthogonal Polynomials*, Amer. Math. Soc. Coll. Publ., Vol. 23, Providence, 1939. §§12.1, 12.2, 12.6

M. Tureckii, A. H., *Theory of Interpolation in Problem Form*, I, II (in Russian), Minsk, Izdat. Vyseisaja Skola, 1968, 1977. §12.1

N. Timan, A. F., *Theory of Approximation of Functions of a Real Variable*, Pergamon Press, Oxford, 1963. §12.5

O. Whittaker, J. M., *Interpolatory Function Theory*, Stechert-Hafner, New York, 1964. §1.5

P. Walsh, J. L., *Interpolation and Approximation by Rational Functions in the Complex Domain*, 5th ed., Amer. Math. Soc. Coll. Publ., Vol. 20, Providence, RI, 1969. §11.6

B. PAPERS ON BIRKHOFF INTERPOLATION AND RELATED TOPICS

1. Ahlberg, J. H., and Nilson, E. N., The approximation of linear functionals, *SIAM J. Numer. Anal.* **3** (1966), 173–182.

2. Atkinson, K., and Sharma, A., A partial characterization of poised Hermite-Birkhoff interpolation problems, *SIAM J. Numer. Anal.* **6** (1969), 230–235, §1.4, 1.5.

3. Balázs, J., and Turán, P., Notes on interpolation, II: Explicit formulae; III: Convergence; IV: Inequalities, *Acta Math. Acad. Sci. Hungar.* **8** (1957), 201–215, **9** (1958), 195–214, 243–258. §§12.1, 12.2, 12.4, 12.6

4. Barrow, D. L., On multiple node Gaussian quadrature formulas, *Math. Comp.* **32** (1978), 431–439. §10.6

5. Beatson, R. K., Jackson type theorems for approximation with Hermite-Birkhoff interpolatory side conditions, *J. Approx. Theory* **22** (1978), 95–104.

6. Beatson, R. K., The degree of monotone approximation, *Pacific J. Math.* **74** (1978), 5–14. §9.7

7. Birkhoff, G. D., General mean value and remainder theorems with applications to mechanical differentiation and quadrature, *Trans. Amer. Math. Soc.* **7** (1906), 107–136. §§1.1, 1.4, 7.1, 7.2, 7.3, 7.5

8. Boas, R. P., Representation of functions by Lidstone series, *Duke Math. J.* **10** (1943), 239–249.

9. de Boor, C., A remark concerning perfect splines, *Bull. Amer. Math. Soc.* **80** (1974), 724–727. §14.6

10. de Boor, C., Total positivity of the spline collocation matrix, *Indiana Univ. Math. J.* **25** (1976), 541–551. §14.1

11. de Boor, C. Splines as linear combinations of B-splines. A survey. In *Approximation Theory*, Vol. II (G. G. Lorentz et al., eds.), pp. 1–47, Academic Press, New York, 1976. §14.6

12. de Boor, C., and Schoenberg, I. J., Cardinal interpolation and spline functions, VIII. The Budan-Fourier theorem and applications. In *Spline Functions* (K. Böhmer et al., eds.) pp. 1–79, Lecture Notes in Math., Vol. 501, Springer-Verlag, 1976. §7.5

13. Cavaretta, A. S., Sharma, A., and Varga, R. S., Lacunary trigonometric

interpolation on equidistant nodes. In *Quantitative Approximation* (R. A. DeVore and K. Scherer, eds.), pp. 63–80, Academic Press, New York, 1980. §11.5, 11.6

14. Cavaretta, A. S., Sharma, A., and Varga, R. S., Hermite–Birkhoff interpolation in the nth roots of unity, *Trans. Amer. Math. Soc.* **259** (1980), 621–628. §§11.1, 11.2

15. Chalmers, B. L., A unified approach to uniform real approximation by polynomials with linear restrictions, *Trans. Amer. Math. Soc.* **166** (1972), 309–316. §§9.3, 9.7

16. Chalmers, B. L., Uniqueness of best approximation of a function and its derivatives, *J. Approx. Theory* **7** (1973), 213–225. §9.5

17. Chalmers, B. L., The Remez exchange algorithm for approximation with linear restrictions, *Trans. Amer. Math. Soc.* **223** (1976), 103–131. §9.7

18. Chalmers, B. L., Johnson, D. J., Metcalf, F. T., and Taylor, G. D., Remarks on the rank of Hermite-Birkhoff interpolation, *SIAM J. Numer. Anal.* **11** (1974), 254–259. §2.1

19. Chalmers, B. L., and Taylor, G. D., Uniform approximation with constraints, *Jahresber. Deutsch. Math.-Verein.* **81** (1978/79), 49–86. §9.7

20. Chui, C. K., Smith, P. W., and Ward, J. D., Degree of L_p-approximation by monotone splines, *SIAM J. Math. Anal.* **11** (1980), 436–447.

21. Chui, C. K., Smith, P. W., and Ward, J. D., Monotone approximation by spline functions. In *Quantitative Approximation* (R. A. DeVore and K. Scherer, eds.), pp. 81–98, Academic Press, New York, 1980.

22. Cimoca, Gh., Über ein Interpolationsschema, *Mathematica (Cluj)* **11** (1969), 61–67.

23. Demko, S. C., Lacunary polynomial spline interpolation, *SIAM J. Numer. Anal.* **13** (1976), 369–381. §14.7

24. DeVore, R. A., Monotone approximation by splines, *SIAM J. Math. Anal.* **8** (1977), 881–905.

25. DeVore, R. A., Monotone approximation by polynomials, *SIAM J. Math. Anal.* **8** (1977), 906–921. §9.7

26. DeVore, R. A., Meir, A., and Sharma, A., Strongly and weakly non-poised H-B interpolation problems, *Canad. J. Math.* **25** (1973), 1040–1050. §1.2, 8.1, 8.2, 8.5, 8.6

27. Drols, W., Zur Hermite–Birkhoff Interpolation, *Resultate Math.* **2** (1979), 225–227.

28. Drols, W., On a problem of DeVore, Meir and Sharma. In *Approximation Theory*, Vol. III (E. W. Cheney, ed.), pp. 361–366, Academic Press, New York, 1980. §8.7

29. Drols, W., Zur Hermite–Birkhoff-Interpolation: DMS-Matrizen, *Math. Z.* **172** (1980), 179–194. §§8.1, 8.3, 8.5, 8.6, 8.7

30. Drols, W., Fasthermitesche Inzidenzmatrizen, *Z. Angew. Math. Mech.* **61**, (1981), T275–276. §8.2

31. Dyn, N., Hermite–Birkhoff quadrature formulas of Gaussian type. In *Approximation Theory*, Vol. III (E. W. Cheney, ed.), pp. 371–376, Academic Press, New York, 1980. §10.5

32. Dyn, N., On the existence of Hermite-Birkhoff quadrature formulas of Gaussian type, *J. Approx. Theory*, **31** (1981), 22–32. §§10.1, 10.5

33. Dyn, N., Lorentz, G. G., and Riemenschneider, S. D., Continuity of the Birkhoff interpolation, *SIAM J. Numer. Anal.* **19** (1982), 507–509. §5.6

34. Elsner, L., Über Birkhoff-Interpolation und Richardson-Extrapolation, *Z. Angew. Math. Mech.* **53** (1973), 57–60.

35. Elsner, L., and Merz, G., Lineare Punktfunktionale und Hermite-Birkhoff-Interpolation, *Beiträge Numer. Math.* **4** (1975), 69–82.

36. Epstein, M. P., and Hamming, R. W., Noninterpolatory quadrature formulas, *SIAM J. Numer. Anal.* **9** (1972), 464–475.

37. Ferguson, D. R., The question of uniqueness for G. D. Birkhoff interpolation problems, *J. Approx. Theory* **2** (1969), 1–28. §§1.4, 1.5, 3.1, 4.3

38. Ferguson, D. R., Some interpolation theorems for polynomials, *J. Approx. Theory* **9** (1973), 327–348.

39. Ferguson, D. R., Sufficient conditions for Peano's kernel to be of one sign, *SIAM J. Numer. Anal.* **10** (1973), 1047–1054.

40. Ferguson, D. R., Sign changes and minimal support properties of Hermite–Birkhoff splines with compact support, *SIAM J. Numer. Anal.* **11** (1974), 769–779. §§7.3, 7.5

41. Fiala, J., Interpolation with prescribed values of derivatives instead of function values, *Apl. Mat.* **16** (1971), 421–430.

42. Fiala, J., An algorithm for Hermite-Birkhoff interpolation, *Apl. Mat.* **18** (1973), 167–175.

43. Fletcher, Y., and Roulier, J. A., A counterexample to strong unicity in monotone approximation, *J. Approx. Theory* **27** (1979), 19–33. §9.7

44. Freud, G., Bemerkung über die Konvergenz eines Interpolationsverfahrens von P. Turán, *Acta Math. Acad. Sci. Hungar.* **9** (1958), 337–341. §12.5

45. Glaeser, G., Prolongement extrémal de fonctions différentiables d'une variable, *J. Approx. Theory* **8** (1973), 249–261. §14.6

46. Goodman, T. N. T., Perfect splines and Hermite–Birkhoff interpolation, *J. Approx. Theory* **26** (1979), 108–118. §14.6

47. Goodman, T. N. T., Hermite Birkhoff interpolation by Hermite–Birkhoff splines, *Proc. Roy. Soc. Edinburgh* **88A** (1981), 195–201. §§13.4, 13.6, 14.1, 14.3

48. Goodman, T. N. T., and Lee, S. L., Another extremal property of perfect splines, *Proc. Amer. Math. Soc.* **70** (1978), 129–135. §14.6

49. Goodman, T. N. T., and Lee, S. L., The Budan–Fourier theorem and Hermite–Birkhoff spline interpolation, *Trans. Amer. Math. Soc.* **271** (1982), 451–467.

50. Greville, T. N. E., Spline functions, interpolation and numerical quadrature. In *Mathematical Methods for Digital Computers*, Vol. 2 (A. Ralston and H. S. Wilf, eds.), pp. 156–168, Wiley, New York, 1967. §14.5

51. Haussmann, W., Hermite-Interpolation mit Cebysev-Unterräumen. In *Numerische Methoden der Approximationstheorie*, Bd. 1 (L. Collatz and G. Meinardus, eds.), pp. 49–55, Birkhäuser-Verlag, Basel, 1972 (ISNM Vol. 16).

52. Ikebe, Y., Hermite–Birkhoff interpolation problems in Haar subspaces, *J. Approx. Theory* **8** (1973), 142–149. §5.5

53. Jaffe, L., Rolle regular Birkhoff matrices. In *Approximation Theory*, Vol. II (G. G. Lorentz et al., eds.), pp. 397–404, Academic Press, New York, 1976. §7.6

54. Jerome, J. W., and Schumaker, L. L., On L_g-splines, *J. Approx. Theory* **2** (1969), 29–49.

55. Jerome, J. W., and Schumaker, L. L., Local bases and computation of g-splines. In *Methoden und Verfahren der mathematischen Physik*, Bd. 5 (B. Brosowski and E. Martensen, eds.), pp. 171–199, Bibliographisches Institut, Mannheim, 1971.

56. Jetter, K., Duale Hermite-Birkhoff-Probleme, *J. Approx. Theory* **17** (1976), 119–134. §§7.5, 13.3, 13.6

57. Jetter, K., Birkhoff interpolation by splines. In *Approximation Theory*, Vol. II (G. G. Lorentz et al., eds.), pp. 405–410, Academic Press, New York, 1976. §§13.4, 13.5, 14.1

58. Jetter, K., Nullstellen von Splines. In *Approximation Theory* (R. Schaback and K. Scherer, eds.), pp. 291–304 (Lecture Notes in Mathematics, Vol. 556), Springer-Verlag, 1976. §§7.5, 10.4

59. Jetter, K., Optimale Quadraturformeln mit semidefiniten Peano-Kernen, *Numer. Math.* **25** (1976), 239–249.

60. Jetter, K., Approximation mit Splinefunktionen und ihre Anwendung auf Quadraturformeln, Habilitationsschrift, Hagen, 1978. §§10.4, 10.6

61. Jetter, K., Lorentz, G. G., and Riemenschneider, S. D., Rolle theorem method in spline interpolation; *Analysis*, to appear. §§7.4, 13.2, 13.3, 14.1, 14.2

62. Johnson, D. J., The trigonometric Hermite–Birkhoff interpolation problem, *Trans. Amer. Math. Soc.* **212** (1975), 365–374. §2.4

63. Johnson, R. S., On monosplines of least deviation, *Trans. Amer. Math. Soc.* **96** (1960), 458–477. §10.4

64. Karlin, S., Best quadrature formulas and interpolation by splines satisfying boundary conditions. In *Approximation with Special Emphasis on Spline Functions* (I. J. Schoenberg, ed.), pp. 447–466, Academic Press, New York, 1969. §§13.1, 14.1

65. Karlin, S., Best quadrature formulas and splines, *J. Approx. Theory* **4** (1971), 59–90. §§10.6, 14.5

66. Karlin, S., Total positivity, interpolation by splines and Green's functions of differential operators, *J. Approx. Theory* **4** (1971), 91–112. §14.1

67. Karlin, S., Some variational problems in certain Sobolev spaces and perfect splines, *Bull. Amer. Math. Soc.* **79** (1973), 124–128. §14.6

68. Karlin, S., Interpolation properties of generalized perfect splines and the solutions of certain extremal problems, I, *Trans. Amer. Math. Soc.* **106** (1975), 25–66. §14.6

69. Karlin, S., Oscillatory perfect splines and related extremal problems. In *Studies in Spline Functions and Approximation Theory* (S. Karlin et al., eds.), pp. 371–460, Academic Press, New York, 1976. §14.6

70. Karlin, S., and Karon, J. M., Poised and non-poised Hermite–Birkhoff interpolation, *Indiana Univ. Math. J.* **21** (1972), 1131–1170. §§3.1, 3.3, 3.5, 4.1, 4.4, 4.5, 6.1, 6.2

71. Karlin, S., and Micchelli, C. A., The fundamental theorem of algebra for monosplines satisfying boundary conditions, *Israel J. Math.* **11** (1972), 405–451. §10.4

72. Karlin, S., and Pinkus, A., Gaussian quadrature formulae with multiple nodes. In *Studies in Spline Functions and Approximation Theory* (S. Karlin

et al., eds.), pp. 113–141, Academic Press, New York, 1976. §§10.5, 10.6

73. Karlin, S., and Schumaker, L. L., The fundamental theorem of algebra for Tchebycheffian monosplines, *J. Analyse Math.* **20** (1967), 233–270.

74. Karlin, S., and Ziegler, Z., Chebyshevian spline functions, *SIAM J. Numer. Anal.* **3** (1966), 514–543. §§13.1, 13.4, 14.1

75. Keener, L. L., Hermite–Birkhoff interpolation and approximation of functions and its derivatives, *J. Approx. Theory* **30** (1980), 129–138. §9.7

76. Kimchi, E., and Richter-Dyn, N., An example of a nonpoised interpolation problem with a constant sign determinant, *J. Approx. Theory* **11** (1974), 361–362. §4.1

77. Kimchi, E., and Richter-Dyn, N., Best uniform approximation with Hermite–Birkhoff interpolatory side conditions, *J. Approx. Theory* **15** (1975), 85–100. §9.7

78. Kimchi, E., and Richter-Dyn, N., Properties of best approximation with interpolatory and restricted range side conditions, *J. Approx. Theory* **15** (1975), 101–115. §9.7

79. Kimchi, E., and Richter-Dyn, N., On the unicity in simultaneous approximation by algebraic polynomials, *J. Approx. Theory* **18** (1976), 388–389. §§9.4, 9.7

80. Kis, O., Notes on interpolation, *Acta Math. Acad. Sci. Hungar.* **11** (1960), 49–64 (in Russian). §§11.2, 11.4

81. Kis, O., On trigonometric $(0,2)$-interpolation, *Acta Math. Acad. Sci. Hungar.* **11** (1960), 255–276 (in Russian). §§11.5, 11.6

82. Kis, O., Remarks on the error of trigonometric $(0,2)$-interpolation, *Acta Math. Acad. Sci. Hungar.* **22** (1971), 81–84 (in Russian).

83. Kloostermann, H. D., Derivatives and finite differences, *Duke Math. J.* **17** (1950), 169–186.

84. Krein, M. G., The ideas of P. L. Chebyshev and A. A. Markov in the theory of limiting values of integrals and their further developments. *Uspehi Mat. Nauk* **6** (1951), no. 4 (44), 3–122 (*Amer. Math. Soc. Transl.*, ser. 2, **12**, 1–122). §10.1

85. Lee, John W., Best quadrature formula and splines, *J. Approx. Theory* **20** (1977), 348–384.

86. Leeming, D. J., Representation of functions by generalized Lidstone series, *J. Approx. Theory* **5** (1972), 123–136.

87. Leeming, D. J., Convergence theorems for $(0, n - 1, n)$ interpolation, *Mathematica* (*Cluj*) **15** (1973), 209–217.

88. Linden, H., Pittnauer, F., and Wyrwich, H., Zur Birkhoff-Interpolation ganzer Funktionen, *Acta Math. Acad. Sci. Hungar.* **31** (1978), 259–268. §12.6

89. Lorentz, G. G., Monotone approximation. In *Inequalities*, Vol. III (O. Shisha, ed.), pp. 201–205, Academic Press, New York, 1972. §9.7

90. Lorentz, G. G., Birkhoff interpolation and the problem of free matrices, *J. Approx. Theory* **6** (1972), 283–290. §§5.1, 5.5, 6.1, 6.5

91. Lorentz, G. G., The Birkhoff interpolation problem: New methods and results. In *Linear Operators and Approximation*, Vol. II (P. L. Butzer and B. Sz.-Nagy, eds.), pp. 481–501, Birkhäuser-Verlag, Basel, 1975. §§3.1, 3.5, 4.1, 6.3, 8.1, 8.2, 8.5, 9.6

92. Lorentz, G. G., Birkhoff interpolation problem, CNA-report 103, The University of Texas at Austin, July, 1975. §§2.4, 4.3, 5.1, 6.3

93. Lorentz, G. G., Zeros of splines and Birkhoff's kernel, *Math. Z.* **142** (1975), 173–180. §§2.1, 7.4, 7.5, 7.6, 14.1

94. Lorentz, G. G., Coalescence of matrices, regularity and singularity of Birkhoff interpolation problems, *J. Approx. Theory* **20** (1977), 178–190. §§3.1, 4.4

95. Lorentz, G. G., Symmetry in Birkhoff matrices. In *Multivariate Approximation* (D. C. Handscomb, ed.), pp. 105–113, Academic Press, New York, 1978. §§2.3, 2.4

96. Lorentz, G. G., Independent sets of knots and singularity of interpolation matrices, *J. Approx. Theory* **30** (1980), 208–225. §§5.1, 5.5, 6.5

97. Lorentz, G. G., Theorem of Budan-Fourier and Birkhoff interpolation, *Turán Memorial Volume*, Hungarian Academy of Sciences, in press. §2.1

98. Lorentz, G. G., The analytic character of the Birkhoff interpolation polynomials, *Canad. J. Math.* **34** (1982), 765–768. §5.6

99. Lorentz, G. G., and Lorentz, R. A., Probability and interpolation, *Trans. Amer. Math. Soc.* **268** (1981), 477–486. §§4.5, 6.5

100. Lorentz, G. G., and Riemenschneider, S. D., Probabilistic approach to Schoenberg's problem in Birkhoff interpolation, *Acta Math. Acad. Sci. Hungar.* **33** (1978), 127–135. §§2.2, 6.4

101. Lorentz, G. G., and Riemenschneider, S. D., Birkhoff quadrature matrices. In *Linear Spaces and Approximation* (P. L. Butzer and B. Sz.-Nagy, eds.), pp. 359–373, ISNM, Vol. 40, Birkhäuser-Verlag, Basel, 1978. §10.1

102. Lorentz, G. G., and Riemenschneider, S. D., Recent progress in Birkhoff interpolation. In *Approximation Theory and Functional Analysis* (J. B. Prolla, ed.), pp. 187–236, North-Holland, Amsterdam, 1979. §4.5

103. Lorentz, G. G., and Riemenschneider, S. D., Birkhoff interpolation: Some applications of coalescence. In *Quantitative Approximation* (R. A. DeVore and K. Scherer, eds.), pp. 197–208, Academic Press, New York, 1980. §§3.1, 4.3, 4.6

104. Lorentz, G. G., Stangler, S. S., and Zeller, K. L., Regularity of some special Birkhoff matrices. In *Approximation Theory*, Vol. II (G. G. Lorentz et al., eds.), pp. 423–436, Academic Press, New York, 1976. §§8.1, 8.2, 8.3, 8.4, 8.5, 8.6, 8.7

105. Lorentz, G. G., and Zeller, K. L., Degree of approximation by monotone polynomials, I, II, *J. Approx. Theory* **1** (1968), 501–504, **2** (1969), 265–269. §9.7

106. Lorentz, G. G., and Zeller, K. L., Monotone approximation by algebraic polynomials, *Trans. Amer. Math. Soc.* **149** (1970), 1–18. §§9.2, 9.3, 9.4, 9.7, 14.7

107. Lorentz, G. G., and Zeller, K. L., Birkhoff interpolation, *SIAM J. Numer. Anal.* **8** (1971), 43–48. §§5.1, 6.1, 6.5

108. Lorentz, G. G., and Zeller, K. L., Birkhoff interpolation problem: coalescence of rows, *Arch. Math.* **26** (1975), 189–192. §§3.1, 3.5, 4.6

109. Lorentz, R. A., Uniqueness of best approximation by monotone polynomials, *J. Approx. Theory* **4** (1971), 401–418. §§9.4, 14.7

110. Lorentz, R. A., Nonuniqueness of simultaneous approximation by algebraic

polynomials, *J. Approx. Theory* **13** (1975), 17–23. §9.5

111. Lorentz, R. A., Probabilistic regularity of Birkhoff matrices with all active knots, *Ges. Math. u. Datenverarbeitung*, Bonn, to appear. §6.5

112. Maekelae, M., Nevanlinna, O., and Sipilae, A. H., On some generalized Hermite-Birkhoff interpolation problems, *Ann. Acad. Sci. Fenn. Ser.* **AI**, 563 (1974).

113. Maekelae, M., Nevanlinna, O., and Sipilae, A. H., Hermite interpolation by generalized rational functions, *Ann. Acad. Sci. Fenn. Ser.* **AI**, 564 (1974).

114. Malozemov, V. N., and Pevnyĭ, A. B., Spline interpolation, *Mat. Zametki* **26** (1979), 817–822 (in Russian).

115. Malozemov, V. N., Pevnyĭ, A. B., and Vasil'ev, A. A., Interpolation and approximation by spline functions of an arbitrary defect, *Vestnik Leningrad. Univ. Mat. Meh. Astronom.* **1979**, vyp. 4, 23–30 (in Russian).

116. Marusciac, I., Sur les polynomes d'interpolation et de préinterpolation lacunaire, *Mathematica (Cluj)* **11** (1969), 111–125.

117. Meir, A., and Sharma, A., Lacunary interpolation by splines, *SIAM J. Numer. Anal.* **10** (1973), 433–442. §§12.6, 14.7

118. Melkman, A. A., The Budan–Fourier theorem for splines, *Israel J. Math.* **19** (1974), 256–263.

119. Melkman, A. A., Interpolation by splines satisfying mixed boundary conditions, *Israel J. Math.* **19** (1974), 369–381. §13.6

120. Melkman, A. A., Hermite–Birkhoff interpolation by splines, *J. Approx. Theory* **19** (1977), 259–279. §13.4, 13.5, 13.6, 14.1

121. Melkman, A. A, Splines with maximal zero sets, *J. Math. Anal. Appl.* **61** (1977), 739–751. §14.6

122. Melkman, A. A., On a Hermite–Birkhoff problem of Passow, *J. Approx. Theory* **33** (1981), 81–84.

123. Micchelli, C. A., The fundamental theorem of algebra for monosplines with multiplicities. In *Linear Operators and Approximation* (P. L. Butzer et al., eds.), pp. 419–430, Birkhäuser-Verlag, Basel, 1972 (ISNM Vol. 20). §10.4

124. Micchelli, C. A., Best quadrature formulas at equally spaced nodes, *J. Math. Anal. Appl.* **47** (1974), 232–249.

125. Micchelli, C. A., Best L^1 approximation by weak Chebyshev systems and the uniqueness of perfect splines, *J. Approx. Theory* **19** (1977), 1–14.

126. Micchelli, C. A., and Rivlin, T. J., Quadrature formulae and Hermite-Birkhoff interpolation, *Adv. in Math.* **11** (1973), 93–112. §§10.1, 10.5, 10.6

127. Mühlbach, G., An algorithmic approach to Hermite–Birkhoff interpolation, *Numer. Math.* **37** (1981), 339–347. §1.6

128. Nemeth, A. B., Transformations of the Chebyshev systems, *Mathematica (Cluj)* **8** (1966), 315–333. §1.4

129. Passow, E., Alternating parity of Tchebycheff systems, *J. Approx. Theory* **9** (1973), 295–298. §9.6

130. Passow, E., Hermite–Birkhoff interpolation: a class of non-poised matrices, *J. Math. Anal. Appl.* **62** (1978), 140–147. §§4.4, 4.5

131. Passow, E., Conditionally poised Birkhoff interpolation problems, *J. Approx. Theory* **28** (1980), 36–44.

132. Passow, E., Extended Chebyshev systems on $(-\infty, \infty)$, *SIAM J. Math. Anal.* **5** (1974), 762–763. §9.6

133. Peano, G., Resto nelle formule di quadratura espresso con un integralo definito, *Atti della reale Acad. dei Lincei, Rendiconti* **5**, Vol. 22 (1913), 562–569. §7.2

134. Pence, D. D., Hermite-Birkhoff interpolation and monotone approximation by splines, *J. Approx. Theory* **25** (1979), 248–257. §§13.5, 13.6, 14.1, 14.7

135. Pence, D. D., Best mean approximation by splines satisfying generalized convexity constraints, *J. Approx. Theory* **28** (1980), 333–348. §14.7

136. Pólya, G., Bemerkung zur Interpolation und zur Näherungstheorie der Balkenbiegung, *Z. Angew. Math. Mech.* **11** (1931), 445–449. §1.5

137. Pólya, G., Sur l'existence de fonctions entières satisfaisant à certaines conditions linéaires, *Trans. Amer. Math. Soc.* **50** (1941), 129–139.

138. Popoviciu, T., Aspura unei generalizari a furmulei de integrare numerica a lui Gauss, *Acad. Republ. Popul. Romine Studii Cerc. Mat.* **6** (1955), 29–57 (in Rumanian). §§10.5, 10.6

139. Poritsky, H., On certain polynomials and their approximation, *Trans. Amer. Math. Soc.* **33** (1931), 274–331. §1.4

140. Riemenschneider, S. D., and Sharma, A., Birkhoff interpolation at the nth roots of unity: convergence, *Canad. J. Math.* **33** (1981), 362–371. §§11.1, 11.2, 11.4

141. Riemenschneider, S. D., Sharma, A., and Smith, P. W., Convergence of lacunary trigonometric interpolation on equidistant knots, *Acta Math. Acad. Sci. Hungar.* **39** (1982), 27–37. §§11.5, 11.6

142. Riemenschneider, S. D., Sharma, A., and Smith, P. W., Lacunary trigonometric interpolation: convergence. In *Approximation Theory*, Vol. III (E. W. Cheney, ed.), pp. 741–746, Academic Press, New York, 1980. §11.6

143. van Rooij, P. L. J., Schurer, F., and van Walt van Praag, C. R., *A Bibliography on Hermite-Birkhoff Interpolation*, Eindhoven, December, 1975.

144. Roulier, J. A., Monotone approximation of certain classes of functions, *J. Approx. Theory* **1** (1968), 319–324.

145. Roulier, J. A., and Taylor, G. D., Approximation by polynomials with restricted ranges of their derivatives, *J. Approx. Theory* **5** (1972), 216–227. §§9.3, 9.4

146. Saxena, R. B., and Sharma, A., On some interpolatory properties of Legendre polynomials, *Acta Math. Acad. Sci. Hungar.* **9** (1958), 345–358. §§11.1, 12.6

147. Saxena, R. B., and Sharma, A., Convergence of interpolatory polynomials, *Acta Math. Acad. Sci. Hungar.* **10** (1959), 157–175. §§11.1, 12.6

148. Scherer, R., and Zeller, K., Der Einsequenzensatz in der Birkhoff-Interpolation, *Resultate Math.* **2** (1979), 186–195. §6.5

149. Schmidt, D., Strong unicity and Lipschitz conditions of order $1/2$ for monotone approximation, *J. Approx. Theory* **27** (1979), 346–354. §§9.4, 9.7

150. Schoenberg, I. J., On monosplines of least square deviation and best quadrature formulae, I, II, *SIAM J. Numer. Anal.* **2** (1965), 144–170, **3** (1966), 321–328. §14.4

151. Schoenberg, I. J., On Hermite–Birkhoff interpolation, *J. Math. Anal. Appl.* **16** (1966), 538–543. §§1.1, 1.2, 1.4

152. Schoenberg, I. J., On the Ahlberg–Nilson extension of spline interpolation: the g-splines and their optimal properties, *J. Math. Anal. Appl.* **21** (1968), 207–231. §§13.1, 14.4, 14.5, 14.6

153. Schoenberg, I. J., A second look at approximate quadrature formulae and spline interpolation, *Adv. in Math.* **4** (1970), 277–300. §10.6

154. Schoenberg, I. J., The perfect B-splines and a time optimal control problem, *Israel J. Math.* **10** (1971), 261–274. §14.6

155. Schoenberg, I. J., and Whitney, A., On Pólya frequency functions, III: The positivity of translation determinants with an application to the interpolation problem by spline curves, *Trans. Amer. Math. Soc.* **74** (1953), 246–259. §§13.1, 13.4, 14.1

156. Schumaker, L. L., Zeros of spline functions and applications, *J. Approx. Theory* **18** (1976), 152–168.

157. Schumaker, L. L., Toward a constructive theory of generalized spline functions. In *Spline Functions* (K. Böhmer et al., eds.), pp. 265–331 (Lecture Notes in Mathematics, Vol. 501), Springer-Verlag, 1976. §7.5

158. Sharma, A., Some remarks on lacunary interpolation in the roots of unity, *Israel J. Math.* **2** (1964), 41–49. §§11.2, 11.4

159. Sharma, A., Lacunary interpolation in the roots of unity, *Z. Angew. Math. Mech.* **46** (1966), 127–133. §§11.2, 11.4

160. Sharma, A., Some poised and nonpoised problems of interpolation, *SIAM Rev.* **14** (1972), 129–151. §12.6

161. Sharma, A., and Leeming, D. J., Lacunary interpolation—$(0, n-1, n)$ case, *Mathematica (Cluj)* **11** (1969), 155–162. §12.6

162. Sharma, A., and Prasad, J., On Abel–Hermite–Birkhoff interpolation, *SIAM J. Numer. Anal.* **5** (1968), 864–881.

163. Sharma, A., Smith, P. W., and Tzimbalario, J., Polynomial interpolation in roots of unity with applications, In *Approximation and Function Spaces* (Z. Ciesielski, ed.), pp. 667–681, North Holland , New York, 1981. §§11.5, 11.6

164. Sharma, A., and Tzimbalario, J., Some strongly nonpoised Hermite–Birkhoff problems, *J. Math. Anal. Appl.* **63** (1978), 521–524. §§4.4, 4.5

165. Sharma, A., and Varma, A. K., Trigonometric interpolation, *Duke Math. J.* **32** (1965), 341–357. §§11.5, 11.6

166. Sharma, A., and Varma, A. K., Trigonometric interpolation $(0,2,3)$ case, *Ann. Polon. Math.* **21** (1968), 51–58. §11.6

167. Sharma, A., and Varma, A. K., Lacunary trigonometric interpolation on equidistant nodes (convergence), *J. Approx. Theory* **35** (1982), 45–63. §11.6

168. Shull, C. S. F., Birkhoff type interpolation of entire functions, *J. Approx. Theory* **28** (1980), 260–266. §12.6

169. Stieglitz, M., Reguläre Hermite–Birkhoff-Funktionale, *Z. Angew. Math. Mech.* **55** (1975), 530–532.

170. Stieglitz, M., Beste Quadraturformeln für Integrale mit einer Gewichtsfunktion, *Monatsh. Math.* **84** (1977), 247–258. §§10.6, 14.5

171. Stieglitz, M., Beste Quadraturformeln für Inzidenzmatrizen ohne ungerade gestützte Sequenzen, *J. Approx. Theory* **25** (1979), 176–185. §§10.1, 10.3, 14.4, 14.5

172. Stieglitz, M., Der Satz von Budan–Fourier für allgemeine polynomiale Monosplines, *Z. Angew. Math. Mech.* **59** (1979), 217–223.

173. Surányi, J., and Turán, P., Notes on interpolation, I, *Acta Math. Acad. Sci. Hungar.* **6** (1955), 67–80. §12.6

174. Suzuki, Chisato, A quasi-Hermite $(0, q)$-interpolation problem and its explicit

solution, *Information Processing in Japan* **17** (1977), 137–143.

175. Švedov, A. S., Comonotone approximation of functions by polynomials, *Dokl. Akad. Nauk SSSR* **250** (1980), 39–42; *Soviet Math. Dokl.* **21** (1980), 34–37. §9.7

176. Swartz, B. K., and Varga, R. S., A note on lacunary interpolation by splines, *SIAM J. Numer. Anal.* **10** (1973), 443–447. §14.7

177. Turán, P., On rational polynomials, *Acta Szeged* **11** (1946), 106–113. §5.3

178. Turán, P., On the theory of mechanical quadrature, *Acta Sci. Math. Szeged* **12** (1950), par. A, 30–37. §10.5

179. Turán, P., On some open problems of approximation theory, *J. Approx. Theory* **29** (1980), 23–85. §12.6

180. Varma, A. K., Some interpolatory properties of Tchebicheff polynomials; $(0,1,3)$ case, *Duke Math. J.* **28** (1961), 449–462. §12.6

181. Varma, A. K., Trigonometric interpolation, *J. Math. Anal. Appl.* **28** (1969), 652–659. §11.6

182. Varma, A. K., An analogue of a problem of J. Balázs and P. Turán, III, *Trans. Amer. Math. Soc.* **146** (1969), 107–120. §§11.1, 12.1, 12.6

183. Varma, A. K., Hermite–Birkhoff trigonometric interpolation in the $(0,1,2,M)$ case, *J. Austral. Math. Soc.* **15** (1973), 228–242. §11.6

184. Varma, A. K., and Sharma, A., Some interpolatory properties of Tchebyscheff polynomials: $(0,2)$ case modified, *Publ. Math. Debrecen* **8** (1961), 336–349. §12.6

185. Vértesi, P. O. H., On the convergence of the trigonometric $(0, M)$ interpolations, *Acta Math. Acad. Sci. Hungar.* **22** (1971), 117–126.

186. Vértesi, P. O. H., Notes on the convergence of $(0,2)$ and $(0,1,3)$ interpolation, *Acta Math. Acad. Sci. Hungar.* **22** (1971), 127–138. §12.6

187. Vértesi, P. O. H., A problem of P. Turán (On the mean convergence of lacunary interpolation), *Acta Math. Acad. Sci. Hungar.* **26** (1975), 153–162. §12.6

188. Windauer, H. Zur symmetrischen Lücken-Interpolation, *Z. Angew. Math. Mech.* **51** (1971), T31–T32.

189. Windauer, H., On Birkhoff interpolation: free Birkhoff nodes, *J. Approx. Theory* **11** (1974), 173–175. §5.5

190. Windauer, H., Symmetrische Birkhoff-Interpolationen, *Acta Math. Acad. Sci. Hungar.* **26** (1975), 29–39. §2.3

191. Zeel, E. O., Trigonometric $(0, p, q)$-interpolation, *Izv. Vyss. Ucebn. Zaved. Mat.* **3**, 94 (1970) 27–35 (in Russian). §11.6

192. Zia-Uddin, Note on an "alternant" with factorial elements, *Proc. Edinburgh Math. Soc.* **3** (1933), 296–299. §4.6

193. Zielke, R., Remarks on a paper of Passow, *J. Approx. Theory* **20** (1977), 162–164.

194. Cavaretta, A. S., Micchelli, C. A., and Sharma, A., Multivariate interpolation and the Radon transform, *Math. Z.* **174** (1980), 263–279. §2.5

195. Cavaretta, A. S., Micchelli, C. A., and Sharma, A., Multivariate interpolation and the Radon transform. II. Some further examples. In *Quantitative Approximation* (R. A. DeVore and K. Scherer, eds.), pp. 49–62, Academic Press, New York, 1980. §2.5

196. G. Glaeser, L'interpolation des fonctions différentiables de plusieurs variables, *Proceedings of the Liverpool Singularities Symposium*, Vol. II (Lecture

Notes in Mathematics, Vol. 209), Springer-Verlag, Berlin, 1971. §2.5

197. Kergin, P., A natural interpolation of C^k functions, *J. Approx. Theory* **29** (1980), 278–293. §2.5

198. Jetter, K., A new class of Gaussian formulas based on Birkhoff type data, *SIAM J. Numer. Anal.*, to appear. §10.6

199. Lorentz, R. A., Simultaneous approximation, interpolation and Birkhoff systems, *J. Approx. Theory*, to appear. §9.7

200. Micchelli, C. A., and Milman, P., A formula for Kergin interpolation in R^k, *J. Approx. Theory* **29** (1980), 294–296. §2.5

201. Micchelli, C. A., and Rivlin, T. J., A survey of optimal recovery. In *Optimal Estimation in Approximation Theory* (C. A. Micchelli and T. J. Rivlin, eds.), pp. 1–54, Plenum Press, New York, 1977. §14.6

202. Nevai, P., *Orthogonal Polynomials* (Memoirs of the American Mathematical Society No. 213), Providence, Rhode Island, 1979. §12.6

203. Cavaretta, A. S. Jr., Sharma, A., and Varga, R. S., Interpolation in the roots of unity: An extension of a theorem of J. L. Walsh, *Resultate der Mathematik* **3** (1980), 155–191. §11.6

204. Saxena, R. B., Sharma, A., and Ziegler, Z., Hermite-Birkhoff interpolation on roots of unity and Walsh equiconvergence, *Linear Algebra Appl.*, to appear. §11.6

205. Zensykbaev, A. A., Monosplines of minimal norm and best quadrature formulas, *Uspehi Mat. Nauk* **36** (1981), 107–159 (in Russian).

Symbol Index

Subject Index

For names of authors quoted and their location in the text see Bibliography and References.